DRYLAND FARMING

A Concept of Future Agriculture

ARUPRATAN GHOSH

P.K. JANA

KALYANI PUBLISHERS

LUDHIANA - NEW DELHI - NOIDA (U.P.) - HYDERABAD
CHENNAI - KOLKATA - CUTTACK - GUWAHATI - KOCHI - BENGALURU

KALYANI PUBLISHERS

Head Office

B-1/1292, Rajinder Nagar, **Ludhiana**-141 008 • Ph : 0161-2760031, 2745756
E-mail : kalyanibooks@yahoo.co.in **Website :** www.kalyanipublishers.co.in

Administration Office

4779/23, Ansari Road, Daryaganj, New Delhi-110 002
Ph : 011-23271469, 23274393 **E-mail :** kalyani_delhi@yahoo.co.in

Works

B-16, Sector-8, **NOIDA** (U.P.)

Branches

1, Mahalakshmi Street, T. Nagar, **Chennai**-600 017 • Ph : 044-24344684

Gopabandhu Lane, Behind Govt. Bus Stand, Badambadi,
Cuttack-753 012 (Odisha) • Ph : 0671-2311391

3-5-1108, Narayanaguda, **Hyderabad**-500 029 • Ph : 040-24750368

10/2B, Ramanath Mazumdar Street, **Kolkata**-700 009 • Ph : 033-22416024

Arunalaya, 1st Floor, Saraswati Road, Pan Bazar, **Guwahati**-781 001 • Ph : 0361-2731274

Koratti Parambil House, Convent Road, **Kochi**-682 035 • Ph : 0484-2367189

No. 24 & 25, 1st Floor, Hameed Shah Complex, Cubbonpet Main Road, **Bengaluru**-560 002

KPP N 24048 05

Third Revised Edition, 2017

Datalink Computer
(DRY-FA17)

ISBN 978-93-272-8197-2

PRINTED IN INDIA
At Diamond Agencies Pvt. Ltd., B-125, Sec. 63, NOIDA

PREFACE

First Edition

According to the demographers every day two hundred thousand is being added to the world population, which is likely to reach 8.2 billion by 2025. World dryland areas will home a good chunk of it. By now (2004) all the productive lands have nearly been exhausted for the production of food crops. In order to maximise the production, such land and its environment has been polluted to a dangerous level, only within last two decade. Under this condition the growing world population can't be fed properly unless the production in dryland is improved in immediate future.

Till today a comprehensive idea for dryland farming is very much lacking in literature, documents and works. This is basically due to multi-disciplinary nature of the subject. Hence, necessity arises for a comprehensive textbook on 'dryland farming'.

Though a lot of research is being carried out to generate knowledge for development of production in dryland areas, reflection of the result in the field is not at all up to the mark. The very low economic status, almost to the brink of starvation, in such areas proves the comment. Lacking of holistic approach of experts, researchers and farmers is possibly the main cause of this.

The teachers, researcher and students, interested to work on dryland, have to refer various books, monographs and periodicals to collect information on dryland farming. But, that collection usually fails to provide a clear conception about the subject. Frequently the N G Os and different developmental agencies, who are engaged in development of dry areas, require a comprehensive idea about the methodology of sustainable agricultural development under water stress situation.

Keeping this in mind, this book has been designed and it may fulfil the need of a comprehensive book on dryland farming. It is indeed a critical task, however, books on various topics, technical bulletins, and periodicals on related subject have been consulted and relevant information or basic principle have been incorporated in the text. This book has been compartmentalized in a fashion to convey a holistic idea on this field to the teachers, students, researchers, developmental workers and farmers and those concerned in dryland development.

Chapter I and II deals with situation and conception of dryland farming. Chapter III provide the basic information required for dryland farming and Chapter IV supplies the information on different technologies. Chapter V, the longest one, describes the management part of dryland farming. It is the most

PREFACE

Third Edition

At present, i.e. at the later part of 2017, the agrarian situation all over the Globe is deep down in serious trouble mainly because of non-availability of water required for agriculture. Hence, technology of crop production in water-scarce situation, i.e. dryland situation, is still the most vital aspect of future agriculture.

Considering the prevailing and coming uncertainty of water supply ominously hovering over the farm production sector, we decided to revise the book; "Dryland Farming - a concept of future agriculture", published in 2005, along this line. This revision work accommodates necessary rectification and addition, required to make this book more comprehensive, exhaustive and palatable to study.

It is indeed a difficult task to work on this revision because of its multidisciplinary nature and diversified character. However, we have decided to lay stress upon elaboration or addition of some basic informations as well as practical aspects of the dryland farming.

As such no new chapter is added, but important additions have been made almost in every chapter of the first edition of the book. We tried to improve the lucidity of the book, both in respect of technology and language, and at the same time rectify the odd mistakes.

The presentation has been improved through addition of 'list of figures', 'list of table', and 'boxes' that may help the reader to conceive the idea of dryland farming technology in a holistic manner.

Dr. R G Maiti and Dr. Pradip Bannerjee helped us a lot in the revision work. The authors are particularly thankful to Smt. Susmita Ghosh, without whose help this work could not see light of the day within this short period.

We still think that there is enough scope to improve the book further. The reader are earnestly requested to send their valuable suggestions to us for considering the same in the next edition.

At the last, but not the least, we pay our gratitude to Mr. Raj Kumar of Kalyani Publishers for the commendable job in respect of publication and marketing of the previous editions. This had provided us the scope to take up the necessary revision work.

B-5/78, Kalyani
1st August, 2017

Arupratan Ghosh
Dr. P.K. Jana
arupratan61@gmail.com
arupratan_g@rediffmail.com

important chapter and has been divided into water, soil, crop, organic matter, mulch, animal and social aspects. The book has ended with Chapter VI that offers a specific model of dryland farming.

The authors acknowledge their indebtedness to the authors of various books, monographs and periodicals, from which information have been consulted frequently to create this book. We express sincere gratitude to Dr. R. G. Maiti, former professor, for his constant help and providing moral boost during preparation of the manuscript. We are also thankful to Dr. B. A. Stewart, Dr. S. Chatterjee, Mr. A. Chakroborty, and Dr. G. Sounda for their constructive criticism and valuable suggestions. Our grateful thanks are due to Dr. M. Pandit, Dr. P. Bandopadhyaya, Dr. B. Mondol, Dr. D Bose and Mr. R. Sengupta for their help in different ways.

Authors are indebted to Mr Raj Kumar and Mrs Usha Raj Kumar of 'Kalyani Publishers' for their interest and pain without which this work might not see the light of the day.

B–5/78
P.O.–Kalyani
Dist. Nadia,
W.B. 741235

Arupratan Ghosh
P.K. Jana

ONE MAY GO THROUGH THE FOLLOWING

A leak of just one drop per second wastes 45,000 litres of water a year, which is sufficient for a person for 15 months.

World health organisation (WHO) has estimated the per day per head requirement of water as 135 litres. However, it includes the requirements of all developed, developing and underdeveloped countries of the world. In India if we become water conscious, 100 litres per day per head can easily serve the purposes.

According to study conducted by international water management Institute, a quarter of world' population or one third population of developing countries including India are living in regions that will experience severe water scarcity within the first quarter of the next century.

In a dense forest area of 2500 mm rainfall about 1000 to 1500 mm rainfall can pass down and on an average one hectare of forest can store about 6 to 7 lakh tones of water after filtering it. Further more the topsoil itself can hold 1.2 lakh tones of water. So to make life sustainable in respect of water, in a particular area, one third of dense undisturbed green cover is recommendable.

Effect of runoff on soil –
According to Central Soil and Water Conservation Research Training Institute (CSWCRTI), Deradun, 5334 million tones of soil (approximately @ 16 tones/hectares as against permissible 4 tones) is being eroded annually and the figure is increasing year after. Of this, 29% is being permanently lost to sea, 10% is deposited in reservoirs as silt and rest 61% displaced from one location to another.

As study by the National Committee, on integrated water resources development, the loss of storage capacity of reservoirs, tanks etc. due to siltation has deprived India a whooping 5000 crore rupees as capital loss.

Revival of a dead well-
Mr. H. G. Lakshminarayana Bhat of Karnataka has two acres of barren land having lateritic soil on the top. To develop the land he dug out a well up to 30 ft. but fail to create any source of water. Next year, inspired by a story of rainwater harvesting, he constructed 1.5 X1.5 feet continuous contour trenches of 60 to 70 feet apart in that land. He spends about 1100 rupee to construct the trenches. It gives an immediate result and a higher degree of moisture content in summer encourages the growth of the coconut sapling planted in the previous year. In third year the well registered a steady 10 to 12 feet of water in the summer and thick grass covers the land completely. The water of that well was utilised to irrigate the tree crops and also for other purposes. On an average daily uptake of water go up to 900 to 1000 litres. Gradually the ecology of the entire land area is changed and creates a sustainable economically profitable farm.

Recharge of well through roof top water harvesting-

Mr. U K B Nambiar, an ex-serviceman of kerala, constructed his house in 1975 and facing water crisis in every summer, which grew year after year. Then in the year 1993-94 he developed his 100 sq. ft. terrace fitted with one 4-inch PVC pipe that leads the collected roof water to the bottom of a coconut plant. The soil of the tree basin is full of decomposed and semi-decomposed plant residues like coconut husks and leaves. It acts as a good soak-pit and infiltrate the entire harvested water into the under ground. As a result the well registered 3, 4, 5, and 6.5 feet of standing water in the month of May in the year 1995, 1996, 1997, and 1998 respectively instead of 1ft. water level earlier. Finally the summer water level has been stabilised at that level.

Water harvesting through Tanks for irrigating in Karnataka-

Karnataka state of India has more than 36,508 big and small tanks of which some of them are 15-16 centuries old. The tanks of Malnad region, which constitute about 25% of these tanks, are generally small and a great number only harvest rain. Frequently the stored water percolates and rainfed paddy is raised in the fields below. Apart form big/ major irrigation projects, 80% (0.65million ha) of states total minor irrigation potential is contributed by tanks. Nearly 38% of the tanks have a command area less than 4 ha and 50% have command areas between 4-20 ha. Command areas between 20-200 ha and more than 200 ha shares 10% and 1.4% of total number of tanks. At present there is very little scope to create new tanks in the state but proper maintenance and modernisation of the existing tanks are the serious task to fought against draught in Karnataka.

Scarcity of drinking water in rural India-

The Drinking Water Mission at the commencement of the Sixth plan designated about 2.31 lakh villages i.e. 40% of total villages as 'problem village'. These villages either had no potable water source or, if present, the source was more than 1.6 km away or was at a depth of 15mt or at an elevation of more than 100mt. Some of these sources of drinking water contaminated chemically (fluorides etc.) or biologically (worm, germ, pathogen etc.)

A comparison of water used (in percent) by various sectors in different locations-

Locations	Household	Industry	Agriculture
Germany	14	83	3
Singapore	45	51	4
Europe	13	54	33
India	3	4	93
Bangladesh	3	1	96
Global	8	23	69

Salvage of a river and a village only by well-executed check dams-
The people of Thunthi Kankasiya village of Dahod district of Gujrat state of India were habituated to face sever water crisis before 1994. In those days more than 75% of them used to migrate at least for 10 months every year. There were no wells in the village. Uncertain mono-cropping within 4 months period is the only farming option for them. The village is situated in the upper catchment area of the river Machhan, which dried up four months after the monsoon and rapidly carried away most of the rainfall as runoff towards the down stream.

In this situation, in the year 1994 with the help of a N G O, the people of the village began the construction of a check dam in April and completed within record 85 days without any govt. help. A series of small and medium sized check dams, starting from the source of the river, were later built along the river Machhan to slow down the runoff and impond about 4.53 lakh cubic metres of runoff water for irrigation and other purposes. In the process the trapped runoff water percolates and recharge ground water, which in association with other water conservation methods like bunding, trenching, and terracing provides good amount of water in 23 numbers of wells of the area. At the same time the river can provide enough water to meet irrigation demands despite the draught in the year 2000. They tried to trap every drop of rainfall in their area and brought their self-made economic development within a span of 7 to 8 years.
The impact of proper implementation of water conservation technology on the Thunthi Kankasiya village is unique and are stated as follows:-

Development occurred	Status before 1991	Status as on 2000
Perennial drinking water in wells	Nil	23
River Dams	Nil	1
Water availability (months)	4	12
Irrigation coverage (hectares)	Nil	135
Number of crops per year	Below 1	2 -3
Annual agricultural production (tonnes/hectare)	0.9	4.0
Average migration rate (Percent)	78	5
Average migration period (months)	10	2
Income per house hold (Rs.)	8590=00	35620=00

LIST OF TABLES

LIST OF FIGURES

CONTENTS

CHAPTER 1

INTRODUCTION

1.1. GLOBAL SITUATION

Out of total world population more than 1.5 billion live in dryland areas. The bulk of this population belongs to sub-Saharan Africa and Asia, the region that is more or less inhabited by the world's poorest people. The present number of about 500 million of these poor people is expected to rise to 800 million in 2020. These dryland areas are under increased pressure due to the prevalence of poor agriculture practices, over-grazing and intensive cropping to meet food and fiber requirement of man and animal. In the process a total of 1.2 billion hectares already experience moderate to severe and extreme degradation chiefly in relation to fertile topsoil.

The technological development of dryland areas is very poor, particularly in agriculture front. Thus this land is unable to meet the pressure of people's need. A severe environmental degradation is unavoidable due to this pressure. In addition, the poor development in the social and economic front often creates serious disturbances. The ultimate result is the persistence of perennial poverty in the dryland areas.

In some countries the dryland covers more than 50% of the total geographical area. Most of these countries are economically very poor. This poor economic status leads to a poor social consciousness,

which is the major cause of excessive population growth in these countries. The final outcome of this situation is the low growth rate of gross domestic production (GDP). Naturally per capita income, expenditure and food/nutrition availability is very low in these countries.

To improve this downgrade situation in these areas, most of which have limited scope for industrial production, technological development in respect of economically viable food and fodder production is the only way. Primarily, it involves the considered use of every available drop of water. When the agricultural development is effected in this way, the economic condition of the people will improve. This will automatically usher in the development in other fields like education, health, nutrition etc.

The economically developed countries like Israel, which is also situated in extremely dry areas, made use of the appropriate agricultural technology for economic production of food and fodder in their country.

1.2. INDIAN SITUATION

1.2.1. Socio-economic Conditions

Closer at home, out of total landmass of 329 million hectares with an average annual rainfall of about 1120 mm, 224 million hectares are moderate to extreme dry. These areas alone provide home to some 40% of our total human population and two third of cattle population. Maximum of these areas is generally mono-cropped and dependent on rain only. No significant technological development in agricultural front, particularly in respect of water conservation to produce a second crop, is found in these areas and only subsistence agriculture predominantly exists. Instead of high value crop these areas are compelled to produce 90% of coarse cereals, 91% of pulses, 86% of oil seeds and 67% of cotton of the total production of our country obviously with low productivity. At the same time, in our country, 90% of people of dry areas depend only on agriculture for their livelihood. Interestingly an estimated 45% of total food grains come from the dry areas and that too with the help of age-old package of practices. Naturally, poverty is a constant companion of the people living in these areas.

In the context of our national economy, the growth rate of country's GDP heavily depends on the agriculture production. In spite of a present degrading trend in comparison to past few decades, still about 20% of our GDP comes from Agriculture sector. The downtrend in this national economic indicator occurs when the food production of dryland areas is adversely affected by drought.

1.2.2. Research Position

In recent years scientists, farmers and others involved in improving farming practices have developed agricultural techniques and practices for dryland areas in different parts of the world including India. However, their widespread application is yet to be found in the poverty-ridden dry areas.

Through the able guidance of Indian council of Agriculture Research (ICAR), India has conducted lot of research and has developed successful field action technology in dryland areas. However, lack of wide multidisciplinary approach and little attempt to put different information together, many of these remain under utilized or unutilized. Even the cultivation practices of coarse cereals in dryland areas, with available technology, between research plot and farmer's plot differs widely. An analysis on production of sorghum by CRIDA shows that when research plot yields 3970 kg/ ha, the farmer's plot yields only 370 kg/ ha (Table 1.1).

Table 1.1. Yield gap in sorghum between research plot and farmer's plot.

Situation	Number of Case	Yield (kg/ha)	% Drop	Production Senario
(1) **Research plot-**	1	3970	NIL	Possible potential
(2) **Farmer's plot-**				
(a) Verification plot	9	2090	47	Realizable
(b) Demonstrations	39	1460	63	Feasible
(c) Assured input supply	26	1060	73	Attainable
(d) Common rural plots	36	370	90.6	Actual

1.2.3. Status of Dry Areas (A Few Relevant Information)

[A] About land degradation –

(1) Out of a total 329 million hectare (m ha), 175 m ha are in various stages of degradation.

(2) About 50 percent of total cultivable land suffers from soil erosion and salinity problem.

(3) About 30 m ha of fragile land now under cultivation is progressively degrading.

[B] Animal status of dryland areas –

(1) Out of 450 million there are 300 million heads of cattle present in dryland areas.

(2) The annual fodder requirement of India is about 932 million ton of green and 750 million ton of dry as against current availability of 250 and 414 million ton respectively.

[C] Rain water –

(1) On an average, India receives 400 m ha meter of rainwater annually.

(2) Out of this rainfall about 160 m ha m. falls on agricultural land.

(3) Around 24 m ha m equivalent rainfall is available for harvesting and can be stored in small-scale water harvesting structures.

(4) Regions with up to 1000 mm rainfall potentially produce 6.32 m ha m. harvestable run-off water.

(5) Nearly one fourth of the total annual rainfall is received during pre- and post rainy season.

Status of Drought

Identification of drought-prone areas started with the appointment of a committee of officials by the Govt. of India in 1970. The criteria laid down for the selection of the areas were:

(i) Level and periodicity of rainfall,

(ii) Intensity and past occurrence of drought, and

(iii) Proportion of irrigated area to total cultivated area.

Accordingly 44 geographical units, covering 72 districts, partly or wholly, were identified as drought prone areas. Similar studies in the year 1972 and 1973 by 'Irrigation Commission' and 'Task force on Integrated Rural Development' identified the same areas. The state-wise geographical area and population represented by the identified drought-prone areas are given in the Table below:

Table 1.2. State-wise Geographical Areas and Population Covered by Drought

States	No. of districts		No. of tehsils/ blocks		Drought-prone areas as % of total areas in the state	Population as % of total population
	Total in the State	Drought-prone	Total in the state	Drought-prone		
Andhra Pradesh	21	7	195	52	30	22
Bihar	17	4	587	57	14	7
Gujarat	19	10	184	41	29	18
Haryana	7	3	32	6	21	16
Jammu & Kashmir	10	2	44	4	5	7
Madhya Pradesh	43	6	190	16	8	9
Maharashtra	26	6	233	33	14	11
Karnataka	19	10	175	25	25	22
Orissa	13	2	103	25	11	6
Rajasthan	26	13	196	50	56	33
Tamil Nadu	14	2	134	16	17	11
Uttar Pradesh	54	6	232	17	9	4
West Bengal	16	3	297	30	13	7
All India	**395**	**74**	**2602**	**382**	**19**	**12**

Source : Government of India, Planning Commission, Report by the Task Force on Integrated Rural Development, June 1973

1.2.4. Effect of Drought on National Economy

In Indian sub-continent, during last 120 years (1870-1990) drought occurred 425 times across 29 out of 35 meteorological sub-stations. About one fifth of total geographical area was affected once in every 5 years and one third once in every ten years. Apart form these general features, the localized drought situation frequently crop up in many parts of our country. These reduced the crop production to about 44% during the affected years and pulled down the GDP and economic growth.

The first five-year plan (1951-56) had a long-term perspective of doubling the per capita income in next 27 years. If we take the figures of last 40 years (1950-51 to 1990-91), we shall find that the GDP increased at a low annual rate to reach the target in 40 years instead of 27 years as envisaged earlier. This delay could have been avoided by improving the cropping in dryland areas. Though we have received the blessing of '**green revolution**' in between, that has mostly been restricted in the non-dryland tracts.

1.2.5. Ways of Improving the Dryland Areas

To improve the farming condition in dryland areas the following steps may be followed.

(1) Comprehensive, multidisciplinary planning and approach –

(2) Adequate dialogue with local beneficiaries –

(3) Suitability of technology developed in research farms confirmed through on-farm trials with farmers' participation —

(4) Interfacing of research and developmental effort –

(5) Adequate linkage between crop farming and livestock farming –

(6) Close relation among the techniques of water conservation, soil management and crop cultivation –

(7) Development of perennial crops/animals and their varieties/breeds suitable for dryland–

(8) Considering the farmer as partner of development either in research or in policy –

1.2.6. Future of Dryland Farming in India

1.2.6.1. Future Projection

Future dryland farming should be considered more important than the 'greening of gray areas' or the expected gene driven 'Second Green Revolution'. In India about 50 percent of the geographical land will remain rain-fed for an indefinite period. According to several studies and projections in this field made by Paroda (1995), CRIDA perspective plan (1995) and DARE / ICAR IX plan document the projected scenario of 2020 is summarized below:

Year	1995	2020
Total Population (million)	900	1400
Population in dry regions (million)	360	560
Animal population (million)	450	650
Land availability per capita (ha)	0.15	0.07
Productivity of food grains in dryland (kg/ha)	770	2000
Productivity of food grains in non-dryland (kg/ha)	4000	
Water needed (million ha m.)	8	24

According to this projection per capita land availability will decrease to 0.07 ha by 2020. This will render conventional crop production economically non-viable resulting in a possible large-scale migration, particularly from conventionally low productive areas. To avoid this debacle provision of food for the additional population should be made by increasing the production in dryland from 770 kg./ha to 2000 kg/ha, which is simply impossible by the existing technology. Hence, improved technologies for dryland farming have to be brought in.

1.2.6.2. Vision and Approach of Future Dryland Farming

Till today dryland farming is focussed on growing a singular traditional food commodity suitable for subsistent agriculture. The future demand is for the market oriented economy taking comparative advantage of the environment. The planning of dryland farming should, therefore, be laid on "micro-approach" instead of present "macro approach". The future need is the "bottom to up" location specific technological development rather than the present "top to down" region specific approach. The single discipline approach should be replaced by true multidisciplinary approach comprehensible to farmers or implementers rather than to technologists or scientists. The economic viability and farmers' acceptability must be the determining factor in future dryland farming. The approach must aim at long-term productivity and sustainability instead of present quick-return approach disregarding the ecological and economic stability.

Earlier dryland farming was not much concerned with the methodology of water application. The future demand will be the best economic use of rainwater stored or harvested *in-situ* and then recycled for improving the productivity.

Till today dryland farming is almost synonymous to the cultivation of coarse grains, rugged pulses and oil seeds. In future it will represent a well-knit composite farming that will include activity of interdependent production of crop, livestock, fodder, fuel or biomass for economic stability of the people of the region.

Past outlook of Dryland Farming	Future outlook of Dryland Farming
1. For subsistence agriculture,	1. For economic and sustainable agriculture,
2. Consider the odds to do something,	2. Consider the advantages to do something and manage the odds accordingly,
3. Mechanical and labour reducing approach,	3. Biotic and labour intensive approach,
4. Short term measures,	4. Long term measures,
5. Plant population based farming,	5. Individual plant based farming,
6. Economic production of animal, fodder and fuel is not generally considered,	6. Economic production of animal, fodder and fuel will be part and parcel of farming,
7. Application of water (irrigation) is not considered,	7. High impetus on water use (irrigation) planning,
8. Whole farm tilling approach,	8. Minimum land disturbance approach,
9. Soil conservation measures dominated farm planning,	9. Water conservation measures dominated farm planning,
10. Importance was not given to mulching,	10. Mulching will be must for farming,
11. Community approach,	11. Individual farm approach, may be on community basis,

CHAPTER 2

WHAT IS DRYLAND FARMING

2.1. HISTORY

With the development of farming activity through crop cultivation human civilization started unfolding. According to the findings of anthropological studies agriculture came to exist nearly ten thousand years ago when hunter gatherers of early days discovered that a plant comes out when a grain falls on the moist ground and in turn it produces much more number of grains. This might appear trifle today, but was a great discovery at that time that gave birth to **agriculture.**

Since a cropland cannot be carried along with other belongings of those nomadic people, their shifting life gradually got settled down near the field, which promised to supply their major food for survival. Thus started the human settlement and along with it the civilization.

As water is essential for human, these settlements developed in places where water supply was assured. Availability of fertile soil was another factor for development of such settlements. Anthropological studies show that many ancient civilizations like the Indus, Mesopotamia, Maya, and Inca etc. were destroyed mainly because of scarcity of water and loss of soil fertility, which occurred due to the changing geological environment. At present all these places are considered as dry areas.

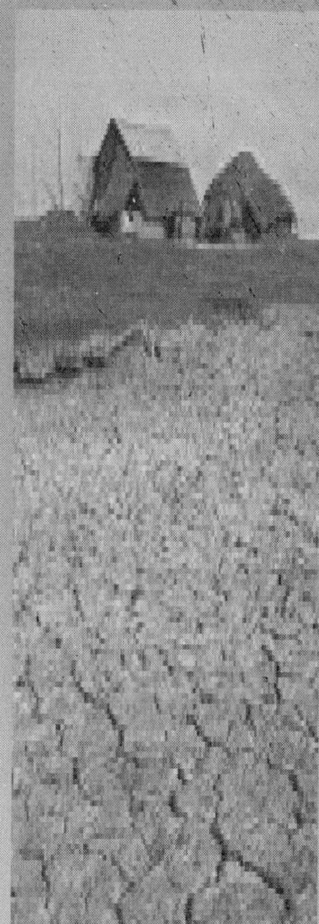

However, the organised crop farming, which we see today, began much later and the systematic cultivation under dryland situation started around 1850. As per the available information, improved technique of dryland farming first started in U.S.A. Wheat was used to be cultivated on the land along the Malad creek. By 1863 the water of that creek turned saline. Under this situation farmers desperately started cultivating this crop without artificial application of water.

In 1894 Colorado State initiated experiments on dry farming and established a sub-station at Cheyenni Wells. The later part of nineteenth century and earlier part of twentieth century saw a big impetus in dry farming in U.S.A. and Canada. Early books like 'Soil Cultural Manual' by Hardy and Champbell (1904), 'Textbook on Dry Farming' by William MacDonald (1909) throw some light on the initial ideas of dry farming.

Up to 1916 farmers paid little attention to water conservation. They used the good rains for their cultivation without irrigation. But soon the farmers of U.S.A. experienced severe drought consecutively for several years when the rainfall turned too little to meet the requirement of the crop. This led them to think about the proper conservation of moisture/ water more efficiently and trigerred the development of many techniques. This process is still going on and yet to reach perfection.

A.K.Y.N. Aiyer (1917) listed the important practices followed by the farmers of dryland areas of India. This included field bunding, fall ploughing, frequent inter-cultivation, drill sowing and growing crops like joar, bajra, ragi etc., which can stand water stress considerably.

Between 1923 and 1943 some package of practices evolved from studies on crop production under dry (no irrigation) situation. These systems were very much similar to those already listed by Aiyer in 1917. Depending on the province where they developed, these packages were named as (1) Bombay (now Mumbai) Dry Farming Method, (2) Madras (now Chennai) Dry Farming Method, (3) Hyderabad Dry Farming Method, (4) Punjab Dry Farming Method etc. These packages consisted of :

(a) Constructing contour bund as basic and essential treatment,

(b) Occasional ploughing of land,

(c) Repeated shallow cultivation of soil to remove weeds as well as to conserve moisture,

(d) Adding moderate quantity of farmyard manure (FYM),

(e) Wider spacing between crop plants,

(f) Repeated inter-cultivation,

(g) Adopting crop rotation and

(h) Keeping the land fallow systematically.

Though technically sound and economically better than the methods followed by the farmers themselves, adoption of Bombay Method was low. It was probably due to the high initial cost on land development and additional inputs in comparison to the low return from crops like joar, bajra, ragi etc. used in this method.

2.2. CONCEPTION OF DRYLAND

2.2.1. What is Dryland ?

Commonly the word 'dry' means absence of moisture. Additional attributes come in when this word is associated with land. To a farmer it refers to moisture stress causing partial or total failure of his crop. To the crop it implies the inability to extract enough moisture to sustain itself.

The land may be dry for certain plants but may not be so for several others. Further, the land may be dry in one season but may not be so in another. Considering all these points, at present, dryland may be defined as "the land which is not able to provide sufficient moisture to produce a commercially viable crop without adopting any water conservation measures".

Apart from areas of very low rainfall (well below 750 mm / year), a large tract of undulating or degraded land with moderate to good but uncertain and erratic annual rainfall (750 mm to 1200 mm) and devoid of a perennial source of water supply, may fall in the category of dryland.

2.2.2. Types of Dryland

Depending on the total rainfall, its distribution and intensity, topography of land and nature of soil, dry areas may be categorised into four types:

1. **Dry Due to Very Low Rainfall.** Areas, which receive less than 50 mm rainfall per month in rainy season or less than 350 mm per year, may be included in this group. Generally, these are desert or desert like land where common agricultural practices are not workable. With the help of composite dry farming techniques, frequently called as 'watershed management', farming in a few selected parts of these areas has been made possible. In India such land constitutes about 13% of the total area.
2. **Dry Due to Low, Erratic Unimodal Rainfall.** These areas receive 50 mm to 200 mm monthly rainfall during rainy season and 350 mm to 750 mm average rainfall annually. Coarse food grains (Jower, Bazra, etc.) can be grown with suitable agro-techniques during the rainy season, but the production is often uneconomic and uncertain. These tracts are generally called 'drought prone areas'. With proper dryland farming techniques parts of these areas can be put to commercial farming. About 22% of the total geographical area of India comes under this group.
3. **Dry due to Undulating Topography With Moderate to Good Rainfall.** There are tracts with as much as 750 mm to 1250 mm rainfall. But due to uneven topography and erratic nature of precipitation these tracts suffer from water scarcity. Further more' high intensity of rainfall on slopy ground causes erosion of fertile topsoil.

   In the past these areas were covered with natural vegetation that prevented loss of rain water and soil erosion. But the spread of human habitation to these areas caused destruction of this vegetative cover thereby enhancing soil degradation and water scarcity. Continuation of this situation over the years finally left behind a barren area where it is now difficult to practice commercial farming. In pockets down the slope some crops like rice are cultivated during rainy season only at subsistent level. This situation prevails on about 36% of total geographical area of India. Proper scientific package of farming practices can lead to commercial farming of these entire areas.

4. **Dry due to other reasons.** Very high altitudes with stiff slopes, rain shadow areas of the hill or mountain, sandy soil with saline ground water etc. come under this group. Specialised methods of cultivation are needed to bring these areas under commercial cultivation. This type of land lies in Leh – Ladakh, parts of Sikkim, rain shadow areas of Western Ghats, eastern coastal belts of South India and some other places.

2.2.3. Advantages of Dryland for Plant Growth

Sunshine is one of the most advantageous aspects of the dryland in respect of plant growth. It is the prime source of energy directly required by plants for photosynthesis. In comparison to the wet areas the dryland areas have more sunny days, and mainly due to this reason the average sunshine hour in dryland areas is high. Utilising this natural blessing one can produce more yields in comparison to the high rainfall areas, if the other management aspects are followed properly. Except in very hot periods when intensity of sunshine is too high for physiological activities, the plants are able to produce good yield throughout the year.

Temperature of about 30°C is always helpful for biological as well as bio-chemical activity of non thermo-sensitive plants. Almost every tropical crop and animal and some acclimatised or hybrid or genetically modified temperate crops and animals will do well in these areas if the farming operations are planned properly with required inputs. During winter the temperature is lower than the corresponding humid regions. This condition is more conducive to the growing of temperate seasonal crops like vegetables, flowers etc.

Aeration in the root zone is vital for any crop (other than paddy and some aquatic crops) for high production. Except vertisol, other soils of dryland areas provide good aeration at the root zone due to higher soil porosity. This condition also prevents water stagnation in root zone that helps better plant growth.

Less Pests and Diseases Problem. High temperature along with dry atmosphere and bright sunshine is deterrent to the development of insects, pests and diseases. Thus, crops grown in dry areas are comparatively free from these biotic maladies.

2.2.4. Disadvantages of Dryland for Plant Growth

Water Scarcity. Scarcity of water is the main factor that limits the growth of plant in dryland areas. This scarcity generates from:

(a) Low rainfall or moderate to good rainfall with unimodal and erratic distribution.

(b) Large scale surface run-off.

(c) Undulating topography with low soil depth that provides little scope to retain rainwater either on the surface or underground.

(d) The water of dry areas sometimes turn alkaline and is not fit for agriculture.

(e) Highly permeable soils or soils with almost no permeability of dryland accelerate the crop related problems.

Soil. Soils in dryland areas are generally not conducive to proper plant growth. The problem is of four types that act against good farming of dryland soil which are described below:

(a) **Poor Soil Structure and Shallow Depth.** Almost all the dryland soils possess poor structure that developed naturally due to various reasons active since long past which has been geared up by the human activity. Shallow soil depth is very common under undulating topography.

(b) **Lack of Plant Nutrients.** Leaching loss of plant nutrients, erosion of top fertile soil, less soil clay-colloids and very low organic matter content lead to a poor plant nutrient status of soils of dryland.

(c) **Erosion.** Erosion of soil is a serious problem in almost all types of dryland areas. The high intensity seasonal rainfall in a dryland area causes severe soil erosion and degrades the land in one season. If not restored it restricts and sometimes prevents the farming activity.

(d) **Salt Accumulation.** High temperature and intense sunshine increases the water loss through evaporation leaving dissolved salts in soil surface. This transforms the soil of the area into saline or alkaline, which is called problem soils.

High Temperature. High temperature (above 40° Celsius) is a common phenomenon in dryland areas. It seriously reduces the growth of crops and animals, particularly in respect of commercial production. It affects adversely the reproductive stage that decreases the yield. Pre-reproductive age also gets disturbed, which becomes a factor for yield reduction. However, the span of such high temperature (above 40°C) is commonly very short generally 1 week to 1 month.

2.3. CONCEPTION OF FARM AND FARMING

2.3.1. What is Farming ?

Farm is a piece of land where something is produced commercially out of any macro- biological activity over a period of time, and **farming** is the package of practice followed in a farm.

2.3.2. Farming Over the Ages

Over millenniums different farming practices have evolved slowly. Farmers of early age identified economic plants suitable for the soil and climate of their locality from the wild stock and adopted them as crops. They developed the package of practices to the best of their capacity to cultivate these plants.

Water is the source of life to both plants and animals including man. Hence, the human settlements developed in places with assured perennial water supply. River valleys not only provided such a facility, these places contained fertile alluvial soil also. Thus, the agro-based civilizations bloomed in and around river valleys since the time farming was taken up by man as a livelihood till the early part of the twentieth century.

However, not all people were lucky to have such prime land as their abode. Some were forced to live in the inhospitable areas in respect of soil and water. Since the package of practices developed in the river valleys was not effective there, these people tried to develop their own technology to suit the situation. These technologies were, however, not able to fight out fully the natural constraints of crop growing in respect of water and soil. Therefore, the production level remained low resulting in poverty and under nourishment in these areas.

Some people had to continue partly hunting and gathering, their old profession. Others took up shift cultivation. Some took up animal grazing to meet their requirement. Till today many of these people are partly dependent on collection, hunting, grazing and shifting cultivation for their food supply. The tribal groups of our country constitute majority of such people.

2.3.3. Future Threat From Existing Farming Practices

During the last 5 to 10 decades the traditional farming has undergone a sea change. In India, in the past, crops were manured entirely from organic sources. Since organic manures contain all the plant nutrients - both major and minor - there was no problem of nutritional imbalance in crop production. Introduction of chemical fertilisers and pesticides during the post-World War and high yielding varieties during the sixties of the past century brought in a revolution in farming practice in India. Food production increased manifold to meet the need of the rapidly growing population. This spectacular event in India was termed as **Green Revolution.**

Though green revolution made the country self sufficient in food it brought many evils along with it. Repeated use of chemical fertilizer without a corresponding supplement of the organic component brought down the inherent productivity level of the soil. The leftover from chemical fertilizers and pesticides spilled over the land and water and polluted our food and drink; it even poisoned mother's milk. **Due to this extensive cultivation, the ecology and environment of intensely cultivated areas are now under serious threat.** The speedy acquisition of fertile farming areas for non-farming activity multiply the threat and finally a chaotic nature-human relationship hover over the mankind. On agricultural front, the beneficial microflora and microfauna have been damaged. The balance between useful and harmful microbiota has been altered giving an upper hand to the latter sometime causing heavy damage to the crop.

Destruction of forest and over grazing of pastureland brought peril to the environment and ecology of relatively non-fertile areas and ultimately converted them into dryland. It has been mentioned earlier that most of the dryland areas of today were covered with natural vegetation, often in the form of dense forest in the past. In India this forestland was partly the property of the landlords and partly of the government. From the fear of loosing control over the landed property after independence landlords sold off their forest trees as timber and fuel wood. Much of the government forest suffered the same fate through illegal felling of trees. This depredation is still continuing unabated. Besides, land was in great demand to grow more food to tackle the famine like situation of post-independence period and for various purposes other than food production. Vast forest areas or fallow pastures were cleared for this purpose. The land that was covered with dense vegetation became barren exposing the surface underneath to the forces of rain, wind etc. People living in these areas faced the consequence of the changing situation.

The main problems of existing farming practices are enlisted below :-

1. **Continuous Cultivation of Specific Crops.** This practice reduces the in-situ bio-diversity by gradual elimination of old, abandoned but potentially important plants/animals/insects/microbes. It reduces micronutrient availability and microbial diversity. It destroys the soil structure. The inherent fertility is gradually reduced as well as the productivity.
2. **Application of chemical fertilisers and other agro-chemicals.** To maintain the productivity and to restrict the intensifying infestation of pests and diseases these applications are essential now. These chemicals and its compounds create mild to severe toxicity to surroundings specifically soil, water, flora and fauna. To accommodate this, many pathogens and pests have changed or mutated themselves and act in a more virulent manner.
3. **No wood and fodder cultivation.** The present farming practices provides no room to produce wood or fodder required for human and livestock. The obvious results are deforestation and over grazing, which causes severe ecological and hydrological imbalance and finally disturb the soil and water status.

2.3.4. Impact of Global warming on Agriculture

Ricardian analysis: It examines the economic impact of climate and other variables on land values and farm revenues. It is a new "Ricardian" approach of analysis in place of traditional 'production-function' approach that estimates the impact of climate changes to predict the damage.

The production-function approach determines the impact of global warming on yields; and often predicts severe yield reduction. This approach has an inherent bias and will tend to overestimate the damage, overruling the indirect and diversified benefits.

Applying the Ricardian approach and considering the climate, farmland price, and other economic and geographical data of United States, it was found that higher temperature in all seasons except autumn reduces the farm values, while more precipitation outside of autumn increases the same.

Actually in 1994 Robert Mendelsohn et. al. (MNS) presented a theory for estimating the impact of global climate change on land values on the basis of "Ricardian" approach. He stated that the relationship between temperature and agricultural productivity, in presence of suitable moisture, is a hill-shaped curve. Here, at 0^0 the productivity will be zero which will gradually increase up to a certain temperature and then gradually decrease, and finally at a specific high temperature the productivity will be very low. However, the model is not fully consistent with this basic Ricardian principle and the omission of an important variable, irrigation, is likely the reason. Actually the value of irrigation capital, legal rights to, and the availability of, irrigation water plays an important role to determine the farmland value.

If the model is applied to a global warming scenario it shows a significantly lower estimated impact of global warming on U.S. agriculture than the traditional production-function approach and, in one case, it suggests that even without CO_2 fertilization, global warming may have economic benefits for agriculture. The Ricardian approach indirectly incorporates production adjustments by focussing on how climate affects land values. That is, 'Ricardian rent' is what accrues residually to the fixed land-climate factor when all other factors are applied optimally.

However, the most comprehensive international study of agricultural impacts on global warming (Rosenzweig et. al. 1993 & Rosenzweig and Iglesias, 1994) finds that, for the case without carbon fertilization (Mendelsohn et. al. 1993), the global average reduction in grain yields would be almost as large as the reduction predicted for United States as a result of declining of U.S. grain belt due to conversion of land use from wheat and corn producing land into non-agricultural one.

2.4. PREVAILING CONCEPT OF DRYLAND FARMING

The practice of dry farming emerged chiefly in the domain of academicians and researchers by combining several technologies to cultivate crops under rainfed condition excluding the consideration for the economic viability of the system. To designate this technology the term, 'rain fed agriculture' was coined. This new terminology gradually infiltrated into the concept of dryland farming. Though literally and conceptually 'dryland farming' and 'rain-fed farming' are not the same, till today many consider these two terms synonymous. However, dryland farming is that part of rainfed farming where water is the most limiting factor.

In the early part of twentieth century Dr. Johan A. Widtsoe, one of the earliest scientists to take up dry farming research, defines the term 'dry farming' as " the **profitable production of useful crops** without irrigation on lands that receive annually a rainfall of 20 inches (508 mm) or less and it may be extended to 30 inches (735 mm) with unfavourable distribution of rainfall ". This is probably the best definition till today, which illustrates the basic concept of dryland farming, particularly under the present economic situation. Accordingly, the phrase 'without irrigation' used in earlier definitions may be deleted. It is because **water harvesting practices and subsequent use of the harvested water, in the form of life saving or drought avoiding or micro type irrigation, is now closely associated with the dryland farming practices.**

Most of the definitions point out to the 'way' or 'method' or 'technology' of cultivation, referred to as **farming** in a water scarce area i.e., in a **dryland**. To understand 'dryland farming' properly it is necessary to have a clear idea about 'dryland' as well as 'farming'.

Steiner *et. al.* (1988), after careful consideration of several definitions, find the 'aridity index concept' of the U N conference on desertification as the appropriate method. This is based on the balance between precipitation (P) and evapotranspiration (EPT). Accordingly, areas with P: EPT ratios of 0.03 to 0.20 are 'arid' i.e. extremely dry and areas with ratio of 0.20 to 0.50 as 'semi-arid' i.e. moderately dry.

As per Indian Agricultural Commission there is another approach to examine and identify the areas in relation to bulk rainfall received in a particular period of crop production. Here the monthly rainfall data are most convenient since the annual or seasonal details do not have much significance to the average time span of most of the crops, which is usually of the order of 90 to 100 days. Keeping this view in mind, all types of land are classified in five groups namely A, B, C, D and E. Out of these groups D (rainfall 50 to 100 mm/month for at least three consecutive months) and E (rainfall>50 mm/month at least three consecutive months) are isolated as dryland. This approach is also not sufficient to create any clear idea about dryland.

In 1970 Indian Council of Agricultural Research (ICAR) used the simple formula of moisture deficiency index of Thornthwaight (1955) to establish the classification of **arid** and **semi-arid** zones of India. The formula of Moisture Deficiency Index or Aridity Index is = (P – PE) / PE X 100, where P = Precipitation and PE = Potential Evaporation. According to this system the limit for semi-arid areas is given by the values of the index varying from –33.3 to –66.7

In recent past Oram (1980) distinguished between rainfed farming and dryland agriculture. According to him successful dryland agriculture is husbandry under conditions of moderate to severe water stress during a substantial part of the year. It requires special cultural techniques and adopted crops and system. Stewart and Burnett (1985) characterized dryland agriculture as rainfed systems that emphasizes on water conservation, sustainable crop yields, limited inputs for soil fertility maintenance and wind and water erosion constraints.

The Food and Agriculture Organization (FAO, 2000) has defined drylands as those regions climatically classified as *arid, semi-arid, or dry sub-humid* based on length of growing period for annual crops. The growing period begins when monthly precipitation exceeds half of the monthly potential evapotranspiration plus a period to use an assumed 100 mm of water from excess precipitation stored in the soil profile. The regions where the monthly rainfall never exceeds half of the potential evapotranspiration have zero growing days and are not included in the drylands. *Arid* regions have 1 to 74 growing days, *semi-arid* have 75 to 119, and *dry sub-humid* regions have between 120 to 179 growing days. Combined, these regions account for 45% of the worlds land area – 7% *arid,* 20% *semi-arid,* and 18% *dry-sub-humid.*

More recently, Australian Center for International Agricultural Research (ACIAR, 2002) stated that dryland cropping refers to those agricultural areas where the average water supply to the crop limits potential yield to less than 40 percent of full (water unlimited) potential.

With these efforts to define 'dryland' and to specify the technologies of farming there, a bunch of new terminology has emerged. From early nineties of the last century attempts were made to define 'dryland farming' and in the process a few new terminology like *rain-fed, arid, semi-arid,* and *watershed* came out. Naturally, nomenclatures like **'Dry Farming'**, **'Rain-fed Agriculture'**, **'Arid Zone Agriculture'**, **'Semi-Arid Agriculture'**, **'Dryland Agriçulture'**, **'Watershed Management'** etc. have emerged and are used by many experts and researchers instead of using any broader terms like 'Dryland Farming'. These different terminologies have actually confused and narrowed the vision and mission of dryland farming.

These different terms gradually shadowed the basic concept of dryland farming propagated by Dr. John A. Widtsoe, which highlighted the concept of 'profitable production of useful crops' for dry areas. Since 'economy' is the heart of any enterprise including any production activity and as the term farming embodies this character the term should be '**dryland farming**'.

2.5. DRYLAND FARMING IN FUTURE

After a detailed discussion about 'dryland', 'farming' and the prevailing ideas of dryland farming the future dryland farming may be defined as **"the package of practices followed in dryland areas to establish and run a commercially viable farm"**. The farming activity will include any kind of biological production having commercial value, suitable for that area. Thus, the activity will vary from area to area depending upon the variable nature of rainfall and land. These two factors also have their own variable that influences the said activity profusely. However, for successful dryland farming attempt should be made to-

1. Conserve and store every raindrop at or around the place of its fall,
2. Utilise every drop of conserved water for farming purposes,
3. Do all sorts of cultivation work along the contour,
4. Disturb the land to a minimum level and select crops that need minimum land for cultivation in respect of the area under canopy cover. Creepers and climbers are good choice for this purpose,
5. Commercially viable deep-rooted perennial crops should be preferred to the shallow rooted perennial crops of shorter duration,
6. Select short duration seasonal crops over long duration seasonal crops to utilise short rainy-season,
7. Follow mixed or inter cropping system to assure more return in unit time and to avoid total loss due to moisture stress or any other reason.
8. Take up integrated and recyclic farming where both crops and animals should be used for production purpose,
9. Prepare separate plans for each farm or farmer involving fuel & fodder production,
10. Consider all possible community works associated with dryland farming through community and social development process.

Sustainability of Dryland Farming

Much of the good cultivable water-enriched land of the past has turned to non cultivable dryland of today by faulty cultivation practices. The examples of the same are scattered all over the globe, especially in and around the places where once the ancient civilizations bloomed. This mindless human activities has to be rectified through proper technology and their adequate application. Hence the hypothesis may be expressed as: If proper sustainable technology is successfully implemented in such a dryland with holistic approach, it may convert the land into non-dry land within a reasonable period. However, the care and attention have to be continued, otherwise the land will return to its dryland form.

Most of the methods used for dryland farming aims at preventing the soil erosion, checking the flow of run-off water and other similar things. In nature these processes are very slow, but human interference increases their speed manifold, particularly in the areas, which are today termed as dryland. The dryland technology, which deal with the items like water, soil, flora and fauna, should be selected judiciously. Slow but steady improvements of the farm environment, made through dryland farming process, are to be monitored and managed meticulously.

If the process of dryland farming is successfully executed over years, the ecological improvement of that area may stay permanently and provide sustainability towards farming. From this angle dryland farming may be redefined as **"the package of practices followed in dryland areas to establish and run a farm in a sustainable and commercially viable way."**

CHAPTER 3

BASIC INFORMATION REQUIRED FOR DRYLAND FARMING

3.1. AGROECOLOGY

The term has come into vogue only during the past two decades. Nonetheless, the concepts of 'agroecology', undefined, dates back to the stage of initiation of agriculture at the onset of human civilization, more than 10,000 years ago. It has roots in both traditional agriculture and ecology.

There are as many definitions of **ecology** as there are ecologists. As per Odum (1971) ecology is " the study of structure and function of nature". Krebs (1985) defined it as "the scientific study of the interactions that determine the distribution and abundance of organisms". Erlich and Roughgarden(1987) described ecology as "the study of the relationship between organisms and their physical and biological environments". However, today this subject includes the interaction between all ranges of organisms, from molecular to complex, at all levels of organisation or environment.

Hence, **agroecology** is the modified ecology that exists in agricultural production systems and surrounding natural resources required to sustain the system. We can explain it as a subject that deals with the interactions among organisms present in an agricultural setting and the interactions between these organisms and their environment. It includes all the

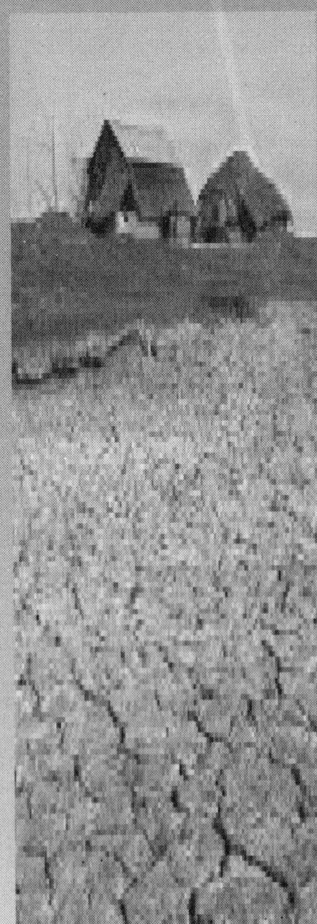

components commonly studied in natural ecological systems but within the well-defined boundaries of an agricultural field, which ranges from only a small (even microscopic) portion of a field to a large farm or entire region or world as a whole.

The main difference between natural-ecology (ecology) and agricultural-ecology (agroecology) is the regulation of these systems; the first one is regulated by nature itself and in case of the second it is regulated by the human requirement. In natural ecology nutrients that are removed from soil by plants are returned to that particular system through decaying of plants or its parts like fruits, leaves etc; otherwise through refuses of the creatures present in that system who feed on these plants. In agricultural ecology the plants remove nutrients from soil but the matured or immature plants or plant parts are taken away from the system for necessary requirements of human and livestock. This means that nutrients of the soil system are not cycled back in the ecological system, and for that matter special management is required to replenish the same and maintain a separate ecology to be necessarily called as agroecology.

Agroecological systems, as well as all other kinds of ecological systems, are made up of both biotic and abiotic components that relate with living and nonliving organisms respectively and termed as **biotic factors** and **abiotic factors** respectively. A third component named **socio-economic factor** is another major component of agroecology, which may not have that much importance for other types of ecological systems.

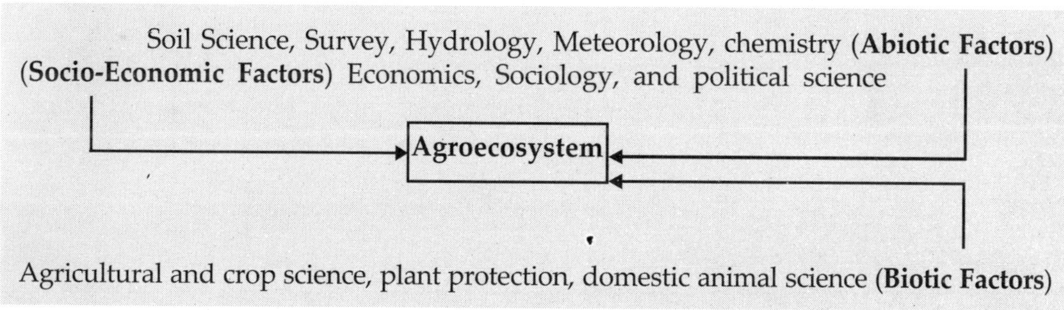

Management is one of the main characteristics of an agro-ecosystem that distinguishes it from natural system and which is predominantly controlled by the said third factor. Though the natural ecosystem is often affected directly or indirectly by human activity, it is not possible for them to manage it. On the other hand agricultural systems are directed by human activity and human choices. Human being decides what plants are to be grown in what area of a farm, when, how much and what types of nutrients, plant protection and other inputs including water has to be applied or replenished in that farm etc. While some of these human choices are influenced by nature, most are directly influenced by socio-economic and political factors.

Population growth is probably the most important defining force of agricultural development. To increase productivity forcefully, different farming strategies (use of hybrid seed, fertiliser, pesticide, fungicide, herbicide and water) are followed without any sustainable approach, as a result the natural ecosystem now-a-days is under serious threat (Table 3.1).

Table 3.1. Present Strategies for Increasing Productivity and their Ecological Consequences.

Present Farming Strategy	Natural Ecological Consequences
Intensive farming;	• Loss of wildlife habitat, • Soil erosion, • Water crisis, • Pollution,
Intensive animal production and goats;	• Desertification, • Denuded land,
Decoupling of animal and agriculture;	• Manure and crop waste are not recycled, • More use of fertiliser change the soil-ecosystem,
No fuel and fodder production;	• Deforestation, • Air pollution through burning more fossil fuel,
Use of plant protection and weed control chemicals;	• Toxic effect on human, livestock, fish, wildlife and all organisms present in natural ecosystem, • Weeds, pests and diseases act differently and more violently leading to long-term problems, • Increase toxicity level of surface and ground water,
Use of fertilisers ;	• Drinking water pollution (nitrates, phosphates), • Reduce inherent productivity of soil, • Disturb microbial ecosystem of soil and reduce their activity,
Irrigation ;	• Salinization,

As population continues to grow, new ways to increase productivity, while maintaining sustainability, must be found. Other-wise the system will be vulnerable to collapse. This new farm strategy should increase the natural enemies of pests, allow nutrients to be recycled and used more efficiently, increase the soil and water conservation capacity of the farm and finally increase the economic productivity. This new ways should lead to stability and sustainability, not only on small farms, but also on larger farms that are production oriented and directed by economics.

3.2. EFFECT OF GEOGRAPHIC FACTORS ON CLIMATIC PATTERN OF ANY DRYLAND AREA

Climatic patterns are determined primarily by geographic factors including *landform, latitude, altitude, distance from large bodies of water, direction and intensity of wind, and the ocean currents.* The type of climate is affected by even relatively small bodies of water through thermal buffering, creating sub-climatic types in the process.

Three geographic factors (altitude, latitude, and longitude) are important for crop production. These are embodied in the **Hopkins Bio-climatic law.** It states that crop production activities (e.g. planting, harvesting) and specific morphological developments are delayed 4 days for each 1^0 latitude, 5^0 of longitude, and 122 meters of altitude. This indicates that successful crop production requires a good understanding of climate and weather pattern of the area.

Large bodies of water or a mountain can modify the meteorological factors in a region. A mountain range in the path of prevailing wind creates moist conditions on the windward side and dry condition on the leeward side. As the air is forced to climb upwards along the mountain, the air gets compressed and cools adiabatically to the dew point and loses its moisture. Hence, it becomes a dry air mass once it goes accross the top of the mountain. For bigger mountain range the leeward side is called **'rain shadow area'.** Some times it is called **orographic effect**. In case of large water bodies it is just opposite and the windward side will be drier than the leeward side.

Topography and type of land surface (presence of rock, vegetation etc.) influences micro-climate and induce modification of temperature and humidity of the area. North facing slopes and south facing slopes may differ significantly in climate and vegetation. The north facing slopes and west facing slopes is relatively warmer in northern hemisphere.

3.3. HYDROLOGY CYCLE OF DRYLAND AREAS

It is the movement of water or water vapour in and around the soil surface and restricted in troposphere of atmosphere and lithosphere of the earth mass. There is a continuous and unending cycle of water between the earth and atmosphere. Its effect on life on earth may be compared with the blood circulation of our body. **The main driving force of this cycle is temperature which is the result of the sun-shine falling on earth.** The process of precipitation, water running into streams, rivers and oceans, infiltration of water into soils, recharge of soil water and ground water, interception of precipitation, evaporation of water from land and ocean, transpiration by vegetation, formation of snow, ice, fog, dew and cloud constitute the hydrology cycle. It is actually the circulation or movement of water in solid, liquid and vapour forms (Figure 3.1).

Before going in to details of Hydrology cycle we can have an idea of movement and position of total global water estimated by M. I. Lovich. Which is narrated below.

Portion of Hydrosphere	Water ($1000km^3$)	Percentage
1. Ocean-	1,370,323	93.96
2. Underground water (Unavailable)-	56,000	3.85
3. Underground water (In Zone of active water circulation)-	4,000	0.27
4. Glaciers-	24,000	1.56
5. Lakes including $5000km^3$ man made reservoirs-	280	0.019
6. Soil moisture-	85	0.006
7. Atmospheric vapour-	14	0.001
8. River water-	1.2	0.0001

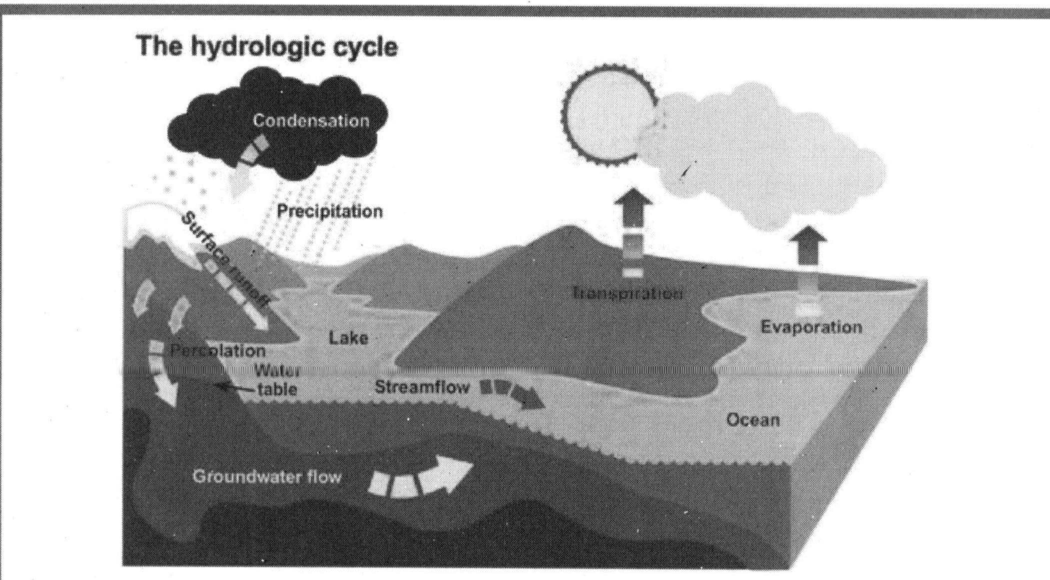

Fig. 3.1. Hydrology Cycle

Out of the total 1,454,703.2 '000km^3 only 525 '000km^3 of water participated in hydrology cycle as total annual precipitation and the same amount evaporates annually from the surface of entire globe.

To study the hydrology cycle, the role of (1) humidity, (2) precipitation, its relation with surface run-off and ground water recharge, (3) evaporation from water bodies and land surface, and (4) transpiration from vegetation, should be known. The general behaviour of these aspects with special reference to the dryland areas is narrated below.

Table 3.2. Annual status of chief components of hydrology cycle of Globe. (After M. I. Lvovich)

Chief components of hydrology cycle	Total volume of water precipitates/year (km^3)	Total depth of water precipitates/year (mm)	Percentage of the components (%)
A. In total land mass (144900 thousands km^2)			
1. Atmospheric precipitation	113500	1148	21.6
2. River runoff to ocean	41000	350	7.8
3. Evaporation & transpiration	72500	789	13.9
B. In total ocean area (361100 thousands km^2)			
1. Precipitation	411600	11400	78.4
2. River water inflow	41000	114	7.8
3. Evaporation	52600	1254	86.1

3.3.1. Air Humidity

Humidity refers to the water content of the atmosphere. Liquid water is converted into water vapour by evaporation for which necessary energy is provided by solar radiation in the form of temperature. Water vapour is colourless, odourless and tasteless and sometime consists of minute droplets of water suspended in the air. The amount of water vapour in the atmosphere depends upon wind and temperature. Wind distributes the water vapour in the atmosphere. Higher the temperature more is the amount of water vapour that can be held by the atmosphere. Commonly atmospheric air is a gaseous mixture of nitrogen (78%), oxygen (21%) and carbon dioxide (0.03%) with a few other gases including water vapour. Water vapour is most unstable among the gases present in the air. Its concentration is higher in lower most part of the atmosphere. The troposphere, which is the lower layer of the atmosphere, contains all the water vapour present in the atmosphere. Condensation of vapour to form clouds and subsequent precipitation is likely to occur in troposphere. In hot humid areas it is up to 4 percent by volume, whereas in areas of inclement climate it is up to 0.01 percent. In general, entire volume of water contained in the atmosphere is about 14,000 cubic km. In the dryland areas, in general, the humidity is very low due to very low moisture. The humidity of a locality depends entirely on the continuous evaporation of moisture from that area.

Status of Humidity : Humidity status of air is generally considered and measured through different ways in relation to temperature. These are narrated below:

(a) **Absolute Humidity (AH) :** It is the actual weight of water vapour per cubic meter of air and expressed as gram/m^3.

(b) **Relative humidity (RH):** It is the water vapour present in Percentage of maximum water vapour that can be retained by air at a given temperature. It is expressed as the ratio of water vapour actually present in the air and maximum water vapour that can be retained by that air at a given temperature. For example per cubic meter of air at 20°C can hold 17.3g water vapour in maximum (100%). So if it holds 13g/m^3, then the relative humidity is 75%. The relative humidity falls with increase in temperature and rises with decrease in temperature.

Conversion of absolute humidity to relative humidity can be done only when the temperature is known. For example 15g/m^3 at 17.5°C equals to 100% relative humidity while at 25°C it equals to 65% and at 30°C it equals to 50% relative humidity

(c) **Vapour Pressure Deficit (VPD):** VPD indicates 'drying effect' of the air. It is the relationship between vapour pressure of a particular gas, here the water vapour, and the normal air pressure (pressure of all gases in the air together) which is expressed as millibar or kilo-pascal (KPa). It normally ranges from 10 to 50 millibar or from 1 to 4 KPa.

(d) **Dew point:** It is the absolute humidity at a particular point of temperature below which the excess water vapour in air condenses. For example air with absolute humidity 10g/m^3 and 20 g/m^3 has dew point of 11°C and 22.5°C respectively.

Effect of Relative Humidity: RH 95% at 20°C means 16.5 gm/m^3 absolute humidity (see table). Relative humidity affects leaf area development and stomatal conductance thereby interfering with the photosynthesis and dry matter production (Jolliet, 1994). High levels of relative humidity can lead to yield loss.

Table 3.3. Comparison of Relative Humidity, Absolute Humidity and VPD.

Temp.	10°C		15°C		20°C		25°C		30°C	
RH (%)	AH (g/m^3)	VPD (kpa)	AH (g/m^3)	VPD (kpa)	AH (g/m^3)	VPD (kpa)	AH (g/m^3)	VPD (kpa)	AH (g/m^3)	VPD (kpa)
100	0.42	0	12.86	0	17.30	0	23.09	0	30.43	0
95	8.94	0.06	2.21	0.09	16.47	0.12	21.94	0.16	28.91	0.21
90	8.41	0.12	11.57	0.17	15.60	0.23	20.79	0.32	27.39	0.42
85	8.00	0.18	10.93	0.26	14.73	0.35	19.63	0.48	25.87	0.64
80	7.53	0.25	10.28	0.34	13.87	0.47	18.84	0.63	24.34	0.85
75	7.06	0.31	9.64	0.43	13.00	0.59	17.32	0.79	22.82	1.06
70	6.59	0.37	9.00	0.51	12.13	0.70	16.17	0.95	21.30	1.27
60	5.65	0.49	7.71	0.68	10.40	0.94	13.86	1.27	18.26	1.70
50	4.21	0.61	6.43	0.85	8.67	1.17	11.55	1.59	15.22	2.12
40	3.77	0.74	5.14	1.02	6.93	1.41	9.24	1.90	12.17	2.55
30	2.82	0.86	3.86	1.20	5.20	1.64	6.93	2.22	9.13	2.97

Measurement of humidity: Humidity can be measured in various ways viz. measurement of RH (in %), Absolute humidity (in g/m^3), moisture VPD (in Pa or KPa), and Dew point (in °C). RH (sometime dew point) is used to monitor infestation of disease and its control while VPD is the best indicator for transpiration. However, conversion from any form of humidity measurement to other is possible.

At present three types of sensors with different measuring principles are commonly used: (1) Dry and Wet bulb instrument psychrometer, (2) Electronic humidity sensor or capacitive sensor and (3) hair-hygrometer.

1. **Dry and wet bulb instrument.** A simple and yet ingenious method to determine air humidity is by an instrument that contains two thermometers placed side by side inside the greenhouse. One measures the normal air temperature and the other, continuously wet, measures the temperature at wet condition. The RH may be determined manually from the data collected from these two thermometers. For automatic data recording these two thermometers may be placed inside a specially designed apparatus that generates the data continually and send them to the computer for analyzing and providing the humidity situation.
2. **Electronic (capacitive) sensor.** Electronic sensors, mostly capacitive sensors produce a signal developed in the presence of water vapour in the air and transfers it to the computer. The computer converts the signal to Relative Humidity reading.

These sensors may not be reliable in high humid condition. After about 6 months it becomes inaccurate, and is required to be calibrated every month particularly in wet conditions.

3. **Hair Hygrometer.** Human hair becomes longer as humidity increases and shortens when becomes dry. This property of hair has been used to form this instrument, which is quite cheap and is available in convenient pocket size. Such instruments are recommended for use in temperature range of 5° to 35°C and relative humidity of 40 to 95%.

3.3.2. Atmospheric Precipitation

When the air gets in touch with the hot earth surface its temperature increases and, consequently, it ascends. As the mass of air goes up, it expands due to low pressure of the upper region and gets cooled down. As a result, the water vapour present in it condenses, which leads to the formation of water droplets or small ice crystals, thus making cloud or mist. When water droplets in a cloud coalesce togather, they become bigger and heavy enough, to fall on earth as precipitation.

Another type of precipitation results from condensation of water vapour, abundantly present in the air near the ground surface, due to lowering of temperature. The said condensation takes place directly on the earth's surface or on certain surface objects. This type of precipitation includes dew, rime, frost etc. Precipitation in the form of dew may produce average 10-30 mm of water annually, while rime in forest, under favourable conditions, may produce maximum of 35 mm annually.

The common forms of precipitation are rain, snow, drizzle and hail. In case of **rain** the size of water drops/droplets are larger than 0.5 mm in diameter. The maximum raindrop size is about 6mm and larger than 6mm tend to break up into smaller drops during their fall. **Drizzle** is a form of rain when numerous water droplets of size less than 0.5 mm with an intensity less than 1 mm/hour falls on ground surface. **Snow** consists of ice crystals that usually combine to form flakes. Fresh snowfall has an initial density varying from 0.06 to 0.15 g/cm^3, with an average value of 0.1g/cm^3 (i.e. 10 mm snow = 1 mm rain). In case of Hail or **hailstorm** the irregular pellets or lumps of frozen rain, of size greater than 8mm, occurrs as precipitation. This results when upward air currents are very strong, blowing the raindrops up to freezing heights. However, in dryland areas rain is the prime kind of precipitation, available for farming purposes.

Adequate data on rainfall of a dry area provide the basic information required for meaningful management of land, water and crop resources of that area. The important rainfall parameters are:-

1. **Rainfall amount.** it is the depth (mm. or cm.) at which rainwater stand on a horizontal surface without infiltration, runoff and evaporation for a specific rainstorm or for a specific period of observation like for a day, week, month or year.
2. **Rainfall Duration.** It is the period during which a particular rainfall occurs is known as duration of rainfall that is expressed in terms of seconds, minutes and hours.

3. **Rainfall Intensity.** The most important parameter that plays a great role in soil and water conservation. It is defined as the rate at which rainfall takes place, that is, the amount of rainfall occurring per unit time. It is expressed as mm/hr or cm/hr. Depending upon the intensity the rainfall is classified as light, moderate and heavy. Rainfall intensity trace to 2.5 mm/hr, 2.5 to 7.5 mm/hr and more than 7.5 mm/hr are generally termed as light, moderate and heavy respectively.
4. **Rainfall Frequency.** Frequency or return period or recurrence interval denotes the period in years during which a storm of a given duration and intensity can be expected to recur. It is nothing but an expectation, so better way to express this is by use of 'per cent chance', e.g. 50 years frequency has a 2% chance. It has an economic significance while performing soil and water conservation related works or projects. Depending on this data the life period of a project is fixed and the structures are designed considering the largest calamity that is expected to occur during that period.

In tropics the layer of cloud is up to 10 km thick. During the rainy season when the droplets from the upper region of the cloud, on their downward journey to earth, penetrate through this thick layer of clouds they coalesce with each other and grow bigger in size. These big drops finally descend on the surface of the land as heavy rain.

The precipitation that has an intensity of 0.5 to 1.0 mm / minute or higher is called very heavy rain or **shower** providing a lot of water in a short period. As the dryland areas have very little vegetative cover to break the speed of this rainfall or to absorb some of this water at least temporarily, such showers cause vigorous surface erosion and run-off. This run-off carries away most of the rainwater from this area, thereby leaving little scope to replenish the water vapour of the atmosphere, This drastically reduces the chance of formation of further cloud in that region. Naturally the duration of rain in that area is short and the gap between two rains is long.

3.3.3. Evaporation and Transpiration

Evaporation is a process of loosing water from a moist or wet surface to the atmosphere due to vapour pressure difference, which is of tapering nature from the ground level upwards. From the moist soil surface the water is lost at about the same rate as from a free water surface having the same exposure and temperature. Water is evaporated continuously from the surface of seas and oceans globally (specially from tropics and subtropics) and from the surface of river, lakes, moist soils, plant leaves, snow and ice locally. Evaporation of water creates water vapour that caters to the atmosphere laterally and vertically by the air current. Temperature of the atmosphere controls the conversion of water into water vapour.

Evaporation

Evaporation requires heat energy, which is normally available from sun shine. For annual global evaporation (525000 m^3 of water) about $12.6 x 10^{23}$ Joules of heat is required, which is equal to 25 % of the solar energy received by the earth annually. Evaporation of water is a complicated and time variable process, and its rate is affected by environmental factors like temperature, air flow or wind speed, humidity or Vapour pressure deficit (VPD), and atmospheric pressure. However, the amount of evaporation from water surface can be estimated from the following expression:

$$Q = k\ (d/P)S$$

Where, Q is the amount of water evaporated per unit time; k is the proportionality factor; d is the vapour pressure deficit; P is atmospheric pressure; and S is the area of evaporating surface.

However, **Penman's equation** is widely used and accepted formula to calculate evaporation rate, which is stated below:

$$E = \frac{\left(\frac{\Delta}{\gamma}\right)(Rn - G) + L\ Ex}{1 + \left(\frac{\Delta}{\gamma}\right)}$$

Where E = evaporation (mm/day);

Δ = relatinship (slope) between saturation (water) vapour pressure and temperature;

γ = psychometric constant (0.66 m bar/°C);

Rn = net radiation (cal / cm^{-2}/day);

G = soil heat flux (cal / cm^{-2}/day);

L = latent heat of evaporation of $0.1m^3$ (= 59 cal);

Ex = isothermal evaporation (mm/day);

Transpiration

Actually transpiration is a passive process of evaporating water from plant body through openings of stomata of the leaf. It is a vital physiological process of plant not only for growth and development but also for its survival. The water enters through root system and moves upwards within the plant as 'transpiration stream' throughout the day hours. Hence continuous evaporation of the said transpiration stream of water through stomata maintains the flow, which is the source of water for maintenance of turgidity and posture of the plant. This process or transpiration of water keeps the transpirating surface (leaves) cool, neutralizing the incident radiant energy, which would otherwise heat the tissue. Transpiration also categorically helps the plant to uptake plant nutrients and despatches the same to leaf through this upward transpiration stream of water.

Evapotranspiration

It is the total amount of water lost due to evaporation from soil surface and transpiration from crop/plant for a particular area during a specific time.

To measure evapotranspiration of crops, Doorenbos and Pruitt (1977) presented a modified version of Penman's equation. The modification known as the Penman-Monteith equation, is valid for a dry crop completely shading the ground and is as follows:

$$ET = \frac{\frac{\Delta (Rn - G)}{L} + \frac{c_p \rho_a (e_a - e_d)}{Lr_a}}{\Delta + \gamma^*}$$

Where, ET = Evapotranspiration from a dry crop surface,

Δ = relationship (slope) between saturation (water) vapour pressure and temperature;

Rn = net radiation,

G = soil heat flux,

L = latent heat of evaporation of $0.1m^3$,

c_p = specific heat of dry air at constant pressure,

ρ_a = density of moist air,

$e_a - e_d$ = vapour pressure deficit,

r_a = aerodynamic resistance of the crop, and

γ^* = modified psychometric constant = $\gamma \left(1 + \frac{r_c}{r_a}\right)$

where, r_c = canopy diffusion resistance of the crop.

Modified Blaney-Criddle method is another way to measure evapotranspiration. According to the basic formula ET_0 is the mean value of evapotranspiration over a specific period (mm), which is derived from the Blaney-Criddle factor (f). This factor f is determined from the mean daily maximum and minimum temperature (T) in °C over the month considered and the mean daily percentage of annual day time hours (p) for the given month and latitude, and the formula is;

$$\mathbf{f = p\ (0.46\ T + 8).}$$

Now, in order to obtain the evapotranspiration values of a particular crop, ET_0 values are multiplied by the crop coefficient (K_c or c) for that particular crop.

Hence the $\mathbf{ET_0 = c\ [\ p\ (0.46\ T + 8)]}$ **mm/day**,

Here $\mathbf{ET_0}$ is defined as the rate of evapotranspiration from an extended surface of 8 to 15 cm tall, green grass cover of uniform height, actively growing, completely shading the ground and not short of water. These figures (to determine f) are developed using values of day time minimum humidity (RH_{min}), ratio of actual to maximum possible sunshine hours (n/N) and day time wind conditions (u) at 2m height.

In arid regions, loss of water due to evaporation from the soil is relatively low due to low moisture content, the main component of loss being transpiration from the plants. Evaporation from plants is of two types. One is the common physical process of evaporation at the plant surface including the water that comes out through diffusion from the cuticle. The other is the physiological process by which the plants evaporate the excess water drawn from the soil by their root system while absorbing the nutrients. This process of water loss from the plant surface is called transpiration. About 99 per cent of water absorbed by the roots is lost to the atmosphere through transpiration. The loss is mainly from the stomata pores present on leaves.

The moisture released through transpiration from plant and evaporation from soil, though quite negligible in relation to the total moisture present in the atmosphere, it is sufficiently significant in respect of localized atmosphere of a dryland areas particularly to create microclimate.

3.3.4. Run-off

Surface Run-off

The atmospheric precipitation provides water for ground water recharge and surface run off. The surface run-off is made up of gravity driven flow of water moving over the earth's surface along the watershed or drainage slope from uplands towards low lands. This flow, initially as numerous small streams join together to form rivulets and rivers, which ultimately travel to the sea. In dryland areas these rivers are small and usually dry except during the short spell of rainy season when they are large and furious causing flash flood.

For determination of surface run-off, the value of flow of water through a given cross section of a channel or stream or river can be calculated with the help of the following basic formula.

$$Q = v.f.$$

Where, Q = run-off (m^3/sec)

v = Velocity of water flow (m/sec) and

f = Cross section area (m)

Run-off of a Specific Area is Defined by the Following Characteristics

(a) **Specific run-off**: It is the amount of water, which flows from 1 sq km of drainage area of a channel or stream or river basin per unit time. It may be stated as -

$$M = Q.10^3/F \text{ basin}$$

Where Q = Annual av. Discharge ($m.^3$ / sec) and

F basin = A drainage area of the basin (km^2)

If the value of M is converted to flow thickness (h) in millimetres then it will be **h =31.5 M.**

(b) **Average run-off:** It is the arithmetic mean value of a river run-off over a large period of observation (40 to 50 years).

(c) **Run-off coefficient**: It is the ratio between run-off over a certain period of time and the rate of atmospheric precipitation over the same period (usually per year). It is expressed as:

$$N = y/x$$

Where N = Run-off coefficient;

y = Run-off/year (mm);

x = Rainfall/year (mm).

There are number of methods used to estimate field runoff. *Rational method, Cooks method, Hydrologic soil cover complex number method* and *Table method* are suitable for prediction of design peak runoff for small watersheds. {For details of these methods see Chapter- 4.3.7.}

Runoff occurs only when the rate of precipitation, i.e. its intensity, exceeds the rate of infiltration. Once the infiltration rate is exceeded by precipitation a small water-head builds up on the soil surface and the water starts to move due to gravity. Initially the water begins to fill the small to micro surface depressions and then it starts flowing as overland flow. If this flow continues, the flowing water immediately create tiny to small unstable channel flows, which finally combines and generate a permanent course or flow path of runoff water. However, runoff in the field is affected by different factors which are stated below:

(1) Intensity and duration of rainfall:
(2) Soil and soil surface condition:
(3) Vegetative cover:
(4) Topography or slope of land surface:
(5) Conservation practices adapted:

By manipulating these factors the infiltration of water into underground, retention of water in and around ground surface can be increased and thus *quantities and rates of runoff* may be reduced significantly. Through this the ***'time of concentration'*** (longest time taken for runoff water to travel from any point in a catchment to its main outlet) may be increased, which provide more time for infiltration of runoff water thus reducing quantity and rate of runoff. *Time of concentration* is the time required by a drop of water to travel from the farthest point to the ultimate out-flow point of a particular catchment or watershed area.

There are many empirical formulae available to estimate 'time of concentration'. Many of them basically estimates the time of concentration for channel flow. Considering the dominance of overland flow Haan et. al. (1982) suggested the following equation on the basis of formula given by Kirpich (1940) to calculate the time of concentration.

$$t_c = 0.02L^{0.77}\, S^{-0.385} + \left[\frac{2L_0{}^n}{\sqrt{S_0}}\right]^{0.467}$$

Where, t_c = time of concentration (min),
L = length of channel reach (m),
S = av. Slope of channel reach (m/m),
L_0 = length of overland flow (m),
S_0 = slope of the land (m/m), and
n = manning type roughness coefficient,

('n' ranges from 0.02 to 0.8 for smooth land surface to dense grass/forest land)

The big catchments or watersheds are generally designed by the nature. In case of small or very small catchments or watersheds, the area can be designed artificially as per requirement of a farm particularly in dry land areas.

Ground Water Run-off

The free water found beneath the ground surface is referred to as **ground water.** A part of the precipitation, which escapes as run-off, generally gets absorbed by soil in the place of rainfall through the surface of flooded land, flowing rivulets, rivers and irrigation system to enrich the amount of groundwater and thus to raise the water table. This water also flows towards the lower static head. The rate of ground water run-off (mt./sec) is generally estimated by the following formula;

$$Qg = Qdn - Qup$$

Where Qg = Ground water run-off,

Qdn = Surface run-off or water discharge at the down stream station or lower static head zone of aquifer ,

Qup = Water discharge at the up stream station or higher static head zone of the same aquifer.

3.4. WATER BUDGET

The quantitative representation of the hydrologic cycle for a large particular area, like river basins or separate regions or the entire globe, is termed as 'water budget'.

Actually this is the quantification of available moving water or component of hydrology cycle of the considered region or globe for a specific period i.e. a year. This may be quantified for smaller periods like seasons or months and finally added to obtain the yearly figure.

The general 'water budget equation' for a closed river basin for an arbitrary period was proposed by M.A.Velikanov (1948):

$$x = y + z \pm u \pm w.$$

Where x is the precipitation over the area (mm); y is the runoff through the river (mm); z is the evaporation minus condensation (mm); u is positive or negative value of the net changes of volume of available water resource (like water level of river, lake, groundwater or height of snow cover etc.); w is the value of water exchange between the basin under consideration and the adjacent basins, which may be positive or negative.

This formula can only be used if (1) very little percolation loss towards groundwater; (2) the underground basin area is somewhat equal to that of surface drainage basin; (3) the influx of ground water from outside of the basin area is negligible and similarly out ward flow of groundwater from the basin is also negligible.

Water budget for an individual farm

This concept can be used to generate a formula to make a water-budget for a farm of specific size of an individual. This can be done annually, seasonally or monthly. The hypothetical formula for a farm may be as follows:

$$X = Y + Z + U + C - W$$

Where, X = Av. Precipitation (mm);

Y = Av. Surface runoff (mm);

Z = Av. Evaporation loss (mm);

U = Av. Storage of rainwater (mm) in farm pond and in the soil up to root zone depth;

C = Av. Water requirement of crop for transpiration (mm);

W = Uptake of groundwater from well;

To make a water-budget for a farm by utilizing this equation, it is required to assume that the deep percolation loss from the farm is negligible.

3.5 WATER ACCOUNT OF A FARM

To quantify actual water accounts of a farm, for a period of a year or a season, on the basis of the water budget equation we may derive the following equation:

$$x = a + p = y + z + u + c - w$$

Where, **x** = Total precipitation (m^3) for the specific period, a rain-gauge data;

a = Area of the farm (m^2);

p = Total rainfall of the period (m);

y = Total surface runoff of the farm (m^3). The amount of runoff may be determined by making bund or trench along the boundary of the farm with one or two outlet fitted with necessary measuring device.

z = Total evaporation (m^3) of the farm determined from pan-evaporation data and area of farm.

u = Total volume of rainwater storage capacity or stored water plus (+) amount of water utilized from these water harvesting structures (m^3);

c = Total water requirement of all the crops (area-wise) cultivated in the farm (m^3);

w = Total water, taken out from a well (dug / bore) or collected from outside (m^3) if any;

Here we assume that negligible amount of rainwater will be lost due to deep percolation and the evaporation loss from the water harvesting structure will be least. This may be a rough estimate but it is a good exercise to have a handy idea about the proper planning of a dryland farm. From this equation one can estimate the requirement of water for a particular purpose like crop cultivation, runoff, amount of water storage etc.

The methods of determination of these data (water requirement of crop, estimation of runoff/flowing water, planning of water harvesting structure, water holding capacity of soil, etc.) are available in the respective chapters or sub-chapters of this book.

3.6. SOIL

Soil is a complex poly-disperse system comprising of solid, liquid and gaseous phases. The *minerals* and *organic matters* in soil together constitute the solid phase, while water forms the liquid phase and the *soil air*, the gaseous phase. Mineral matters comprise the largest fraction of soil and exist in the form of particles of different sizes and shapes encompassing the void space called *soil pore space*. Organic matters derived from plant and animal residue at various stages of decomposition remain interspersed in soil minerals; pore space is occupied by air or water or both.

Plant roots exist and grow in pore space. *Soil water* is present in these pore spaces or arround the soil particles in the form of a thin layer, which directly control the amount of water and air present in the soil and thus indirectly controls plant growth and crop production. The *soil air* is totally expelled from soil where water is present in excess of total pore space. On the other hand water in liquid form may be almost absent in a soil which is termed as dry.

The pores of soil are classified into micro pores and macro pores depending on the diameter of the pore. Pores with a diameter of 2 to 20 micron are called micropores and those with a diameter of 200 micron to 0.3mm are called macro pores. When the water of macropores are emptied by gradual evaporation & percolation, the soil solution (plant nutrients dissolved in soil water) spills over from the micropores and absorbed by the roots. Water holding capacity of soil depends on the physical properties of soil like texture, structure and expressed differently by related characters like porosity, plastisity etc.

3.6.1. Soil Texture

Soil texture refers to the relative proportion of sand, silt and clay. Particles larger than sand are called gravel and stone. Generally, rock fragments larger than 2cm. in diameter are stones, materials between 2cm. and 2mm. are gravels, and the soil mineral matter smaller than 2mm in diameter is the fine earth constituted by sand, silt and clay and are called *soil separates*. The relative sizes of sand, silt and clay as proposed by the United States Department of Agriculture (USDA) and International Soil Science Society (ISSS) are given in the table (3.4) below.

Table 3.4. Size limits of Soil Separates.

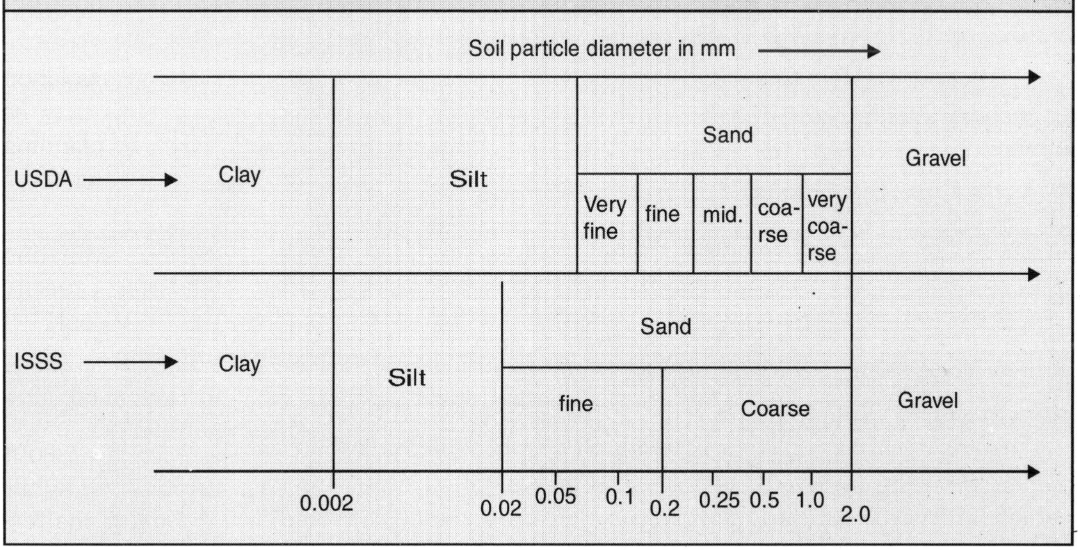

As per ISSS the sand particles of 0.2 to 2mm in diameter are classified as coarse sand and of 0.02 to 0.2 mm in diameter as fine sand. Because of the large size of sand particles macropores exist between them that facilitate free movement of air and water. The sand group includes all soils comprising 70 percent or more of sand fraction by weight. These soils have more number of macropores; hence their water holding capacity is very poor.

The size of clay particles is less than 0.002 mm in diameter. When the clay fraction is more than 50 percent, the soil is termed as clay soil. Due to very small size of the clay particles, these soils have greater number of micropores that can hold more water. Silt particles are medium in size ranging from 0.02 to 0.002 mm in diameter.

The loamy soil contains sand, silt and clay fraction almost in equal proportions. It has a good water holding capacity. A loam in which sand is dominant, is classified as sandy loam and in the same way a soil is classified as silt loam or clay loam in which either silt or clay is dominant (Figure: 3.2).

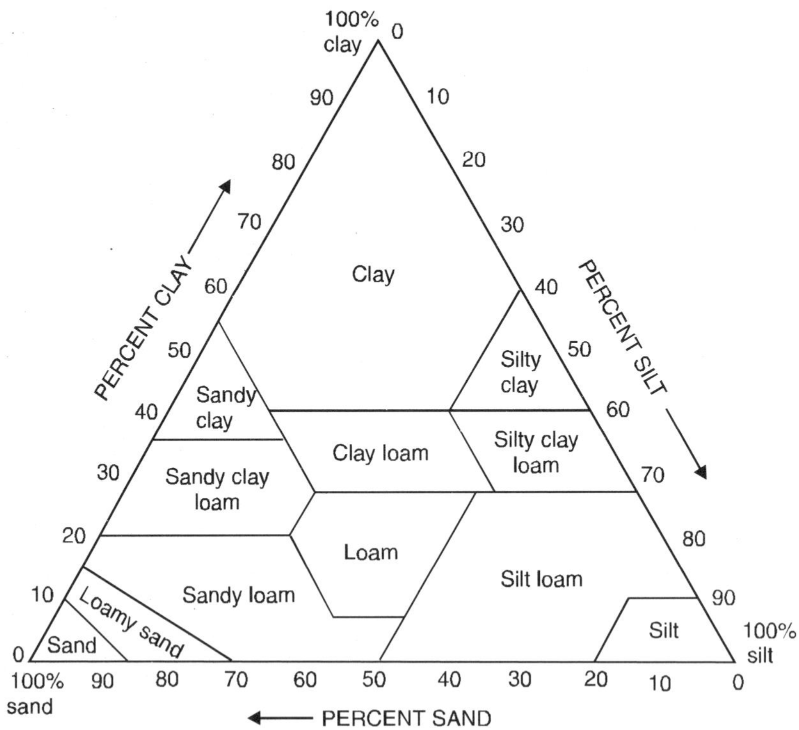

Fig. 3.2. Textural Classification of Soil

Plasticity of Soil. It is the nature of soil by which it tends to be sticky on wetting and become hard and cloddy when dry unless handled properly. This characteristic depends on the textural class of soil. Plasticity increases with increase of clay content which are non-flocculating in nature.

3.6.2. Porosity of Soil

Porosity, i.e. the percentage of pore space, is determined largely by the arrangement of the solid particles (sand, silt, clay, humus, etc.) of soil. The following formula may be used for calculating the porosity of a given soil.

$$\text{Porosity} = \frac{1-d_v}{d} \times 100$$

Where – d_v = bulk density or volumetric mass (mg/cm^3)

d = partical density or specific mass (mg/cm^3)

The porosity or total pore space of sandy and medium to fine textured surface soils generally ranges from 35 to 50 percent and 40 to 60 percent respectively. Presence of high organic matter and granulation increases the porosity more than 60 percent. Porosity varies with soil depth and in compact subsoil it sometimes drops to 25 to 30 percent.

The macro and micropore space, which sometime called as non-capillary and capillary pore space, have a separate and specific influence on porosity. The flow of water in soil and water holding capacity of soil generally increases with the increase of macropores and micropores respectively. Hence, due to high percentage of macropores the movement of air and water is quite high in sandy soils despite having low porosity. Therefore, the size of the pore space (macro or micropore space), rather than total pore space (porosity) is an important consideration for plant growth.

On agriculture point of view the best soil is the well-granulated silt loam that contains about 50 percent pore space, equally shared by both macro (for air) and micro (for water) pores and the rest 50 percent is soil solid of which 10 percent is generally organic matter.

Continuous cropping i.e. cultivation often reduce the macropore space of a soil. Depth of soil also affects the porosity of a soil. The following table (3.5) illustrates the same:

Table 3.5. Effect of cropping, soil depth and organic matter on pore space.

Sampling Depth (inch)	Soil Treatment	Organic Matter (%)	Macropores (%)	Micropores (%)	Porosity (%)
0 to 6	Virgin	5.6	32.7	25.6	58.3
	Cultivated	2.9	16.0	34.2	50.2
6 to 12	Virgin	4.2	27.0	29.1	56.1
	Cultivated	2.8	14.7	36.0	50.7

Source : W. D. Laws, and D. D. Evans (1949)

Structural Pore Space. In field soils the pore space in between secondary soil particles are normally created by plant roots and other soil fauna. This type of pores may be termed as structural pore space, which is highly unstable on wetting. These pores can be stable on wetting if some agent can hold the soil particles apart against the surface tension set up between particles as the soil wets. Clay, humus, soil conditioners, and action of different soil flora and fauna can help to stabilize the structural pore space on wetting.

Clay particles help to keep sand particles apart on wetting, provided they retain their individuality, that is, provided they are flocculated. The polysaccharides in the humus are the most active fraction in stabilising soil structure, thus stabilising the structural pore space.

3.6.3. Soil Structure

Under natural conditions, primary soil particles (sand, silt and clay) remain clustered with bonding agents as aggregates and those are referred to as secondary particles. The arrangement of primary and secondary particles into certain defined patterns or in closed packing is called soil structure. Soil structure is strictly a field term descriptive of the gross, over-all aggregation or arrangement of the primary soil separates. However the other aspect of soil structure is concerned with the size, shape and appearance of the individual lumps of soils, or peds, or crumbs, which form the body of the soil.

Platelike, prismlike, blocklike, and spheroidal are the four primary types of soil structure generally recognized. The last three have two subtypes each and the details of all are narrated below in tabular form:

Type	Subtype	Occurrence
Platelike	Platey - also leafy & flaky	May occur in any part of profile. At times Inherited from the soil material.
Prismlike	Prismatic - level tops Coulmnar - rounded tops	Both basically subsoil manifestations. Common in soils of arid and semiarid regions.
Blocklike	Blocky - cubelike Blocky - subangular	Common in heavy subsoils, particularly those of humid regions.
Spheroidal	Granular - porous Crumb - very porous	Characteristics of the furrow slice. Subject to wide and rapid changes.

Soil structure has a profound effect on soil properties like erodibility, porosity, hydraulic conductivity, infiltration and water holding capacity. Platey structures normally hinder free drainage. Crumb and granular structures provide the most favourable physical properties for plant growth. Common methods of soil structure management include (1) addition of organic matter, (2) adoption of suitable tillage, (3) soil conservation and (4) cropping practices. The quality of the soil structure may be expressed in terms of porosity, aggregation, cohesiveness or permeability for water or air.

3.6.4. Soil Aggregates

As per Martin et. al., (1955) soil aggregate is naturally occurring cluster of soil particles in which the forces holding the particles are much stronger than the forces between adjacent soil particles or their groups. This process may be called granulation. Good aggregation leads to good structure formation. To deal with the soil aggregation the factors that is responsible for aggregate formation and those which give the aggregates stability once they are formed are to be considered properly.

Genesis of aggregation or granulation is influenced by several factors such as (a) Wetting and drying, (b) freezing and thawing, (c) activity of roots and soil animals,

(d) organic matter, (e) decaying flora and fauna including microorganisms, and (f) physical disturbances like tillage and other intercultural operations. Any action that will develop line of weakness, shift the particles back and forth, that force contacts and encourages aggregation. The major reason that encourages formation of granular-type aggregates in surface soil is organic matter. Organic matter not only binds but also lightens and expands, making porosity a characteristic in individual soil aggregates. The electrochemical properties of humus and clay are probably effective in the organisation and the later facilitates stabilization of the aggregates. It may be assumed that granulation or aggregation is a highly biological aspect.

Stability of aggregates is of great practical importance. Some granules readily disintegrate by any simple action like beating of rain and few others resist disintegration. Apparently three major factors influence aggregate stability which are (1) the mechanical binding action of microorganisms like mycelia of fungi etc., (2) the cementing action of intermediate products of microbial synthesis and decay of gum like substances and certain polysaccharides which are sometimes called "pre-humus constituents", and (3) also the cementing action of more resistant stable humus components that provide most of the long-term aggregate stability. {For role of soil organic matter in soil aggregates see 5.4.1.5.}.

3.6.5. Mineralogical and Chemical Composition of Soil

The fragments of rocks as well as minerals often create coarsest sand particles. Quartz (SiO_2) usually dominates the finer grades of sand as well as silt. Various feldspars and micas frequently occur with it in addition. The sand grains are sometimes coated with gibbsite, hematite and limonite. The coarser clay fractions are composed of minerals such as quartz, hydrous oxides of iron and aluminium and the complex aluminosilicates. **Kaolinite, illite,** and **montmorillonite** are the three main mineral types, which alongwith a few others, generally occur in clay fraction. These groups vary markedly in the characters like plasticity, cohesion and adsorption, kaolinite being the lowest in each case and montmorillonite highest.

Since quartz (SiO_2) dominates, sand and silt are generally inactive chemically. Even the primary minerals which may contain nutrient elements in their chemical makeup are generally so insoluble that their nutrient supplying ability is almost nil. Chemically, kaolinites and montmorillonites are aluminium silicates but in some cases montmorillonite may contain sodium, iron, or magnesium in adition. Illite, often referred to as hydrous mica, is a potassium aluminium silicate. Its high potash content gives it a special nutrient significance. In well-weathered soils, especially those in hot, humid tropics, oxides of iron and aluminium are prominent if not dominant, even in the clay-sized fraction.

Reaction of Soil Solution. It is an important aspect of a soil that actually monitors growth of plant. Reaction of soil solution i.e. whether the solution is acid or alkaline or neutral, is usually expressed in terms of pH value. It is the comparative position of hydrogen ions (H^+) and hydroxyl ions (OH^-). The acid solution possesses a predominance of hydrogen over hydroxyl ions, the alkaline solutions shows the reverse situation and the concentration of both the ions are equal in neutral solutions. The **pH value** is the logarithm of the reciprocal of the hydrogen ion concentration.

For mineral soils the extreme range in pH value extends from 3.5 to 10 or above. The pH value 7 of a soil solution is the point where reaction of that solution is neutral. As it approaches to 3.5 the acidity of the solution increases and as the value moves towards 10 from 7 the alkalinity increases. Several essential elements tend to become less available as the pH is raised from 5 to 7.5 or 8. However, 6 to 7 is he best range of pH value for a soil solution because in this state it can provide almost all essential elements to the plants in available form.

3.6.6. Supply and Availability of Plant Nutrient in Soil

The study of soil from the standpoint of higher plants or crops has clearly identified certain elements as essential for growth of plants/crops. According to eminent German chemist Julius von Liebig, the growth of plant/crop is directly related to certain 'minerals' at a minimum level in soil and by the deficiency or absence of that necessary constituent, all others being present, the soil is rendered barren for all those plants/crops. The capacity of soils to supply the said essential elements is a fundamental aspact of plant growth.

Investigation and research shows that there are sixteen (16) elements which are necessary for normal plant growth. Out of which nine (9) elements are used in relatively large quantities and the rest are used in a very small quantity. The list of these elements are given as follows:

Required in Large Amounts

(a) Carbon, Hydrogen and Oxygen - —available in Air and Water;

(b) Nitrogen, Phosphorous, Potassium, Calcium, Magnesium and Sulfur —available in Soil Solids;

Required in Small amounts

(c) Iron, Manganese, Boron, Molybdenum, Copper, Zinc and Chlorine- —available in Soil Solids.

Essential Elements of the Soil. The group of six elements mentioned in 'b' is generally called **macronutrients** and the group mentioned in 'c' is called as **micronutrients**. Among the macronutrients nitrogen(N), phosphorus(P) and potassium(K) are commonly supplied to the soil either as manure or as commercial fertilizers and for that reason they are termed as ***primary*** or ***fertilizer elements***. In the same way, as calcium(Ca), magnesium(Mg), and sulfur(S) are not required to be supplied as fertilizer, they are referred to as ***secondary elements***.

The chemical composition of representative surface soils shows that nitrogen and phosphorus are almost always present in comparatively small amounts. Moreover, a large proportion of these elements at any one time are held in combinations unavailable to plants. For example, simple compounds of phosphorus are relatively insoluble in most soils. In contrast, total quantity of potassium is usually plentiful in soils except in sandy soils, but the main problem is its unavailability to plant.

The nutrient elements generally exist in two conditions: (a) complex and rather insoluble compounds, and (b) simple, more soluble forms readily available to higher plants. The latter forms are commonly present in the soil solution. The forms of macronutrients, especially the primary nutrients, occur in soil are as follows:

Nitrogen - Ammonium salts (NH_4^+), Nitrite salts (NO_2^-) & Nitrate salts (NO_3^-);

Phosphorus - Phosphate of Ca, K, Mg, etc & soluble organic forms {HPO_4^-, $H_2PO_4^{--}$};

Potassium - Potassium ions & potassium salts e.g. sulphates, carbonates etc.(K^+);

Fertility and Productivity of Soil. The term "fertility" refers to the inherent capacity of a soil to supply nutrients to plants/crops in adequate amounts and in suitable proportions. It has no relation with growth of plant or crop yield. The term "productivity" refers to the ability of a soil to produce crop. It is directly related with yield of crop. Productivity is a broader term since fertility is only one of a number of factors that determine the magnitude of crop yields.

3.6.7. Classification of Soil

Soil as a concept is evolved and nurtured on the basis of two basic ideas :-

1. The first group, the pedologist, considers the geological aspects of soils i.e. its origin, formation, description and classification. They consider the soil as a natural body and give less emphasis on its immediate practical utilisation. They classify soils as it occurs in their natural environment.
2. The other group, the edaphologist, consider biological aspects of soil i.e. its relation with growth of plant. They consider various properties of soil, which are related to plant production.

In dryland farming the dominant viewpoint will be that of the edaphologist. However, knowledge of pedology is required for general understanding of soils as they occur in nature and are classified.

Soil Profile. Soil profile is a vertical section of in-situ soil mass. The vertical section of soil column in most cases is composed of a series of distinct layers that are approximately parallel with the soil surface and are called soil horizons. Each soil is characterised by a given sequence of these horizons. Collectively the soil horizons make up the soil profile of a particular soil. So attention will be given to the major horizons making up soil profiles and the terminology used to describe them.

Three distinct horizons O, A, B and a fourth one C above bedrock are usually recognized in a common profile (Figure: 3.3). The uppermost O- horizon is the organic horizon formed above the mineral soil. This un-decomposed (O_1) or well-decomposed (O_2) organic matter layers are produced out of litters collected from plant and animals. A -horizon, having uniform and well-accepted characteristics, is the zone of maximum biological activity. This horizon lie at or near the surface characterised as maximum zones of leaching (eluviation) and is divided as A_1, A_2, etc. The B –horizon is the region of maximum accumulation of materials where illuviation (washed in) from above or even from below has taken place. The specific B-horizons are referred as B_1, B_2, etc. C-horizon is the unconsolidated parent material underlying the 'Solum' (A and B horizons) and is outside the zone of major biological activities.

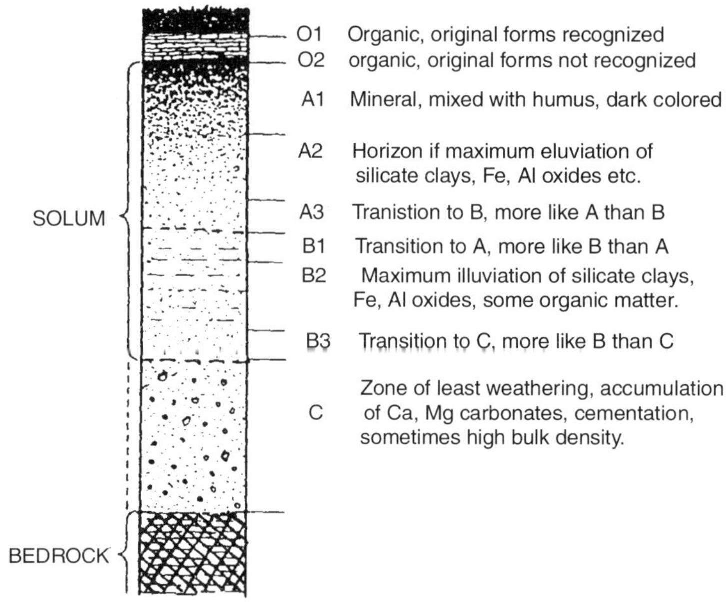

Fig. 3.3. Soil profile.

Classification. 'Individual specific soil' (a soil) and 'soil as a whole' (the soil) are the two extreme concepts which represent opposite ends to elaborate the idea of soil classification. In 1883, Russian soil scientist V. V. Dokuchaev and his associates classified soil in the concept of 'soil as a natural body'. Chernozem (black earth) and Podzol (under ash) are the common examples of that classification. With the same concept and with the able guidance of Mr. C. F. Marbut, the U. S. Department of Agriculture (USDA) developed a system of soil classification in the year 1938 and improved the same in 1949. This system also had a strong genetic base i.e., based on process of soil formation.

The presently used comprehensive system of soil classification is based on the properties of soils as they are found in the field. Published in 1960 by USDA it maintained natural body concept but is less dependent upon genetic influences. There are six categories of classification in the new system: *order* (the broadest category), *suborder, great group, sub group, family* and *series* (the most specific category).

In this soil classification system there are ten recognized orders of soil, which, along with their approximate equivalents in old classification systems, are as follows:

1. Entisol Azonal, some Low-Humic Gley soils
2. Vertisol Grumusols
3. Inceptisol Ando, Sol Brun Acide, some Brown forest, Low-Humic & Humic Gley soils
4. Aridisol Desert, Reddish Desert, Sierozem, Solonchak, some reddish brown soils
5. Mollisol Chestnut, Chernozem, Brunizem(Prairie), Rendzinas, some brown soils
6. Spodosol Podzols,Brown podzolic soils and groundwater podzols

7. Alfisol	Grey-Brown podzolic. Degraded Chernozems, associated planosol
8. Ultisol	Red-Yellow Podzolic soils, Associated planosols and Half-Bog soils
9. Oxisol	Lateritic soils, Latosols
10. Histosol	Bog-soils

This is a vast subject and all these ten orders have separate suborders, great groups and subgroups. Then depending upon the variations of soil properties and locality the corresponding family and series are classified and named.

3.7. SOIL WATER

Water may exist as solid, liquid or vapour in a soil, but the liquid form is most commonly found. In unsaturated soils, water occurs in vapour form in the soil air. Where only a small amount of water is present in soil, it exists between crystal lattices and on the external surface of soil particles as a very thin film, which is not available to plant. With addition of more water slowly, the thin film grows in thickness around the soil solid. The soil air present in pore spaces is slowly driven out making space for water to occupy. A soil is fully saturated when the entire pore spaces, macro and micro, are filled with water excepting some entrapped air occupying some small space in the pore space. The water content of soil of which all the soil pores are filled with water is referred to as **maximum retentive capacity**. At this point the thickness of the water around soil particles is increased, water molecules of the water surface are held less tightly. Finally they are pulled down into more deeper layers by gravitational forces.

In a very simple way, for plants, the soil acts as a water tank of finite capacity. At field capacity the tank is full, at saturation it is overflowing, while at permanent wilting point (PWP) it is empty (in fact at PWP the soil still contains an appreciable amount of water that is unavailable to plants).

The moisture content of soil after drainage of gravitational water is called ***field capacity*** (FC). It is the upper limit of soil water available to plants. The soil moisture tension at field capacity varies from soil to soil but it generally ranges from 1/10 to 1/3 atmospheres (atm.) At field capacity, the micropores are filled with water, and almost all the macro-pores are devoid of water, so any further drainage due to gravitational force is very slow.

Solar energy removes soil water through evaporation. Plants also exhaust this water by transpiration. The soil dries gradually and ultimately plants can no longer obtain enough moisture to meet transpiration requirements and remain wilted unless water is added to the soil. The soil moisture content at which plant wilts and does not regain its turgidity unless water is added to the soil is known as ***permanent wilting point*** (PWP). Soil moisture tension at PWP is generally taken as 15 atm, though it may range from 7 to 32 atm. It is the lower limit of ***available water*** in soil for plants. Therefore, the available water is the water held in soil between FC and PWP. When the soil water content falls below PWP and is held at a tension of 15 atmospheres and above, it is considered as ***unavailable water***. Similarly when the soil water is above the FC and is held at a tension between zero to 1/3rd atmosphere, this water is also unavailable to most of the plants. However, depending upon the textural class the presence of soil-water differs in different stages (Table 3.6).

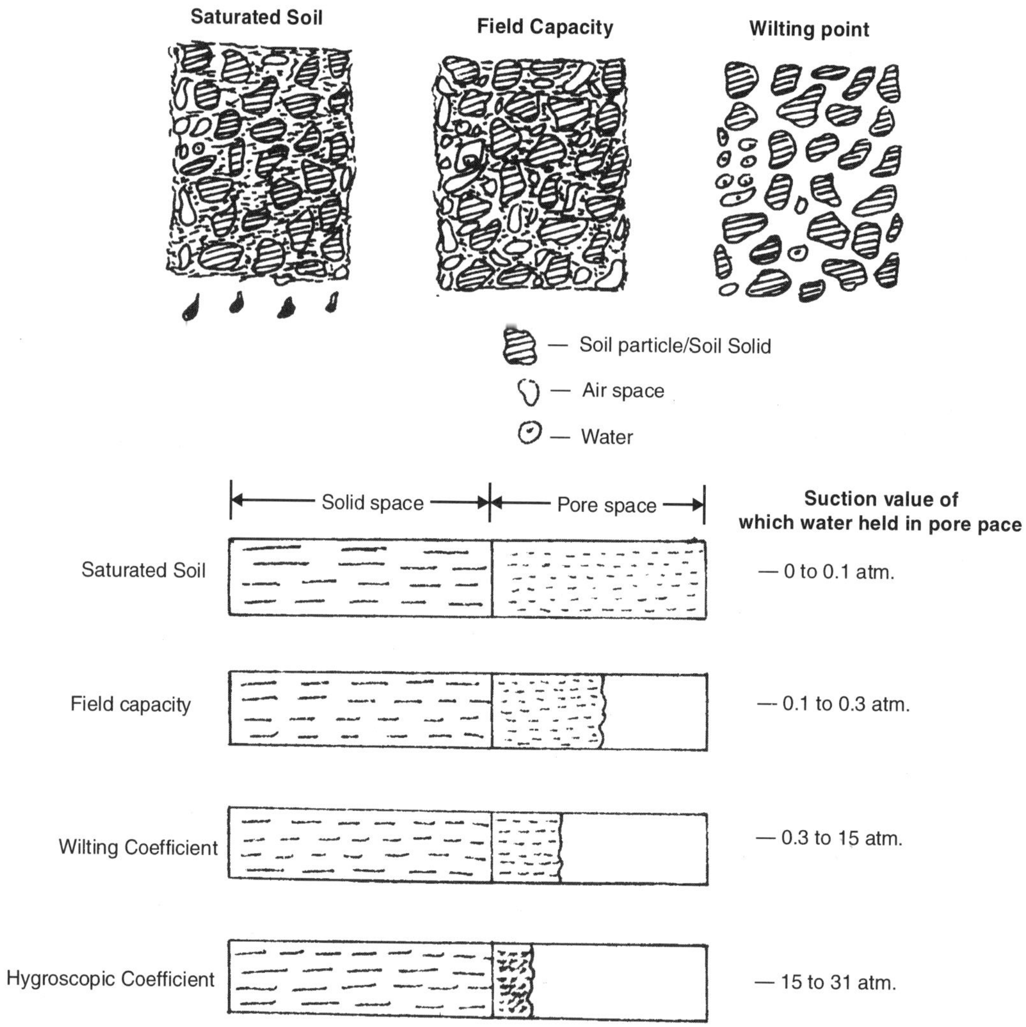

Fig. 3.4. Physical and Volumetric Position of Soil Water in a well granulated Soil. (Modified from USDA publication)

Table 3.6. Range of moisture (%) in different type of soils on dry weight basis.

Textural class of soil	Field capacity	Permanent wilting point	Available water
Sand	6–12	2–6	4–6
Sandy loam	10–18	4–8	6–10
Loam	18–26	8–12	10–14
Silt loam	23–31	11–15	12–16
Clay loam	27–35	13–17	14–18
Clay	31–39	15–19	16–20

Further drying of soil leaves only a very thin film of water around soil particles. Water in the soil is then mainly in the form of vapour. Due to hygroscopic nature of soil particles, water is held very tightly. ***Hygroscopic-coefficient*** of water is defined as the amount of water that the soil contains when brought to equilibrium with air of one standard atmosphere with 98 percent relative humidity at room temperature or 25 degree centigrade. At this point, the water is at a tension of 31atmospheres and is referred to as ***hygroscopic water.*** This water is not available to plant, but may be available to certain bacteria. A soil rich in silt, clay fractions, organic and colloidal matters, has a high hygroscopic coefficient.

With increasing supply of water, the water film held around soil particles thickens and occupies the pore spaces. Due to force of gravity, water from macropores is removed. The water retained by soil at this stage or after cessation of the downward movement is the ***capillary water.*** This water is held at a tension of 1/3 to 31 atmospheres and much of it is in the liquid state. Capillary water remains between FC and hygroscopic coefficient. The capillary water supplies the largest part of the water needed by plants. However, the whole of capillary water is not usually available to plants. This water also provides an upward movement through soil by capillary force created by the surface tension. Naturally, thicker the capillary space less is the height of upward

Table 3.7. Guideline of Determining soil moisture by Judging appearance and feeling of soil.

Available soil moisture remaining	Light texture	Medium texture	Heavy texture
0 to 25 per cent	Dry, loose, flows through fingers.	Powdery, dry, some times slightly crusted but easily broken down into powdery condition.	Hard, baked, cracked, sometimes has loose crumbs on surface.
25 to 50 per cent	Appears to be dry, will not form a ball	Somewhat crumby but holds together from pressure.	Somewhat pliable, forms ball under pressure*.
50 to 75 per cent	Tends to ball under pressure but seldom holds together.	Forms a ball some what plastic, will some-times stick slightly with pressure.	Forms a ball, ribbons out between thumb and forefinger.
75 per cent to field capacity (100 per cent)	Forms weak ball, breaks easily will not stick	Forms a ball, is very pliable, sticks readily if relatively high in clay.	Easily ribbons out between fingers has.
At field capacity	Upon squeezing no free water appears on soil but wet outline of ball is left on hand.	Upon squeezing no free water appears on soil but we outline is left on hand.	Upon squeezing, no free water appears on soil but wet outline of ball is left on hand.
Saturated	Water appears on ball and hand.	Water appears on ball and hand.	Water appears on ball and hands.

(*Ball is formed by squeezing a handful of soil very firmly)

movement and vice versa. In case of plain water the height (h) of upward movement may be expressed by the formula-

$$h = 0.15/r$$

Where, r = radius of the capillary space.

In the dryland area with sandy or coarse soil, it can't provide much water through capillary, for plant growth, from below the root zone area. In dryland areas sometimes the soils are of sandy type i.e. contain more of larger particles. Naturally its macropores are more in comparison and consequently the water present in the micropores of that type of soil is less. The water applied in this type of soil makes downward movement fast and goes out of root zone quickly. The quantity of this water is also quite high, the lateral movement being much less.

An idea about soil moisture is essential to farm a land. There are many methods for measurement or determine the soil moisture, some of them are laboratory methods and some are in-situ methods. At the same time some are accurate and some are approximate. However, **appearance and feel of the soil** is the oldest method of determining soil moisture, present in the top layer of the soil, and this type of assessment requires sufficient experience. Though there is a guideline of assessing soil moisture by visual observation and hand feelings but, the accuracy of approximation improves with experience.

In relation to plant growth, the quantity of available water in this type of soil is also very meagre. In sandy to sandy-loam soil, the available water is between 8% to 15% when the moisture content ranges between 1.5 to 2.8 inches/foot depth of soil. Similarly, the wilting point is also low. Whereas, in clay loam, the wilting point starts at a higher water content, than that of sandy loam, because more number of finer particles of clay loam soil retain more water which is unavailable to plants.

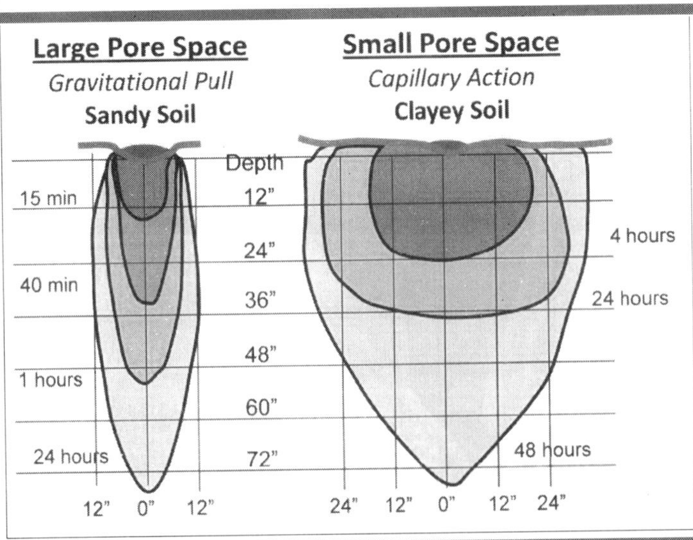

Fig. 3.5. Pattern of downward movement of water with time into a sandy loam and a clay loam soil.

3.7.1. Measurement of Soil Moisture/Water

The soil moisture content is commonly expressed as the weight percentage i.e. grams of water associated with 100 grams of dry soil. Moisture content is also sometimes expressed in volume percentage. It is the volume of soil water as a percentage of the volume of the soil sample.

Methods. In the laboratory soil moisture content in samples can be determined by methods like (a) gravimetric method, (b) rapid moisture meter using calcium carbide, (c) soil moisture gauge. In field condition or in-situ moisture content can be estimated by using (a) resistance blocks, (b) Tensiometers (c) neutron moisture meter, (d) Indicator plant etc.

Out of these methods **gravimetric method** is the one most commonly used, where a sample of moist soil from the field, usually taken in the form of a core, is dried in an oven at a temperature of 100 to 110° C and weighed again. The moisture lost by heating represents the soil moisture in the field sample. This method is the most accurate, but is not practical for farm use.

The **resistance blocks** i.e. the porous blocks made up of gypsum or nylon or fibreglass measures the electrical resistance of these blocks in different moisture regime. The resistance in the blocks is determined by the moisture content and in turn by the tension or suction of water in the nearby soil. The relationship between the resistance reading and the soil moisture percentage can be determined by calibration. This method gives reasonably accurate moisture readings over the range of 1 to 15 atmospheres suction.

Tensiometers measure the tension with which water is held in soils. Here the water in the **tensiometer** equilibrates through a porous cup with adjacent soil water and the suction in the soil is same as the suction in the tentiometer. The water moves out or in the cup depending upon the tension of the soil water decrease or increase respectively. A curve for each soil is needed to determine the moisture percentage. Their range of usefulness is between 0 to 0.8 bar suction and is used for determining the need for irrigation. It can measure soil moisture of different depths by placing porous cup fitted at the end of the tensiometer at desired depths.

The **Neutron Moisture Meter** is another instrument to determine soil moisture which has become popular in recent times due to its convenience in using and accuracy. It contains scalar or rate meter and a probe. The probe contains a fast neutron source, which emit γ-radiation, and a detector of slow neutrons. This probe unit is placed in a metal access tube that is installed in the soil. The fast neutrons are emitted radially in to the soil and collide elastically with soil containing water and slowed down in the process. The detector counts the slow neutrons and transmits information to the scalar or meter placed outside. The volumetric water content of the soil can be determined by the count rate of calibrated instruments of that particular soil. However, it is not suitable for changing water front in a soil and very near ground surface.

Since long past **indicator plants** are used to determine position of soil moisture present in root zone and it helps to determine the time of life saving irrigation. To determine severe water stress condition and moderate water stress condition *Croton sp.* and sunflower plants or seedlings are used respectively. Depending upon the requirements

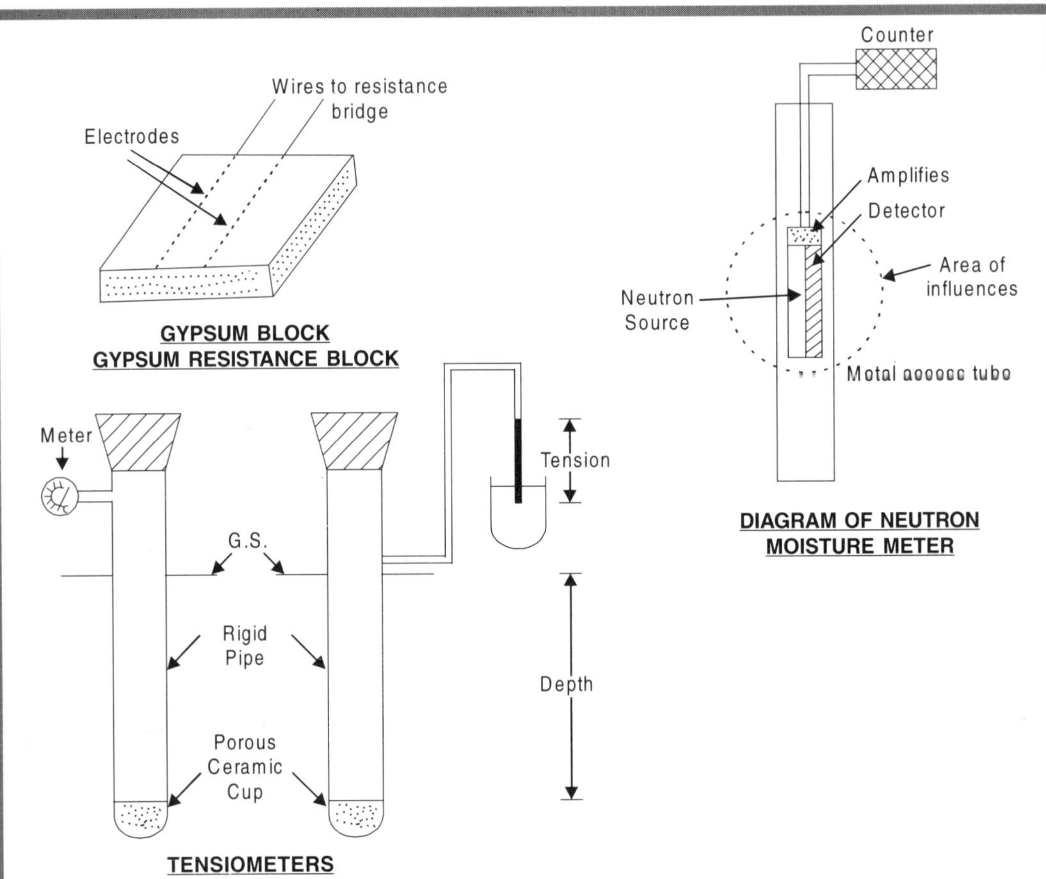

Fig. 3.6. Diagram of different kinds of instruments to measure in-situ moisture.

of crops the selection of indicator plants is done. Few of these indicator plants are planted alongside the crops through out the field, and when most of them shows symptoms of temporary wilting, irrigation is required to be supplied to the crop.

3.8. GROUND WATER

The study of the vertical distribution of water in the subsurface geologic formations reveals a clear picture of the occurrence of the ground water below soil surface. The earth's crust, which is called the lithosphere, is composed predominantly of rocks. The outer part of the earth's crust is normally porous to a greater or lesser depth. The pores or openings in this portion of the lithosphere may be partially or completely filled with water. The upper stratum, where the openings are partly filled with water, is called the ***zone of aeration*** and is divided into three belts viz. (1) Soil water belt, (2) The intermediate belt, (3) The capillary fringe belt.

Immediately below this zone of aeration, where all the openings are completely filled with water, is called the ***zone of saturation.***

After complete saturation of any soil either by rainwater or by irrigation water or by the water of streams, river and any kind of water body, the excess water percolates through soil macro or larger pore spaces to the deeper zone of soil by gravitational force. That entire zone of percolated water is generally called underground hydrosphere. When the water collected from different sources touches soil surface, it infiltrates into soil. A portion of the percolated water is absorbed in the zone of aeration, which starts from the earth's surface and extends up to the ground water table. The rest amount is added into the ground water, leading to the rise of the water table. Though the materials in the earth's crust present below the water table are generally in a saturated state, only a few of them have got sufficient porosity and permeability as well as discharge capacity so as to yield sufficient quantities of ground water. Here the yield refers to specific yield i.e. the volume of water release per unit of the strata per unit change in the component of head.

Actually the permeability and capacity to yield ground water from earth's crust, not only depends on the percentage of pore space, but also size of the pores. The underground earth's crust having more clay and silt particles have more total pore spaces than coarser one but due to more micropores they yield less in comparison to coarser particles. On the other hand a coarse material such as sand contains larger pores, through which water can move easily giving rise to more yield. Therefore, presence of macropores in the water bearing strata is the main factor for getting satisfactory yield of ground water. Depending upon the geologic terrains and character of the earth's crusts and rocks in respect of porosity and permeability the underground strata is characterized as either water bearing or water impermeable one. Actually the flow of water through soil strata i.e. ***hydraulic conductivity*** depends upon the material present in those strata. The hydraulic conductivity of different materials is given below.

Table 3.8. Hydraulic Conductivity of Different Materials of Soil

Materials	Hydraulic Conductivity (meter/day)
Limestone	0.0000004
Sandstone	0.00012
Clay	0.00004
Silt	0.4 to 0.01
Very fine Sand	0,4 to 4
Fine Sand	4 to 40
Medium Sand	40 to 180
Coarse Sand	180 to 260
Very coarse sand	260 to 320
Very fine gravel	320 to 450
Fine gravel	450 to 640
Medium gravel	640 to 900
Coarse gravel	900 to 1200
Very coarse gravel	1200 to 1600

Water bearing strata are the, underground layers which can be expected to yield sufficient ground water. These layers are composed of (1) pebble gravels, (2) gravels, (3) sands, (4) unconsolidated or loosely consolidated conglomerates and sand stones, (5) silt-stone, (6) lime stone, (7) fractured rocks of metamorphic or igneous origin. These water bearing strata are present below the ground surface, from which groundwater can be obtained for beneficial use, are called ***Aquifers***. The water impermeable strata are those, which hardly allow water through or are absolutely incapable of letting it through and yielding it under natural conditions. The compositions of these types of strata are (1) heavy clay to heavy loam, (2) marls, (3) tight and non-fractured rocks of igneous origin and few metamorphic rocks. These impermeable strata are called ***Aquitards***.

3.8.1. Aquifers

The underground strata, from which it is feasible to obtain water is called aquifer. In order to qualify as aquifer, a stratum should not only have a substantial percentage of pore space, but these pore spaces must be large enough to permit the water to move through them towards the suction pressure or high pressure zone. Impervious layer generally floors these strata. It is a relatively continuous (both laterally and vertically) layer saturated with free gravitational water, which alone represents a hydrodynamic entity.

Types of Aquifers. According to geologic setting and hydrologic pressure the aquifers are divided into three types, (1) Un-confined aquifers, (2) Confined un-pressured aquifers and (3) Confined pressured aquifers or '*Artesian*'.

Un-confined Aquifer. It is the zone of water bearing stratum extended from open water table or the bottom line of zone of aeration to any impermeable layer below it. It is very much unstable depending upon the supply of water through soil from rainwater, stream water, or any kind of water present in the surface. This supply is highly, in fact completely dependent on season. This unstability of unconfined aquifer is generally expressed through fluctuation of water table. The heavy drawing of ground water and ground water flow towards the low-pressure zone and subsequent addition of percolated water creates the said fluctuation. Generally in rainy season it is on higher side which gradually goes down to the lowest position towards the end of the summer season.

Vedose/Perched Aquifer. It is a unique diversion of common un-confined aquifer, having all characters of un-confined aquifer but very small and temporary in nature. This situation can occur where an impervious stratum within the zone of aeration intercepts percolation and causes ground water to accumulate in a limited area above that stratum. It is found in the zone of aeration over a non-continuous impermeable stratum and carries a limited amount of water bearing strata. This water wholly depends on the volume of infiltrating rainwater and loss is through evaporation only. This water may be utilized in dry areas if located properly.

Confined Un-pressured Aquifer. It is the zone of water-bearing strata which is confined between two impermeable layers where it has its own water level below the upper impermeable layer or stratum. Actually this water level will be static if these aquifers are tapped from the surface by tube wells. Its hydrodynamic behaviour in and

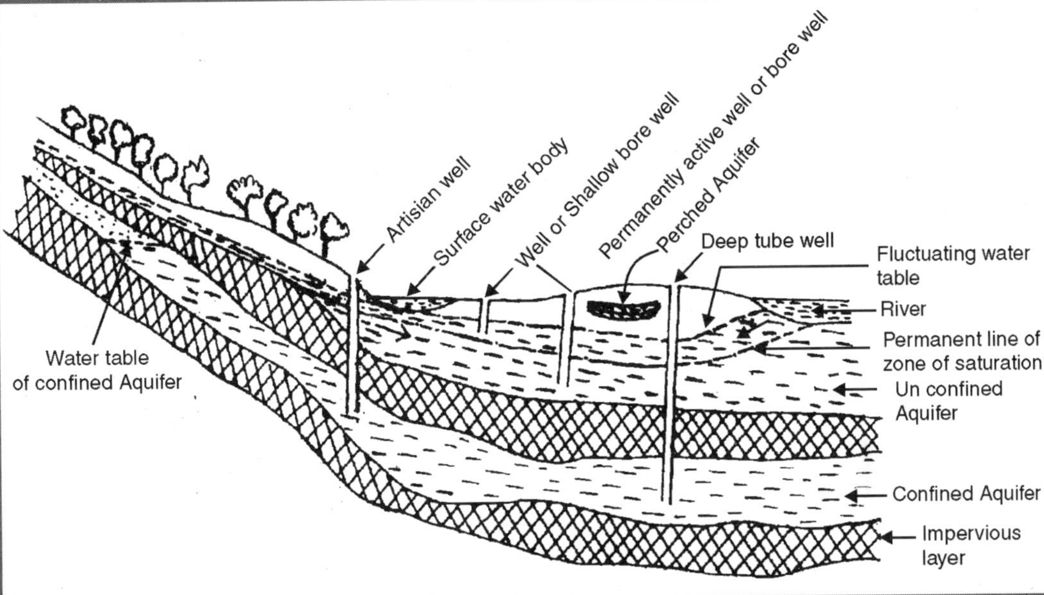

Fig. 3.7. Different Types of Aquifers and Wells related to such Aquifers.

around tube well is more or less same as the un-confined aquifer. The nature and site of recharge is different than that of un-confined aquifers.

Confined Pressured Aquifer. It is the zone of water bearing strata which is also confined between two impervious layers where the water is present in a compressed state due to hydraulic pressure. Because of the presence of upper confining layer, the water of the aquifer is not open to atmospheric pressure. It thus occurs within the pores of the aquifer at pressure greater than one atmosphere and thus the water in the well stands above the top of the aquifer. If a well is drilled into such aquifers then the water would generally rise above the top level of the aquifer to its own water level or piezometric level. This height of water represents the pressure head of the aquifer. Sometimes this height or piezometric level exceeds the ground surface, then after completion of drilling the water flows out automatically above the ground surface with some pressure. This type of well is called **Artesian well.**

Flow of Aquifers. The water present in the aquifers generally flows downward by gravitational force or towards the low pressure or high suction zone. The slope of bottom layer of a aquifer controls the flow in general. If there is any depression in that layer then the underground water may be stored in that depression and is available for use over a prolonged period.

In respect of direction and quantity the gravitational flow of water in aquifer may be influenced by (1) recharge, (2) position of impermeable layer, (3) position of stream or river or any water body and (4) impervious barriers (if any) in the aquifer.

After knowing fully well the above items and managing a few of them properly the outflow of ground water may be controlled. On the other hand, the efficient management of withdrawing ground water may also reduce the fluctuation of water table.

3.8.2. Recharge of Ground Water or Aquifer

The process of addition of water into the ground water or aquifer is generally called recharge.

1. **Recharge From Rainwater.** Though only a small portion of the annual precipitation percolates down to the aquifer but this represents the main source of ground water recharge. It occurs while rain water is absorbed by ground surface, or flows through any waterways or it is stored in tanks, ponds and natural depressions. There are also direct and indirect methods of ground water recharge.

   First the surface soil is soaked by rainwater and gradually percolates downward till it reaches the water table, if it is raining continuously. The amount of recharge or downward percolation to ground water reservoir depends upon the intensity, duration and distribution of rainfall, topography and vegetation cover or roughness of ground surface, and temperature and humidity of atmosphere. Sometimes recharge takes place by the transported rain water flowing through rivers, streams, springs etc. The total amount of annual recharge in India comes to 67 million ha meter.

2. **Recharge From Aquifers.** The natural flow of ground water from higher hydraulic pressure areas to low pressure areas is a very common natural recharge process. Sometime the water from a confined aquifer may directly or indirectly recharge an unconfined aquifer.

3. **Artificial recharge.** Artificial recharge may be required for augmenting the quantity of ground water, disposal of excess water, and for reduction of salt water intrusion. It is the recharge of ground water using rainfall through the effort of man. Sometime the run-off water is retained or stored or its flow is barred temporarily to provide extra time for infiltration or percolation downwards. This is generally done by taking advantage of topography, vegetative cover, and creation of structures like, dams, bunds, percolation tanks, trenches, terraces etc. Pumping of fresh water back into the tubewells (Walton 1970) may do it.

3.8.3. Discharge or Loss of Ground Water

It is the gradual depletion of ground water resulting in the fall of water table. This happens due to various reasons of which the important ones are as follows.

1. **Evapotranspiration.** It is the combined loss of ground water by evaporation from the surface when the water comes up from the aquifer through capillary action and by transpiration through the leaves and other green parts of crops/plants.

2. **Natural Discharge.** It is the flow of water from aquifer towards the down side or low-pressure zone of the aquifer. The loss of ground water through leakage of bottom impervious layer to other aquifer is also a kind of natural discharge.

3. **Artificial Discharge.** It is predominantly the drawing out of the ground water from the aquifer through different kinds of wells. This discharge or yield

depend on human need and recharge of aquifer depend on rainfall. Satish Chandra & Saksena (1975) obtained the following values of this parameter for the river Ganges.

(Ramganga Doab)

	Monsoon period June'71 to Oct'71	Non-Monsoon period Nov'71 to May '72
(a) Recharge		
(I) Recharge due to Rain fall	418	NIL
(II) Natural Recharge	221	51
(III) Artificial Recharge	11	33
(IV) Groundwater in-flow	5	7
	655	91
(b) Discharge		
(I) Evapotranspiration	166	106
(II) Natural Discharge	97	107
(III) Artificial Discharge	40	143
(IV) Ground water outflow	NIL	NIL
(V) Ground water leakage	6	8
	309	364
(c) Change in storage	(+)346	(-)273

3.8.4. Investigation of ground water

For dryland farming a clear knowledge/information about occurrence and movement of ground water is essential. There are several methods of ground water exploration, which are divided in two group e.g. **surface method** and **sub-surface method**.

[A] Surface Methods

1. Geological field reconnaissance. It understands the geology of the area in relation to the possible occurrence of ground water. A study of structural geology in conjunction with stratigraphy to locate possible thickness of water bearing formation and confining beds is required. Gravel, sand, sand stone and lime stone are best water bearing strata and clay, shales and crystalline rocks does not yield enough water. For detail information professional geologists may be consulted.

2. Geophysical method. Sometimes it is possible to estimate the groundwater occurrence by using the geophysical methods. Among these methods (a) electrical resistivity and (b) seismic refraction methods are used for ground water investigation. A brief discussion on these two methods is narrated below:

(a) Electrical resistivity method: The procedure involves measuring the potential difference between two electrodes (potential) resulting from an applied current through two other electrodes (current) outside but in the line with the potential electrodes. The current electrodes consist of metal rods driven into the ground. The potential electrodes are porous cups filled with saturated solution of copper sulphate ($CuSo_4$) to prevent electrical fields forming around them. When the spacing between electrodes is increased, a deeper penetration of the electrical field occurs and a different apparent resistivity is obtained. The apparent resistivity is plotted against electrode spaced with different spacing at the same location and as a result a smooth curve is obtained. Now comparing the character of this curve the nature of the strata and also the depth of the water table can be obtained.

(b) Seismic Refraction Method: - In this method, a small wave is created at the earth's surface, either by using a dynamic charge or by the impact of a heavy instrument. Then the time required for the sound or shock wave to travel a known distance is measured. These measurements are interpreted to determine the nature of substrata. Seismometers or geophones or detectors are used to record the shock waves and convert the vibrations into electrical impulses, which are subsequently recorded on an oscillograph along with time of initial explosion.

Good results are obtained up to 75 to 100 meters, even though, results are reported up to 700 meter and in this method the depth of water table is calculated by using the following formula.

$$D = \frac{S}{2}\sqrt{\frac{V_2 - V_1}{V_2 + V_1}}$$

Where, D = depth to the water table,

S = intercept point in time-distance graph,

V_1 = velocity of wave in the upper layer, and

V_2 = velocity of the aquifer,

[B] Sub-surface Method

Subsurface Geophysical Method: - In this method test drillings (trial bores) are conducted for the purpose of obtaining geological and other information relating to water bearing strata or ground water status. This test drilling is the drilling of small diameter bores up to a certain depth in different location of a farm that provides the information about depth & yield of water bearing strata, and status of ground water level. Persisting with this test drilling it may be utilized as observation well to record the fluctuation of ground water and to conduct pumping tests.

Geophysical measurements from surface to the bottom of the bore will provide a continuous log in an accurate and convenient way to obtain many sub-surface data. There are several methods of well logging such as: (1) Resistivity logs, (2) Spontaneous potential logs, (3) Radioactive logs, (4) Acoustic logs, (5) Temperature logs, and (6) Calliper logs.

3.9. SOIL EROSION

Soil erosion is a mechanism by which the soil particles are detached from the place of their origin and are transported from one place to another through an external agency such as wind or water in motion.

Erosion gradually removes the topsoil of an area. *Water erosion,* caused by rain water, carries down this soil from its place along the slope, forming sheet, rill or gully. Removal of soil particles by high velocity wind is called *wind erosion.*

The major factors affecting soil erosion are the precipitation, wind velocity, soil type, topography, and most importantly the vegetation. Out of these factors, intensity and amount of rainfall is the prime factor that influences soil erosion. However, presence and type of vegetation inversely influences soil erosion. Commonly with the increase of rainfall, soil erosion increases steadily up to 750mm as in low rainfall area presence of natural vegetation is comparatively less. With bare soil the soil erosion increases with increase of rainfall but, with increase of natural vegetation the soil erosion reduces. If the rainfall exceeds 1500 mm, the dense vegetation occurs naturally and thus, the soil erosion due to rainfall reduces radically. This is illustrated in the following figure.

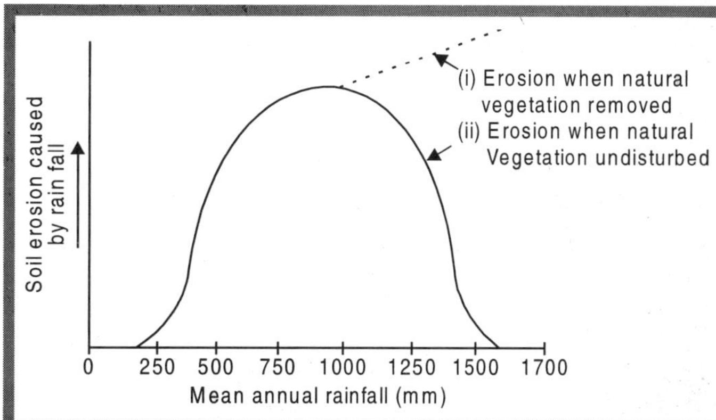

Fig. 3.8. Relation between rainfall & soil erosion in two extreme vegetation situation (from Hudson 1971)

Erosivity and Erodibility. Soil erosion occurs when any erosive agent, particularly the rainfall, acts upon soil. The amount of erosion, depends upon the combination of erosion causing ability of the erosive agent and ability of the land/soil to withstand erosion. Thus erosion is the combined function of erosivity of rainfall and erodibility of soil.

Erosivity of rain can be defined as the potential ability of the rain to cause erosion, which depends on the characteristics (type, intensity, duration, and amount) of rainfall. Technically the kinetic energy of rain is chiefly responsible for the erosivity. The rain has nearly 256 times more kinetic energy than that of runoff. Erosivity of rainfall can be measured by using the rainfall information. However, it can be measured by using experimental data or rainfall simulator.

Using rainfall and soil loss information from experimental plots, Weischmeier et. al.,(1978) suggested the erosivity index (**EL_{30} Index**) for a individual storm having 30 minute maximum intensity of rain during that storm. This can be calculated by dividing maximum amount of rainfall collected in a 30 minute period of that storm to get maximum intensity in mm/hour. There is another system to evaluate erosivity index (**KE > 25 Index**), which is based on low intensity of rainfall (non-erosive rain). It determines a threshold value of rainfall intensity (established as 25mm/hr or 1 inch/hr) at which rain becomes erosive. This method is reported to be not very accurate under tropical & sub-tropical conditions.

On the other hand, **erodibility of soil** depends upon the physical characteristics of land (shape, slope, land use practice, and vegetation/crop) and soil (structure, texture, and organic matter content). Bouyouces, in 1935 suggested that erodibility of soil depends upon its mechanical composition of soil and its proportional ratio as **(% sand + % silt) / % clay.**

3.9.1. Rain Splash and Detachment of Soil

Impact of raindrops has three important influences- (1) it detaches soil, (2) its beating tends to destroy granulation and (3) its splash, under certain conditions, causes appreciable transportation of soil. The rain splash is an important detachment agent. Single raindrops, with the kinetic energy (KE) at the time of fall on the bare soil, lifts the soil particles present at the point of impact, and throw them over a certain distance. This kinetic energy may be expressed as

$$KE = \frac{1}{2} mv^2$$

Where, m = Mass of raindrop (kg),

V = Velocity of raindrop (m/sec).

A continuous splash of raindrops over a specific area detaches large number of particles and makes them loose enough to be carried by the erosion agents. So great is the force exerted by the dashing rain that granules not only get loosened and detached, they may be beaten to pieces also. On a soil of easy detachment type, a torrential rain may splash out as much as 250 tons of soil/ha, some of the drops raising particles as high as 60cm and throwing them horizontally perhaps 120 or150 cm away.

Sometimes raindrops, intercepted by the canopy of trees, coalesce into big drops. These big drops fall from the leaves and hit the ground surface with their potential energy (PE) that detaches the soil particles at the place of hitting. Potential energy is measured by the following formula:

$$PE = m\ h\ g.$$

Where m = mass of the water drop (kg);

h = height difference (m);

g = gravity(m/see);

3.9.2. Sheet or Overland Erosion

When the high intensity of rainfall prolongs and the rate of rainwater fallen on the ground surface exceeds the infiltration as well as moisture holding capacity of the soil the water starts accumulating on the soil surface. With the development of its head or due to land slope, the accumulated water begins to flow. The flow is rarely in the form of a sheet of water of uniform depth and more commonly is a mass of braided watercourses with no pronounced channels. Sometimes large stones and pebbles or vegetation breaks up the flow, often swirling around tufts of grass and small shrubs.

During runoff the loosened soil particles are also transported downwards. The velocity of the flow is commonly expressed by *Manning's equation.* The hydraulic characteristics of a flow generally described by its Reynolds number (Re), which is an index of turbulence of flow. The greater the turbulence, the greater is the erosive power generated by the flow.

Sheet erosion carries down only the detached soil particles, more or less uniformly from every part of the slope.

3.9.3. Rill Erosion

When the sheet or overland flow gains further momentum along the slope, the silt-laden water moves along well-defined miniature channels that spread out in a finger-like pattern over the entire field. Such thin channelling is called rill-erosion. Rills may also start at the bottom, where a small cut is first formed and then rapidly extends headwords up the slope as a channel.

However, rills are usually of temporary nature and often get obliterated by the next storm of sufficient intensity. The fresh channels are of different forms and come up at positions unrelated to the previous rills. Most rills are generally discontinuous and have no permanent connection with the main outlet. However, occasionally a master rill develops and makes a permanent outlet to the stream or river directly or through gully.

Studies (Moss et al., 1982) on the break up of shallow overland flow into small channels or micro rills show that in addition to the main flow path down slope, secondary flow paths develop with a lateral component where these converge to cut small channels or trenches. As expected, the greater erosive power of concentrated overland flow of rill erosion is responsible for the bulk of the sediment removed from the uplands.

3.9.4. Gully Erosion

Once it was thought that gullies are developed from enlarged rills, but studies revealed that their initiation is a more complex process. Several of the small depressions formed initially coalesce to make an incipient channel later. In gully the detachment of soil particles and the erosion are mostly associated with scouring at the base which results in deepening of the channel and undermining the head wall leading to collapse and retreat of the channel up slope (Ologe, 1972). Sediment is also produced further down the gully by stream bank erosion. The reasons that favour gully formation are- (1) high intensity rainfall creating heavy run-off and (2) the lack of vegetation cover due to low average rainfall or heavy deforestation or over grazing. If the velocity or destructive force of the run-off exceeds a critical or threshold value, gullies will occur.

However, gullies are relatively permanent watercourses, which experience ephemeral flows during rainstorm. Unlike river, a head cut and various steps or knick-points along their course characterize gullies. They have relatively greater depth and smaller width, carry large sediment loads and display erratic behaviour depending on rainstorm. Gulling frequently-transforms a cultivable land to a non-cultivable one within the time in which they are stabilizing themselves into that field.

3.9.5. Wind Erosion

The main factor of the wind erosion is the velocity of moving air and the relative roughness of the ground surface. Higher velocity and lesser roughness will cause more erosion. The entrainment of soil particle by wind is effected by the application of a sufficiently large shear force and by the bombardment of the soil by grains already in motion. The movement of soil particles takes place in three stages – (1) Saltation; (2) Surface creep and (3) Suspension.

Saltation: Saltation is the process of movement of soil particles in a series of jumps. The movement is explained by considering the theory of the 'Bernoulli Effect'. Due to wind velocity and its lifting force the medium sized grain (0.1 to 0.5 mm in diameter) rises vertically and then falls with a trajectory of 6 to 12 degree. On striking the ground surface the impact energy of a saltating grain is distributed into a disruptive part that causes disintegrations of the soil particles and lanches them into the air. Generally 55 to 72 percent of the soil lost by a wind erosion is carried in saltation.

Surface creep: It is the rolling of detached coarse grains of soil particles (larger than 0.5 to 3 mm in diameter) along the ground surface obviously driven by wind velocity. About 5 to 25 % of the soil lost by wind erosion occurs by surface creep.

Suspension: The process of suspension describes the movement of fine particles, usually less then 0.1 mm diameter, high in the air and over a long distance, even several thousands of kilometres. The dust comes down to the earth only with rain or when wind velocity is decreased.

To understand the mechanism of soil erosion is vital for proper designing of suitable system of erosion control. Few additional information is necessary for the successful planning of soil conservation work such as status of the affected area, severity of erosion and frequency at which each process operates.

3.9.6. Universal Soil Loss Equation

Out of several efforts, to estimate long time average soil loss from a specific land in a specific cropping or any other vegetative system, the model developed by Weischmeier and Smith has adaptability to a wide range of conditions. This model is termed as Universal Soil Loss Equation and is expressed as –

$$A = R\ K\ L\ S\ C\ P$$

Where, A = average soil loss for a given period (commonly a year) per unit area,

R = rainfall erosivity index (number of rainfall erosion index units for a particular location).

K = soil erodibility factor (soil loss rate per erosion index unit for land of fixed length & slope).

L = slope length factor (ratio of soil loss figures of actual length to that of 21.13m or 72.6 ft).

S = slope steepness factor (ratio of soil loss figures of actual slope to that of 9 % slope).

C = crop factor (ratio of soil loss figures of the land with specific crop cover and management practice to that of tilled non-cropped or fallow land of identical area).

P = soil conservation practice factor (ratio of soil loss figures of the land with specific soil conservation practice like contouring, terracing, strip-cropping etc. to that of conventional farming practices in a land of identical area).

3.10. SURVEY

Survey of a farm or an area is essential for designing the farming activities including planning of soil and water conservation measures. It provides detail knowledge of topography, soil and water situation of the farm. The survey report of a farm is the only instrument by which one could translate the ideas of recommendable technologies properly, particularly the soil and water conservation methods like contouring, terracing, water harvesting and finally the crop selection.

Survey report of a farm generally consists of contour map, soil profile and soil testing data and ground water report. The study should be made step by step as given below:

First Step. It is the preparation of ***contour map***-of a farm. Contour map is a paper where the lines of same elevation, with a specific elevation difference, are plotted inside a boundaried area. The first job is to locate a reference point in the field, which can be used as the basis of the survey operations. The general practice is to establish a grid

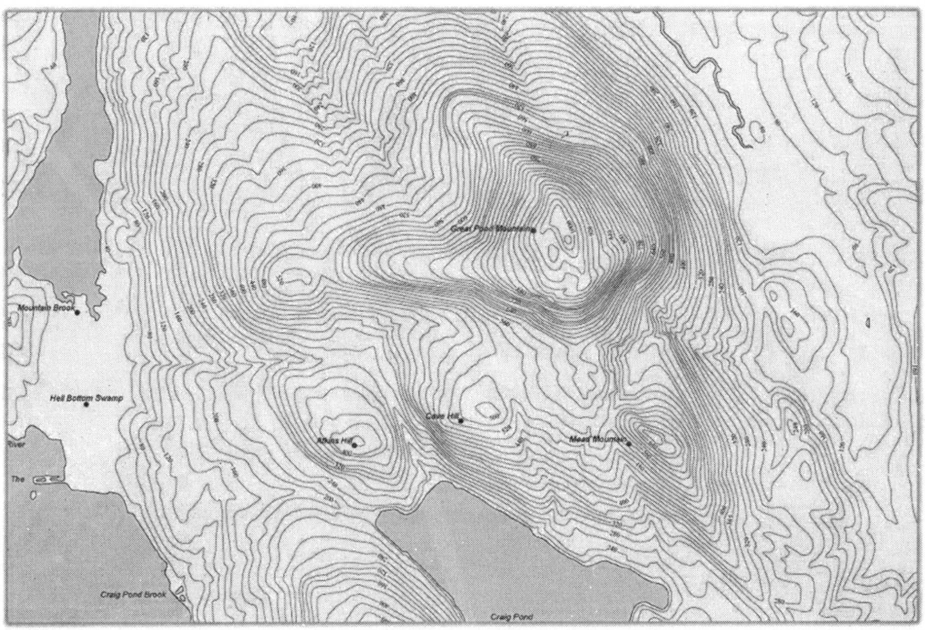

Fig. 3.9. Contour Map.

system over the field and set stakes at the grid points. Next the heights of all the grid points (against the reference point) are to be plotted to prepare the contour map. The usual grid spacing is 30 m, 25 m, 20 m, and 15 m depending upon the nature of surface relief of the area and precision required for the survey. On the basis of said grid points a contour map is to be drawn (Figure: 3.7). Through this map one should make a clear idea about the position of ground surface in respect of ups and downs as well as the slope. On that basis the land development structures like bunds etc., water-harvesting structures and waterways are to be located and designed.

To create contour lines and graded contour lines at farm level the common tools like measuring tape, 2 to 5 metres long *measuring rod, plumb bob, carpenter level, poles & pegs* is required. Home-made equipment namely *N-Frame Level* or *flexible tube level* etc. may be used to determine the contour lines. Farmers can do the job of their own after a little training and practice. {For more detail see the Chapter- 4.3.3.}

Second Step. It is the preparation of **soil profile** ***report***. The information about position, depth and condition of A-horizon is an essential part of farm designing or planning activity particularly in dryland areas. At the same time the position of bedrock or C-horizon (if any) is also an important information for dryland farming.

Third Step. It is the preparation of ***soil testing*** report of different sub-groups of A-horizon or of average root zone depth. The soil testing report should consist of physical and chemical properties of the soil in details including the status of its water holding capacity. The physical property of soil means its groups or class in respect of structure and texture {it is discussed in Chapter-3.6.}. The chemical properties provide us the information or status of available plant nutrients (N P K), organic matter, cation exchange capacity (CEC) etc. These data will determine the future course of action about nutritional management of crops. Both the physical and chemical tests are done in the laboratory after collection of proper samples from the field {the detail of soil testing is discussed in Chapter – 5.2.2.2.}.

Fourth Step. It is the collection of data or information about ***ground water***. It is tough to prepare a realistic report of ground water in dryland areas because of the wide fluctuation of water table or its seasonal existence. This job is also an expensive one. However, test boring in dry season or sinking of pizometers along the slope can provide lot of information about ground water and its fluctuation in a particular area or a farm. On the basis of these information pointing of ground water source for positioning of well or other structures is to be determined.

Out of the common ground water survey methods Test boring or Test drilling is the most reliable one, because it can explore the ground water resources of the area and provide information on the geologic and ground water condition of the site.

The other methods are geophysical in nature. They provide indirect evidence of the presence of possible aquifers through the study of porosity and permeability of subsurface layers. The geophysical methods may be either surface type or bore hole type depending on whether the measurements are taken at the surface or in drilled hole. The commonly used geophysical methods of ground water prospecting are the **Electrical Resistively method** and **Seismic Refraction method**. The seismic refraction method involves creating seismic waves at the surface (sometimes in boreholes) by any sources (called 'shot') like a hammer or accelerated weight drops or tiny explosives etc. The seismic waves travel through the subsurface at a velocity dependent on the density of the soil / rock. The calculation or analysis made by this method is based on the measurement of the portion of the wave front returned to the surface. This provides information on the depth and character of the refractor present below the ground. Only the earliest-arriving reflected wave energy, called "first-breaks" is needed in seismic refraction data analysis. The data are processed and interpreted using a software package.

3.10.1. Remote Sensing for Survey

The satellite-based remote sensing technology is a powerful tool for scientists, planners, and technologists.

What is Remote Sensing. It is the acquisition of the spectral reflectance measurements made from a distant location. Various features of earth surface exhibit typical spectral reflectance patterns, which form a basis for obtaining information on various resources in the earth like vegetation, water, undulation of land surface etc. Basic and applied researches have constantly led to improvements in the existing capability to acquire, process, analyse and interpret the remotely sensed data.

In spectral response of plants, pigmentation dominates in the visible wavelengths of 1.55 —2.80 m. The spectral reflectance of soil is generally less complex in appearance than those of vegetation. The reflectance of soil unit generally increases with increasing wave lengths, particularly in the visible and near infrared portion of the spectrum. The reflectance of soil is influenced by various properties of soil viz. colour, organic matter and iron oxides, texture, structure, surface roughness and wetness.

Remotely sensed data are analysed in analogue or digital format for extracting geological features. In analog format the hard copy photography prints (black and white or colour) of remotely sensed data are visually interpreted to identify and classify different features based on image characteristics such as size, shape, tone, texture, location, association, shadow and aspect. The digital data are analysed and interpreted using special image processing softwares.

The resource maps required for planning are 1:5000000 scale at the national level, 1:50000 scale at the district level and 1:8000 scale at the specific area level.

Use of Remote Sensing Method. Through remote sensing one can gather detail information about ; **soil-** its texture, structure, type, depth, organic matter, moisture etc., **land-** its topography, vegetation, status of degradation, etc., **crop-** its distribution, corresponding area, health, assessment of yield etc. and **water-** its availability and resources like surface, sub-surface and underground. Remote sensing can also help to assess the drought situation of a particular area.

3.10.2. Air-photo interpretation

Air-photo interpretation is the process of studying and gathering required information from photo taken overhead to identify the various cultural and natural features. With careful interpretation, air photos are excellent source of spatial data for studying the Earth's environment. Actually air-photo interpretation is a skill that can only be developed with practice.

Basically this is a unique remote sensing tool taken from Aero-plane maintaining a fixed flight altitude in order to determine its scale in the following manner:

$$\text{Scale of air-photo} = \frac{\text{Altitude of flight}}{\text{Focal length of the lens}}$$

Example: If the flying altitude of the plane is 25000 ft and focal length of the camera is 0.5 ft, then the scale of air-photo will be 1.50000 (Generally these flight-altitude and focal length is available in the margin of any standard air-photo)

The following factors are assessed when trying to identify a feature:

Shape: the form of an object on an air photo helps to identify the object. Regular uniform shapes often indicate a human involvement;

Pattern: similar to shape, the spatial arrangement of objects (e.g. row crops vs. pasture) is also useful to identify an object and its usage;

Size: a measure of the object's surface area (e.g. single-lane vs. multi-lane highways);

Tone/Colour: the colour characteristics of an object, relative to other objects in the photo, are used to identify the feature (e.g. sand has a bright tone, while water usually has a dark tone; tree species can be determined by the colour of their leaves at certain times of the year);

Shadow: a shadow provides information about the object's height, shape, and orientation (e.g. tree species);

Texture: the physical characteristics of an object will change the way they appear on a photo (e.g. calm water has a smooth texture; a forest canopy has a rough texture);

Association/Site: associating the presence of one object with another, or relating it to its environment, can help identify the object (e.g. industrial buildings often have access to railway sidings; nuclear power plants are often located beside large bodies of water).

Stereo perspective

Seeing an area in stereo, or 3-D, is important for determining the topographical relief of an area, as well as the height of objects such as trees and building. Stereoscopic imagery is the result of overlap, which is the amount by which one photograph includes an area covered by a neighboring photograph.

Air photo coverage is generally designed to provide about 60 percent forward overlap between photographs. This allows stereoscopic, or 3D, viewing when the two overlapping photos are used with a stereoscope. In addition, from 20 to 40 percent lateral (side) overlap is allowed when complete coverage of an area is required. For mapping, inventory and vegetation studies, for example, a survey is flown in a series of to-and-from parallel strips with side overlaps between strips over the entire area.

For non-stereoscopic coverage, used in crop sampling or pollution detection, the photographer may choose a 20 percent forward overlap.

Measuring Areas on Maps and Images

Usual print size of air-photo is 23 × 23 cm^2 (9 × 9 sq. inch). The working area of an air-photo is that covered by a single print after omitting overlap with adjacent runs and with next but one print in a run, i.e. area on which boundaries are drawn on alternate prints.

Conventional air-photos always contain scale and distortion errors like optical errors, relief errors and camera tilt errors. In hilly areas, since the scale varies across the photo, terrain elevation should be obtained from a topographic map in which the elevation of the principal point should be measured so the corrected scale is Focal length/ Flight altitude– Terrain elevation. Hence, photograph taken in the scale of 1:50000 of an area situated at 1300 ft above mean sea level will be considered as 1:47400.

Area sampling with point grid overlays: This manual technique for determining the area of features on air-photos and maps involves the use of dot grids. Here the method of calculating the scale of photo is to measure the distance between two widely separated points in the form of a grid, both on photos and on the ground or on accurate map. Hence, the print scale is Distance on print / Distance on the ground a transparent overlay with a grid cell pattern is overlain on the air-photo. By counting the number of cells or dots which occur within a feature it is possible to estimate its area, knowing the area for each dot or cell.

Advantage of Air-photo in Agricultural survey

1. The topography of a particular area is shown in exeserated (Showing higher gradient than actual) form through air-photo survey. So, this is immensely useful in delineation of micro watersheds as the ridge lines can be accurately demarcated.
2. This will help to determine the soil and water conservation structures required to be installed in farmland as well as in non-farm lands.
3. In case of flatter terrains where demarcation of different landform units is extremely difficult by field traversing, the unique advantage of aforesaid survey helps to identify different landform units. These units are essential for soil classification at series level as well as in land use planning.
4. Progress of survey by air-photo is much faster than grid survey by field traversing.
5. It may also help to have a primafacie idea about the character of soil and status of ground water.

3.11. SOIL–WATER–PLANT RELATIONSHIP

3.11.1. Supply of Water to the Root in Soil

The passive movement of water into roots from the immediate vicinity of individual rootlets occurs due to transpiration demand of the foliage of that plant and decreases the soil-water potential. This automatically sets up a gradient, which initiates a flow from surrounding soil towards the root system. There is substantial resistance to this flow, which increases sharply as the soil moisture content falls. There is a tendency for water stress to arise at a rate, which increases with increasing transpiration demand. This supply system is such that water flows into the roots present in the wettest soil, so that the system tends to be self-equilibrating. Experiments shows that a herbaceous plant like lettuce can obtain its full nutrient and water requirements from only a part of its root system as long as the transpiration demand is moderate. When exposed to situations of high transpiration demand, it could obtain its full requirements only when the whole root system was supplied with water and nutrients.

It is easy to demonstrate that roots may be hydrotropic and geotropic but the extent to which either of these tropisms controls growth and conformation in the field is debatable. In experiments it was observed incidentally that actively growing roots would sometimes go away from the wet soil into the non-wet wax. However, in water stress situation there will be no doubt about the proliferation of roots towards wet environment and they can extend their root system fast enough to exploit untapped soil and try to meet their water requirements without replenishment from outside. The efficiency of exploitation of available soil water depends on the size and rate of expansion of the root system rather than the movement of water through the soil towards root.

3.11.2. Entry of Water into the Plant

Transfer of water from the soil into the root takes place through the root hairs, through the relatively undifferentiated tissues near the root tip and through the cortex. The ultimate organs of the roots connecting with the soil are the root hairs; each is a

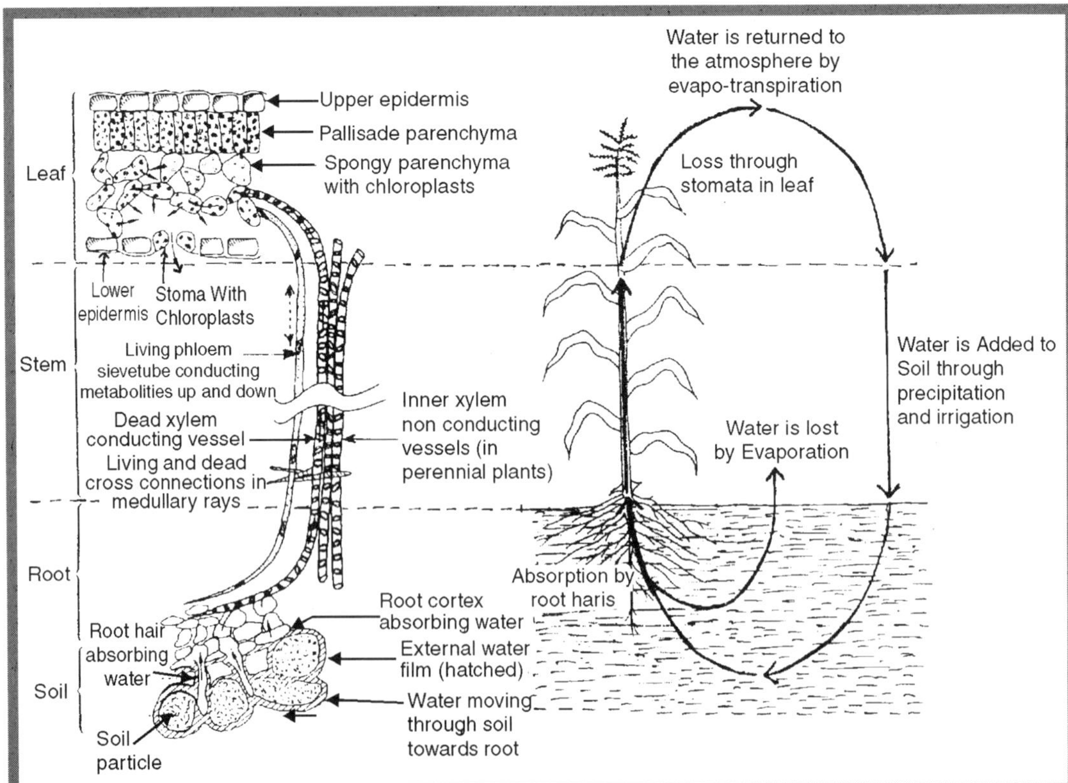

Fig. 3.10. Soil-water-plant Relationship.

protuberance from a single cell of the epidermis. The root hairs are in intimate contact with soil particles and penetrate crevices within them; they are short-lived and disappear after a few days to be replaced by new ones formed immediately behind the advancing root tip. It was formerly supposed that roots did not absorb water from the soil in its older regions but there is now evidence that absorption can and does take place almost anywhere in the root system. In some plants (like pine), it has been estimated that the mature roots absorb three quarters of the total water needs of the plant.

Osmotic pressure within root cells normally exceeds that in the soil and hence water tends to enter, but it is generally considered that more water enters the root in response to pressure or potential gradient between the soil and the plant tissues than by osmotic uptake.

Osmotic pressure of the external medium may sometimes exceed that of cell contents. This can happen naturally as in case of marram grass where it helps the transpiration-reducing mechanisms to cope with the drought condition. In commercial production of dryland areas the high osmotic stress is induced by high application rates of inorganic fertilisers. Sporadic rainstorms or irrigation and subsequent drying of the surface soil can cause violent fluctuation in osmotic potential in the surface soil. This is especially harmful to young seedlings, which have not had time to develop root system deep enough. In dryland, for cultivated plants, evaporation concentrates solutes in the surface soil and increases the osmotic stress.

3.12. AGRICULTURAL DROUGHT

A major constraint to economic production of crops in dryland areas is the 'drought', the acute water stress during the cultivation period, which usually occur in the rainy season. The crop production in dry areas is affected by five distinct categories of drought viz. (1) early season, (2) mid season, (3) late season, (4) chronic and (5) apparent drought.

Early Season Drought generally occurs either due to delayed onset of monsoon or prolonged dry spell soon after the on-set of monsoon ; of which the second one is the most loss making event. The first one delays the sowing. To avoid these two problems, proper seed sowing practices (off-field seeding) clubbed with suitable water conservation technology should be adopted.

Mid season drought occurs due to inadequate soil moisture availability between two successive rainfall events during the crop-growing period. However, the impact of mid season drought varies with the affected growth stage of the crop and the duration of drought spell. Broadly, in annual crops, out of the two crop-growth stages i.e. vegetative and reproductive, yield loss is more when the latter stage is affected by mid season drought.

Late season or terminal droughts occur as a result of early cessation of monsoon and it can be anticipated to occur with greater certainty. Late season droughts are more critical, especially for annual field crops as the grain yield is strongly related to water availability during the mid to late reproductive stages. Further, these conditions are often associated with an increase in ambient temperature leading to forced maturity.

Chronic droughts content are common in extreme dry areas where rainfall and subsequent soil moisture are inadequate to meet the water requirements of common field crops. Apart from these extreme dry areas several other regions of low to medium rainfall, having undulating topography and shallow soils of very low moisture holding capacity, also show the same chronic drought features. These areas are called as 'drought prone areas'.

Apparent drought conditions are seen in some areas with low to medium rainfall, in case of the crops which are selected for economic or other considerations, but do not match with the existing rainfall pattern i.e. intensity and distribution of the place.

3.13. PHYSIOLOGICAL ADAPTATION OF CROP/PLANT TO DROUGHT

Extensive study has been carried out all over the world to understand the response of crops to drought. However, much remains to be understood on the mechanisms of plant adaptation when encountered with severe water stress. Actually plants adapt to water stress by (a) maintenance of higher water and turgor potential, (b) higher root to shoot ratio, (c) tolerance to water stress and higher temperature, and (d) higher crop water use efficiency.

On the basis of 'stability index' of crop in different moisture regime (Eberhart & Russel, 1966) a 'drought susceptibility index' (Frscher & Maurer, 1978) of crop is being developed. Actually drought resistance capacity of a crop is dependent on a few of its own physiological criteria that directly or indirectly relates to dry matter production (including yield) in water stress condition. Some of the criteria, evaluated for drought resistance character, are as follows:

1. Maintenance of higher plant water and turgor potential;
2. Better root characteristics;
3. Maintenance of growth during water deficiency and acceleration on resumption of water supply;
4. Water retention in excised leaves at specified temperature;
5. Effect of crop canopy in respect of temperature and humidity under it;
6. Leaf reflectance and transmittance properties;
7. Osmotic adjustment;
8. Accumulation of proline and glycine betains;
9. Higher water use efficiency;
10. Matching the phenology with the water availability periods.

Few other characteristics such as leaf rolling, conservation of messenger ribonucleic-acid (RNA), membrane permeability, character of chlorophyll has been considered as useful criteria for drought tolerance. However, generalization of all these criteria is not possible and they are useful under a set of conditions. Proper utilisation of these criteria for breeding purposes is not possible unless plant breeders and physiologists work together or either of them learn the language of the other, which generally do not happen in India.

When we speak of tolerance of dry conditions we are talking mostly about the drought tolerance of the growing or mature plant. In the following tables, plants that are useful in arid regions are considered. These plants vary in ability to tolerate aridity and in yields under such dry conditions.

Table 3.9. Relative Draught Tolerance of Food Plants for Dry Regions

Scientific Name	Common Name	Degree of Tolerance (0=none to 3=high)
CEREAL GRAINS		
Zea mays	Corn	1
Sorghum bicolor	Sorghum	1.5
Pennisetum americanum	Pearl Millet	2.5
GRAIN LEGUMES		
Phaseolus vulgaris	Common Bean	1
Vigna unguiculata	Cowpea	1.5
Cajanus cajan	Pigeon Pea	2
Dolichos lablab	Lablab Bean	2
Vigna radiata	Mung Bean	2
Phaseolus acutifolius	Tepary Bean	2.5
Vigna aconitifolia	Mat Bean	2.5
Tylosema esculentum	Marama Bean	3
LEAFY VEGETABLES		
Symphytum officinale	Comfrey	1
Manihot esculenta	Cassava	1
Cnidoscolus chayamansa	Chaya	2
Moringa oleifera	Moringa	2
Leucaena leucocephala	Leucaena	2.5
ROOT CROPS		
Dioscorea rotundata	White Yam	1
Manihot esculenta	Cassava	2
Sphenostylis stenocarpa	African Yam Bean	2
VEGETABLES		
Citrullus lanatus	Watermelon	1
Cucurbita mixta	Mixta Squash	1.5
Cynara scolymus	Globe Artichoke	1
Abelmoschus esculentus	Okra	1.5

Scientific Name	Common Name	Degree of Tolerance (0=none to 3=high)
FRUIT TREES		
Carica papaya	Papaya	1
Psidium guajava	Guava	1
Spondias cytherea	Golden Apple	1
Olea europaea	Olive	1.5
Tamarindus indica	Tamarind	1.5
Zizyphus jujuba	Jujube	1.5
Carissa carandus	Karanda	2
Dovyalis abyssinica	Dove Plum	2
Punica granatum	Pomegranate	2
Anacardium occidentale	Cashew	2.5
Opuntia spp.	Prickly Pear Cactus	2.5
Phoenix dactylifera	Date	3
OIL PLANTS		
Pentaclethra macrophylla	Owala Oil	1
Helianthus annuus	Sunflower	1
Butyrospermum paradoxum	Shea Butter	2

Table 3.10. Relative Drought Tolerance of Non-Food Plants for Dry Regions

Scientific Name	Common Name	Degree of Tolerance (0=none to 3=high)
FEED LEGUMES:		
Gliricidia sepium	Mother of Cacao	1.5
Ceratonia siliqua	St. John's Bread	1.5
Prosopis spp.	Mesquite	2
Leucaena leucocephala	Leucaena	2.5
Acacia albida	Apple Ring Acacia	2.5
Acacia tortilis	Umbrella thorn	2.5
Parkinsonia aculeata	Jerusalem Thorn	3
FEED GRASSES:		
Cynodon dactylon	Bermuda Grass	1
Digitaria decumbens	Pangola Grass	1
Sorghum sudanense	Sudan	1

Scientific Name	Common Name	Degree of Tolerance (0=none to 3=high)
FIBER PLANTS:		
Gossypium barbadense	Sea Island Cotton	1
Agave fourcroydes	Henequen	2
Agave sisalana	Sisal	2.5
TIMBER PLANTS:		
Swietenia mahogoni	Mahogany	1
Acacia tortilis	Umbrella Thorn	2.5
PLANTS FOR ALLEY CROPPING:		
Gliricidia sepium	Mother of Cacao	1.5
Cajanus cajan	Pigeon Pea	2
Leucaena leucocephala	Leucaena	2.5
GROUND COVER:		
Dolichos lablab	Lablab Bean	2
Canavalia ensiformis	Jack Bean	1.5
WINDBREAK:		
Casuarina spp.	Casuarina	2
Eucalyptus spp.	Eucalyptus	1.5
Tamarisk spp.	Tamarisk	2.5
LIVING FENCE:		
Gliricidia sepium	Mother of Cacao	1.5
Bursera simaruba	Gumbo Limbo	1.5
Acacia nilotica	Babul Acacia	2.5
Euphorbia tirucali	Pencil Euphorbias	3

CHAPTER 4

TECHNOLOGY REQUIRED FOR DRYLAND FARMING

4.1. WATER CONSERVATION

Though the terminology 'conservation' and 'management' are different, in respect of dry land farming they are nearly synonymous. All the conservation practices in drylands have almost equal impact on both Soil and Water. In dry areas most of the soil conservation mechanisms are directly or indirectly related to water conservation practices giving rise to the term 'Soil and Water Conservation'. In dry land farming water is the prime item to be conserved and managed for better plant growth. Hence, water conservation should be discussed separately to give due importance to water.

As discussed earlier, in any kind of farming system, including dryland farming, the primary source of water is precipitation and the secondary supply coming from its different storage forms above or below the ground. Thus the principal objective of water conservation in dryland areas should, therefore, be to conserve the rainwater in such a way that the maximum possible part of it may be used for plant growth. Here 'plant growth' refers to 'Economic growth of crop'.

4.1.1. Principles of Water Conservation Technology

The basic principle of water conservation technology is to prolong the 'time of concentration' of

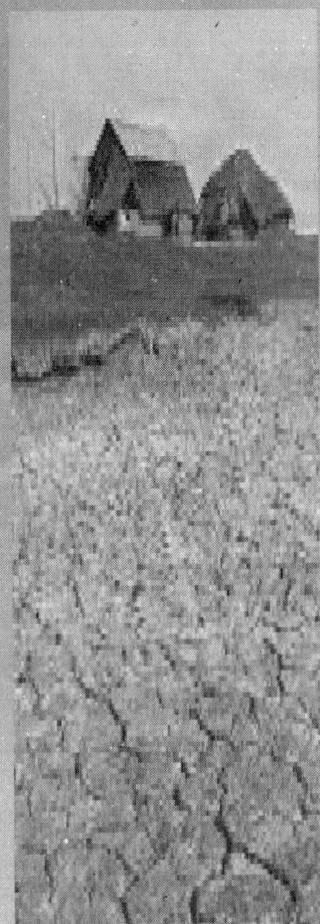

rainwater while flowing on the land surface and induce surface storage. *Time of concentration* is the time required by a drop of water to travel from the farthest point to the ultimate out-flow point of a particular catchment or watershed area. The big catchments or watersheds are generally designed by nature. In case of small or very small catchments or watersheds, the area can be designed artificially as per requirement, particularly in dry land areas. So at the time of managing the big ones or creating the small ones, the said basic principles should be followed. This can be done by increasing the land surface roughness, creating larger obstructions like bunds, vagitation, dams etc. to reduce the velocity of flow and storing a portion of run-off water over the surface for longer period are the basics to be followed while designing water conservation measures or structures.

The loss of water from a field generally occurs by the way of (1) surface run-off, (2) evapotranspiration and (3) ground water outflow. Out of these three processes the third one is a very slow and does not create any immediate loss of rainwater. Direct conservation of this water is more or less impossible. Careful utilization of this ground water for growth and production of crops is the best way to conserve this water indirectly.

The second process of loss i.e. evapotranspiration loss is also a secondary loss and not an immediate loss of rainwater, but it has great significance in respect of plant growth. In hot and dry areas the quantum of this loss is quite high. This loss is very much crucial because the entire amount of such water is 'available water'to the crop. The first process of loss i.e. surface run-off is the prime and immediate loss of rainwater; on which the farming practices generally depend, specially in dryland areas.

The several ways to reduce this loss or to conserve such water, covering of soil by any means is the best way to reduce evaporation loss. However, reduction of transpiration is a very difficult job.

Water Flow-chart of a Particular Dry Land Area

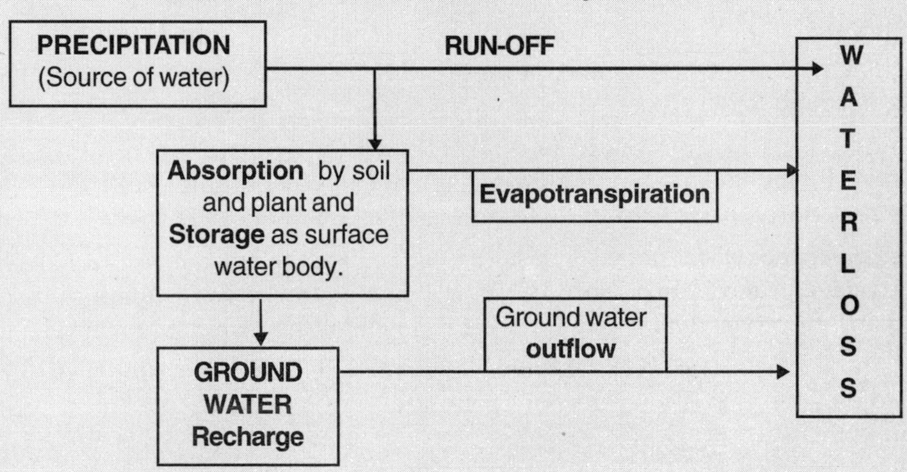

As a natural hydraulic process, it is not possible to stop the natural run-off, nor it is wise to reduce it to a great extent. The conservation technology should be designed first to check the process of run-off up to a certain level and divert that portion of run-off water to different storage spaces over ground, under ground or even in soil itself. Though temporary, this method of storage helps to increase the supply of soil moisture to the crop during dry period, thereby helping different farming activities like crop production, animal husbandry and also human consumption.

History of Water conservation

As early as 3rd millennium B C, farming communities of Baluchisthan and Kutch impounded rainwater in dam built of stone rubble and used it for irrigation. Dholavira, a major site of Indus Valley Civilization (3000-1500 BC), had several reservoirs to collect monsoon runoff. Small size wells are probably a 'Harappan' invention. In the period of Emperor Chandragupta Maurya (321-297 BC) construction of dam, lakes, and irrigation system including sluices are made and supervision by the officers were done regularly. Many similar references and evidences are available in the history that established the existence of developed water conservation technologies in ancient India.

Apart from historical evidences there are several age old practices still prevailing in rural India, which also provide substantial information about useful water conservation technologies.

In North-East Himalayan states, an old system of tapping and carrying of stream and spring water by using bamboo (as drain and pipe) is still prevalent. This bamboo drain (longitudinally half bamboo) can collect subsurface water from steep slopes of the hill sometime in the form of a thin flow. This collected water flow is carried through the bamboo pipes and transported over hundreds of meters, towards a tank/container/small reservoir or directly to the site of the plant/crop.

The 'kundis' of Rajasthan are unique structure, which look like a huge saucer on the landscape having a slope towards center. The surface of the slope is prepared artificially to increase runoff and this circular area acts as catchment of the covered well placed at the center.

The 'ahar-pine' system of irrigation is still used in south Bihar. Ahars are rectangular catchment basins, and pines are channels constructed to utilize the water flowing through hilly rivers.

Karnataka and Tamilnadu is the area with highest concentration of water harvesting/storage tanks (in Karnataka 40000 tanks exist till date) built in earlier days for different purposes with different names.

Underground water harvesting storage structures, below the residential places, to collect harvested runoff water from roof as well as from rocky land surface in desert areas of Rajsthan, are very common and still exist in small village houses to towns like Jodhpur.

Many such structures and systems prevailed in different corners of India and surrounding countries. These are now getting importance by the technical people and experts of this sector.

4.1.2. Basics of Rainwater Conservation

The basics of rainwater conservation or management are (1) to increase the rainwater retention capacity of an area by reducing the velocity of rain drop before it touches the ground, (2) to reduce the velocity of initial flow of runoff water on the ground and to lengthen the path of the flow and (3) to obstruct the well identified flow of run-off in such a way that it increases the ponding time and ponded volume of water.

The rainwater interception capacity of dry land can be improved through *vegetation cover*. In general the dryland areas have very poor vegetation, which leads to low rainwater retention capacity. The canopy of vegetation breaks the speed of the raindrop thus reducing its velocity. Both the dry and the fresh vegetative cover on the ground impede the movement of rainwater on the surface thus improving the chance of absorption of more water. It is useful for high intensity short duration rains, frequently found in dryland areas. In such situation, the soil gets little time to absorb the rainwater which rapidly flows towards down slope and whatever amount absorps by the soil surface gets evaporated. Apart from vegetative cover, ditches, depressions etc. also physically retain water above the soil surface reducing the immediate run-off loss.

By increasing the retention of rain water, run-off may be reduced but cannot be totally stopped; the run-off loss from the field will still contanue. Different land shaping methods should be adapted to utilize the residual run-off water as much as possible for growing of crop.

The speed of movement of run-off water is chiefly dependent on the slope or gradient of the land surface i.e. the topography of the area. The land shaping methods should aim at reducing the velocity of flow of run-off as well as to increases the flow time and the length of path of run-off water, thereby allowing more time for absorption into the soil. There are many methods, which can be suitably utilized for different land situations. The methods are contour tillage, contour bunds, contour trenches, staggered trenches, terrace, and diversion etc. {the details of these methods are described in chapter- 4.3}

These structures can effectively obstruct the flow of run-off water and capture a portion of it for farming activity as well as for recharging the ground water. The run-off water that flows through a specified or non-specified path can be checked by a few other structures like nala bunds, vegetative or sod strip checks and different other check dams, creation of percolation tanks/water harvesting tanks/farm ponds etc. {The details of these methods are given in the chapter 4. 4.}

4.1.2.1. Storing of Rainwater

The rain water can be effectively captured in: 1) Large reservoirs created by large dams having large catchment and command areas, and 2) Small storage areas created by low earthen dams or tanks or ponds with small catchment and command areas; or storing it temporarily in a way (Percolation tanks) that it percolates down into the ground and is stored as ground water.

The study and scientific evidences on **Large and medium reservoirs verses small and mini water harvesting bodies** shows that at village scale, rather at farm scale, small rain water harvesting structure are much more effective than big or medium dams. The small structures are extremely cost effective, more efficient collector of run-off water and high yielding in respect of different human needs. The work done in

Negev desert (annual rainfall is 105mm) by Israeli scientist Michael Evanari shows (Table:4.1)that all other factors like slope etc. remaining the same, the larger size of catchment collects lesser runoff water in comparison. The difference can be quite high. As the Table shows, 3000 microcatchment of 0.1 hectare each will give five times more water together than one catchment of 300 hectares. It means 10 small dams with one hectare catchment each will give substantially more water than one large dam with a 10 hectare catchment.

Table 4.1. Effect of Size of Catchment on the Quantity of Water Harvested.

Sl. No.	Size of catchment (hectares)	Quantity of water harvested (cubic meters/hectares)	Percentage of annual rainfall collected
1	Microcatchment(upto0.1)	160	15.21 %
2	20	100	9.52 %
3	300	50	3.33 %

Source: Michael Evenari *et al* 1971

4.1.2.2. Rooftop Water Harvesting (RWH)

It is the process of harvesting of rain-water from roof, using it as a micro-catchment area, for any specific water storage structure/system. Predominantly this harvested water is used for drinking and other household purposes. However this process of rainwater harvesting is utilized for crop production purposes directly or indirectly. Recharge of ground water by rainwater harvested from roof is by far the best in-situ indirect method.

Roof-water harvesting for crop production

The erratic and unpredictable rainfalls introduce an element of risk, uncertainty and instability in commercial crop production due to water scarcity in vital growth stages. Even the sustainable rain fed relatively high value crop production is facing partial or total crop failure due to such kind of water scarcity between two rains. Hence the idea of micro-watershed, created above the ground designed for an individual plant or a plot or few numbers of plants is followed in case of roof-water harvesting for crop production.

Now-a-days the roof of greenhouse is utilized to harvest and store rainwater for production of crop in the green house.

History & present situation

It is an ancient technology practiced by people of many water scarce areas for drinking and other household purposes in particular. It has been in practice since centuries in many dry tracts of the globe including Gujrat and Rajasthan in India. Unfortunately with the availability of pipe water, mostly taken out from underground, the traditional rooftop water harvesting practices has began to be discontenued and gradually deserted particularly in urban areas. Nowadays with the ever-increasing water crisis there is a new interest to revive this system, sometimes in an improved manner. Most interestingly not only the traditional water scarce (dry) areas, but also many of the water-rich (high rainfall) areas are forced to adopt rooftop water harvesting technology to mitigate the water shortage usually faced in non-rainy days. The classic example is Aizwal, capital city of Mizoram in India, a very high rainfall area where tap water supply is not practical due to its steep hilly terrain, where more than 25000 houses harvest roof-water for their daily use.

Planning

The roof of a house is treated as catchment area for that house. The quantity of rainfall on that roof is trapped and collected for using the same for desired purposes. This collected rainwater may be stored in tanks for domestic use or may be directly utilised for irrigation or for recharging ground water. However it is wise to plan the RWH in such a way that the harvested water can be used both for the storage as well as recharge of ground water.

Before making any plan of a 'Rooftop water harvesting system' the collection of information is must regurding 1) monthly average annual rainfall of the area, 2) peak rainfall in a day, 3) type of roof of the house, and 3) utilisation of harvested water. On the basis of this information the following steps may be adopted.

1. The boundary of referred roof area should be guarded or adjusted in such a way that the entire rainwater will be trapped automatically.
2. In a flat roof the outlets should be correctly placed to dispose the peak rainfall properly. The number and size of the outlets are to be determined on the basis of roof size and peak rainfall. In case of slanting roof, appropriate size of gutter, determined on the basis of slope and length of roof besides peak rainfall, should be placed in a way that it can collect as well as despatch the rainwater fallen on that roof.
3. The delivery pipes will be attached with the outlets or gutters that finally lead towards storage tank for future use or towards infiltration pit/trench/open well/bore well for recharge of ground water.
4. The storage structure or ground water recharging structure (soak-pit) is to be designed properly in relation to the requirement of each and individual RWH system. (a) The size and type of storage structure will be determined in relation to the size of the family, purpose of use, approximate time of the storage, and distribution and intensity of rainfall. (b) The soak-pit or direct delivery to open well/bore well should be designed in accordance with the texture and depth of the soil besides intensity and distribution of rainfall. However, both types of structures should be provided with outlet to tackle overflow at the time of peak-rain hours that will help to avoid problems like soil-erosion etc.
5. In industrial areas where pollution in air is high due to emission of lots of harmful chemical vapours and suspended particles in air, the RWH may be avoided. If performed, it should be made operative after initial rainy days and in case of each rain the initial harvested water should not be collected and

stored. To accommodate these precautionary measures the RWH structure should be designed accordingly.

Advantage

- As the rainwater contains less total dissolved solid (TDS) it remains sweet and is preferred for different household purposes like drinking, cooking, washing etc.
- For irrigation purposes rainwater is far better because of less dissolved salt present in it.
- In dry salt affected areas, instead of soak-pit or well, the recharging of underground water can be done through flooding of agricultural field by the harvested water. This reduces the salinity through leaching.

Methodology

1. For flat roof requisite boundary with proper flooring and for slanting roof correct size of gutter is to be made or purchased to develop the roof to create a perfect micro-watershed for the outlet/s.
2. The delivery pipe system should be properly laid down to carryout the harvested rainwater smoothly.
3. The roof, outlet/gutter, and delivery pipes have to be maintained round the year and thoroughly cleaned once in onset of the monsoon. If the water is used for drinking purpose then cleaning may be done more frequently and more specifically.
4. For drinking and other household purposes, the storage structure (tanks etc.) should be clean thoroughly just before onset of the monsoon and should be provided with appropriate cleaning facilities.
5. For drinking purpose above ground and for other purposes under ground storage structure is recommended. For longer storage chlorination @ 1 gram of bleaching powder per 200 litre water is advised.
6. For drinking purpose growth of micro flora and fauna should be restricted as far as practicable. The storage tank should be completely closed to cut-off the sunlight that stop the growth of algae, prevent the entry of mosquito, cockroaches etc.

Calculations

1. Average gross amount of water that can be collected from a particular roof in a year.

$$Q_t = 10\ (A \times R)$$

Where Q_t = Total rain water collected/year (litre)

A = Area of roof (sq. meter)

R = Average yearly rainfall from last 10 years data (centimetre).

2. Amount of water that can be collected from a particular roof in one-hour effective time @ average of peak rainfall. It may help to determine the size of the tank or soak-pit etc.

$$Q = 10\ (A \times iR)\ C$$

Q = Amount of water harvested per hour of continuous rain (Litre).

A = Open area of the roof (Sq. Meter).

iR = Average intensity of rainfall out of 50% high rainy-days for last 10 years data (Cm./hour).

C = Constant (as per following conditions).

1.1 — If the av. highest intensity of rainy day/year for the last 10 years is 25% or less than that of iR value.

1.2 — If the av. highest intensity of rainy day/year for the last 10 years is between 25 to 50% of iR value.

1.3 — If the av. highest intensity of rainy day/year for the last 10 years is between 50 to 100% of iR value.

1.4 — If the av. highest intensity of rainy day/year for the last 10 years is 100% or more tan that of iR value.

3. (a) The size (in volume) of the tank will be minimum of Q or its any multiplication.

(b) The size (in area) of soak-pit will be $S = (Q/I)/10^4$

S = Total surface area of the soak-pit (sq. meter)

Q = Taken from equation mentioned above (litre).

I = Infiltration rate of the soil (centimetre/ hour).

4.1.2.3 Collection of mist and dew

There is some disagreement about the quantity of dew formed in the tropics and its value to the plant/crop (UNEP 1983:188). Much of this confusion arises because of the difficulties in measuring actual dewfall. According to Evenri et al (1971), the collected amount of dew in Negev desert was on an average 35mm/night. However with better vegetation this amount might be six or seven folds because of the increase of surface area on which it could condense. In UK's uplands, simple structures known as dew pond lined with straw and clay were fairly common. In recent times use of glossy metal sheet or plastic sheet are fairly common to trap and collect dew water for using the same in different household purposes including drinking and direct application to a plant.

4.1.3. Reduction of Evapotranspiration Loss

The methods to conserve the evapotranspiration loss can be done by controlling the evaporation loss and the transpiration loss. Out of these two losses the first one increases with reduction of vegetation, while the second one with increase in vegetation. The evaporation loss is directly related with temperature and transpiration loss chiefly with the sunshine. Air humidity and wind velocity also have an impact on both these losses.

4.1.3.1. Reduction of Evaporation Losses

Evaporation generally occurs from the surface of soil, water and plant out of these three, evaporation loss from soil is most important factor perticularly for plant/ crop growth. Evaporation loss from water reduces the storage of water and evaporation from plant is insignificants.

1. To reduce the high evaporative loss of water from soil surface particularly in dryland areas, principally, we have to obstruct the exchange of heat between soil-air and atmosphere. To achieve this, a technology called '*Mulching*' may be effectively adopted. Continuous evaporation from the fine textured soil is possible due to the sub-surface rise of soil water through capillary action. In caurse-grained soil the process of diffusion takes the place of capillary action. This can be checked principlly by different methods of mulching activity including tillage mulch in heavy soils. In field situation mulching is the best and efficient method of reducing evaporation loss of water. {the detailed of 'Mulching' is described separately in 5.5 }
2. In dryland areas evaporative loss from open water surface is very high due to high temperature, low humidity and hot wind. To reduce this loss no research based recommendations are available at the farmer level till today. However, the following practices may be adopted to reduce this evaporation loss: (i) planting of trees around the water body to control the effect of hot wind and maintain a relatively high humidity and low temperature (effective for small water body), (ii) increasing the width : depth ratio of smaller water body and (iii) covering the surface of small water bodies by non hygroscopic substances like plastics, foamed rubber sheet, asphalt-concrete blocks etc., or dry plant residues e.g. bamboo, *Saccharam munja*, straw, wood dust or bark, sticks etc., or oil mulches, silicon-gum mixture, and fatty alcohols (till today no field level success has been reported with these water covering technology).

   However, experiments conducted at Central Arid Zone Research Institute (CAZRI), Jodhpur, shows that 82%, 74%, 66%, 53%, and 37% reduction of evaporation loss, compared to open tank, may be achieved when polythene sheet, foamed-rubber sheet, polyethylene sheet, bamboo and *s. munja* respectively were used as covering material. Interestingly they found (in 1989) the cost of reduction of evaporation loss was about ₹ 2.54 per 1000 litre saved water using white polyethylene. In case of polystyrene and polyethylene sheet the cost was ₹ 3.07 and ₹ 4.44 respectively.

   In case of tanks and small reservoirs, low-cost, durable, easy to apply thin film of harmless compounds can be applied to reduce evaporation. This will act to reduce evaporation up to 40 per cent without hindering the passage of sunlight and gases, thereby allowing fishes, plants and other aquatic organisms to survive. However, this thin layer is not suitable for tank or small reservoir over 40 ha size as it does not work due to breaking up of the film by normal wave action.

These evaporation suppressants are not commonly used in the field, but some simple methods are standardised and practiced since long past in India. **In Rajasthan**, the driest region of India, underground water bodies are created usually below the buildings and residential houses to store rainwater. For this purpose the precipitation falling on the roof is drawn in to the underground storage tank through a pipeline. Evaporation loss of this storage water is negligible despite of the high temperature generally prevailing there.

3. In general the evaporation loss from plant surface is negligible. Just after rainfall or by dew, a thin water film is created on the plant surface, which subsequently evaporates into air. This water vapour sometimes increases the humidity under the canopy and creates a microenvironment that reduces further evaporation loss thus helping the dryland farming.

4.1.3.2. Reduction of Transpiration Loss

Transpiration is basically a physiological process in which the plants transpire water in the form of vapour through its stomatal openings. It is very difficult to control. Approximately 99 percent of water absorbed by plant roots is transpired into the atmosphere mostly through stomatal pores of leaves. With few measured methods like pruning, use of anti-transpirants we may reduce the transpiration losses. Actually the plants, of their own, use different physical and physiological methods, separately or in combination, to reduce the transpiration losses.

Diminished water supply reduces the water supply to the cells resulting in 'temporary wilting' of plants. During this phase the turgidity of all cells including guard cells fall. This causes the closure of stomatal opening and subsequent reduction of osmotic potential, which virtually stop the transpiration. This greatly reduces further loss of moisture from the plant body. In dry season many plants reduce the size of leaf and other green parts thus reducing the number of stomata in comparison.

Many plants have an inherent capacity to adapt or get acclimatised in dry and water stress conditions by virtue of some of their special physiological modifications like less number and smaller size of stomata, sunken stomata, opening of stomata during night, presence of stomata only on the under surface of the leaf, transformation of leaf to reduce its surface area, having a thick waxy epidermal layer in their body and the like. Such modified characters to withstand dry conditions are called 'xerophytic' character. However, due to the reduction of transpiration the diffusive resistance to carbon-dioxide exchange increases thus decreasing the photosynthesis process, which usually reduces the yield. But to avoid the permanent injury due to water stress the mechanisms for reduction of transpiration loss is essential. Sometimes we may help to reduce transpiration losses by defoliation or shoot pruning and application of anti-transparent chemicals.

Pruning. Systematic and purposeful pruning of green parts, mainly leaves and young shoots reduces the transpiration loss by reducing number of leaves particularly in case of trees. But pruning should be done in such a way that it does not disturb the flowering or fruiting. This is effectively done where dry season matches with the post harvest period. In perennial plants, this pruning is helpful to induce new healthy shoots, which leads to a higher yield in the next season. Certain plants like guava, ber (*Zizyphus jujuba*) etc produce good flowering in new shoots only.

Pruning for Transplants. Pruning also increases the chance of survival of the transplants. When seedlings are lifted from the nursery for transplanting, some injury to the roots invariably occurs, adversely affecting the water uptake. But the transpiration loss of water continues unabated in day time. Being unable to replenish the lost water, seedlings are subjected to severe wilting which may even result in mortality. To save this situation older leaves of dicot and the leafy tops of monocot seedling may be pruned to reduce the transpiration loss. This practice improves water balance in plants and increases the chance of survival. Though this treatment reduces the photosynthesis and slows down the initial growth, this is done when the very survival of the plant is at stake.

Use of Anti-Transpirants. In order to minimize the loss of water through transpiration use of different chemicals, collectively called 'anti-transpirants' or 'Transpiration suppressants' have received much attention in recent years. These chemicals reduce the transpiration loss principally through the reduction of size and number of stomata and also by covering the stomatal openings with a thin film of the substance. An ideal anti-transpirant should restrict the moisture loss from leaves without reducing the photosynthetic activity. Unfortunately this is very difficult to achieve.

Chemicals like phenyl mercuric acetate (PMA), abscisic acid (ABA) and certain alkenyl succinic acids acts by inhibiting stomatal opening. Waxy or plastic emulsions (film forming anti-transpirants) produce an external physical barrier for the escaping water vapour from stomatal pore; silicon penetrates the leaf and act directly on wet cell walls. Reflectants (Kolin) reduce leaf temperature by reflecting solar radiation thereby reducing vapour pressure gradient between stomatal cavity and the nearby atmosphere.

Several experiments using different anti-transpirants on a number of crops have shown that they can be used economically on high value crops, especially in rain-fed dryland areas and for increased survival of transplants under all situations. However, all crop species do not respond similarly to anti-transpirants; even the varieties differ in their response. Season, region, stage of growth and management practices also influence the effect of anti-transpirant, both qualitatively and quantitatively.

4.1.4. Conservation of Ground Water

Satisfactory conservation of rainwater on the surface increases the recharge of ground water. There is a possibility of this ground water to flow towards the low-pressure zone or downward slope along the free water table from the percolating zone to a distant place. There is no established method to restrict this out-flow. However, under favourable situation in a farm, few surface run-off conservation methods may indirectly slow down out-flow of shallow ground water. For example, the field recharge systems like trenches or rectangular tanks across the slope may slow down out flows of ground water through rising of water table.

Underground Dams. This structure can be applied successfully in the beds of seasonally or apparently dry rivers or rivulets or streams. Here an underground barrier (may be called as underground dam) stops as well as collects the flow of ground water and the collected groundwater can be lifted through tube wells or dug wells. Utilizing the locational advantages under favourable situation in a small valley or in a shallow seasonal riverbed, a underground dam can obstruct the flow of ground water and hold it below the ground surface behind the dam. It can raise the water table of an aquifer by restricting the outflow of groundwater, thus making it accessible or available for pumping. There are basically two types of underground Dams: *sub-surface dams* and *sand-storage dams.*

Sub-surface dam is constructed below the ground level, which arrests the flow of an un-confined aquifer. If an aquifer is in a small valley with narrow outlets or extensive beds of waterways, the aquifer gets drained during the dry season through them. A trench may be dug across the outlet reaching down to the bedrock or impermeable sub-soil, which is filled up by any impermeable substance to create a sub-surface dam. This dam will restrict profuse drainage from the groundwater reservoir and can be used throughout the dry season.

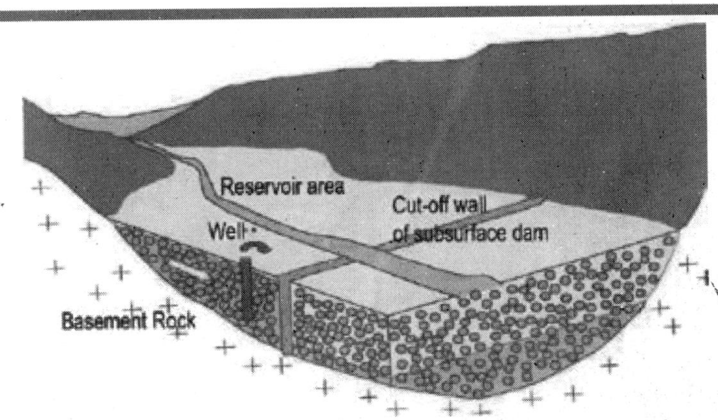

Fig. 4.1. Under Ground Bund/Dam

In some regions valleys have been eroded and then in-filled with alluvial deposits. Ground water may flow in these buried valleys and may be conserved and used by constructing sub-surface dams or barrages by installing waterproof barriers of concrete or plastic sheet in a trench perpendicular to the slope. This system of water conservation methods are widely used in Morocco and have been tried in north-east Brazil (Gischler & Jauregui 1984).

The disadvantage of this method, particularly when this water is directly utilized by the crops cultivated above the sand reservoir, is that, unless great care is taken, the water table may come up too near to the field and salinization may result.

A **sand-storage dam** is generally constructed in stage by stage across a shallow stream bed to collect coarse sand in the up-stream side. This dam impounds water in the stored sediment, which creates an artificial aquifer that is replenished every year. This stored water will be available in the sand reservoir during the dry season.

Although the storage capacity of a sand or gravel filled reservoir is only 25 to 35 per cent of that of an equivalent un-silted reservoir, it has the advantage that as soon as the water level falls more than 60 cm below the surface, evaporation losses virtually cease (Furon 1963). This type of sand/gravel storage dam can withstand flash flood better than normal reservoirs. It offers secure and clean less-contaminated water for drinking and other household and small scale irrigation purposes in a more efficient manner.

The main problem in constructing or creating sand-storage reservoir is to ensure the deposition of coarser sediment/sand. To overcome this problem, construction of check-dam of boulders or brushwood or stone filled gabions, in the stream flow is a recommended method. This dam partially obstructed the stream flow allowing the fine sediments to escape but helps the coarser sediments to deposit (Dixey 1950). Such techniques have been used in Botswana, Namibia, Ethiopia, Mexico, southern USA and many other countries.

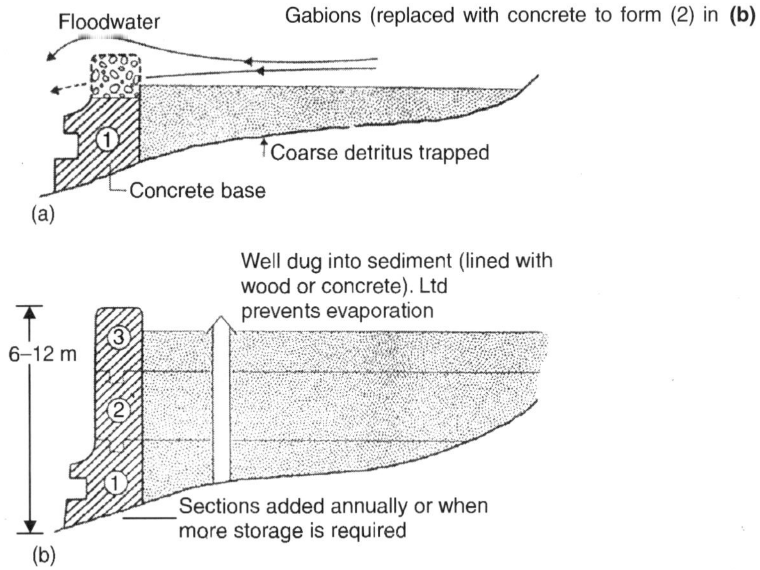

Fig. 4.2. Intermediate technology check-dam and sand filled reservoir (a) Flow partially obstructed by gabions on concrete base (1) and coarser sedment deposited. (b) Gabions replaced or sealed with concrete (2) and gabions placed on top to repeat sediment trapping, the gabions are then replaced with concrete (3) redrawn from pacey et.al. 1977.

Sub-Surface Bunds. In on-farm situation, in an area of shallow soil depth, the shallow underground seepage flow may be collected below the ground by creating underground or *sub-surface bunds* across its flow path. If any impervious layer is found within 1 to 1.5 mt. of ground surface with a reasonable slope, then a contour trench, across the slope of bedrock, filled with relatively impervious material or clay (in the form of a underground contour bund) may store or hold a good amount of water in up slope below the ground. This may be used by the plants planted above it. It is a miniature but elongated form of sub-surface dam.

4.1.5. Conservation of Surface Water

For conservation and storage of water, supplied by rain, different techniques are adopted. These techniques are chiefly related with re-shaping of soil surface considering the crops to be grown and rainfall and topography of that area. But it is always wise not to disturb the original land surface too much and try to utilise the natural shapes of land surface like contours, depressions, waterways etc. to apply the water conservation techniques and for subsequent farm planning.

The conservation techniques like ***broad bed and furrow (BBF) system, tide ridging, check dams, contour cultivation, seepage control measures etc.*** may be suitably and combinedly utilised for conservation of surface water. {For detail of these items please see Chapter- 5.1. & 5.2.}.

4.1.6. Water Harvesting Structures

To mitigate drought and develop water resources in dry farming regions, it would be best to harvest the runoff water, which is otherwise wasted and sometime creates flood in downstream areas. For collection of run-off water from a relatively small piece of catchment area, two types of water harvesting structures/ ponds – embankment type and excavated (dugout) type – are used. Embankment types of ponds are constructed in hilly and undulating topography, by creating a small length of dam across a watercourse. A good amount of run-off water can be stored through this structure {see Chapter- 4.1.2.1 for advantage of small over big dam}. Where the topography does not permit the construction of embankment, excavated ponds can be made in natural depressions. The dugout ponds can be designed and constructed according to the requirement of water and shall be much deeper than usual ponds or tanks. {The details of water harvesting structures are given in 4.3.7 & 8 and 4.4. of this chapter}

The main two causes of loss of harvested water from storage structure are seepage through water harvesting structures and evaporation from open water surface {For details of reduction of evaporation loss please see 4.1.3}. Seepage losses occur due to the high permeability of the beds and walls of the structures. High permeability denotes the lighter texture of the soil and the porous nature of substrata. So suitable seepage control methods can be employed to overcome this problem. {For details of seepage control methods see Chapter- 5.1.2.5.}.

4.1.7. Runoff farming

When the harvested runoff water from un-cropped areas is directed to a cropped area, this technique is called runoff farming. It is identical with "Water Harvesting but for Irrigation Purposes". The rainwater harvesting mentioned here is not similar to that of floodwater harvesting.

Here soil profile acts as a water storage container, but storage in ponds or cisterns is also feasible. Factors affecting the capacity of soil storage are: depth of the soil profile, depth of plant roots, texture, structure, infiltration r nd the water holding capacity of the soil. The catchment-to-field ratio can rang 1:1 to higher ratios . The higher the aridity of an area, the larger is the requ tchment area in relation to the cropping area for the same water yield.

This can be divided into 1. Micro-catchment, Macro-catchment runoff farming. Micro-catchment runoff farming is a method of collecting surface runoff from a small catchment area and storing it in the root zone or in farm ponds of an adjacent area. Macro-catchment runoff farming (catchment area being 1,000 m^2 - 200 ha) system is referred to by some authors as "runoff farming water harvesting from long slopes", as "medium-sized catchment water harvesting" or as "harvesting from external catchment systems".

Runoff farming has proved to be a valuable tool especially in dry marginal areas to increase crop yields and reduce cropping risk, to improve pasture growth, to boost re-aforestation, to allow a higher degree of food production, to fight soil erosion, to make best use of available water resources, to suppress soil salinity and, in a few cases to recharge the local groundwater.

Development of runoff farming is increasing specially in semi-arid and arid areas after 1950, partly due to the successful reconstruction of ancient water harvesting farms in the Negev. Low- cost efficient use of runoff farming in arid zones for food and fuel production could help to restore self sufficiency in food production for local populations in many African, South-west Asian and other dry regions.

4.1.8. Agronomical Methods to Conserve Moisture/Water

Presently Agronomieal methods are considered as most effective technique for conservation of water as well as soil. At the same time these methods have an inherent capacity to provide more economic return to the farmer of dry land areas. {The detail of all these methods are described in soil and crop management i.e. in Chapter no.— 5.2. & 5.3.}

4.2. SOIL CONSERVATION

4.2.1. The Approach

In dryland farming, soil, besides water, is another important factor for crop production. So the aim of soil conservation is the maximization of sustained level of production from a piece of dryland by maintaining soil erosion below certain level. Actually this level permits the natural rate of soil formation which can keep pace with the rate of soil loss. The soil conservation methods not only conserve the soil but at the same time reduce the loss of nutrients, prevent water loss, decrease rate of sedimentation in the down stream and reservoirs thus decreasing the magnitude of devastation of flood.

Though soil erosion is a natural process, it is aggravated manifold by human interference like deforestation, unplanned construction, indiscriminate grazing and afforestation with non-suitable species of our choice. In this process the balance of nature is getting disturbed resulting in severe erosion of the soil. It is now necessary to adapt suitable measures to bring down the erosion process to an acceptable rate, which can match our requirement for sustainable high agricultural production.

4.2.2. Conception of Soil Conservation Technique

The basic strategy of controlling soil erosion should be done on the basis of following principals, (1) covering of the soil to protect it from the impact of raindrop and (2) increasing its infiltration rate to reduce the erosive action of the run-off water. Now with the knowledge of mechanism of erosion of soil particles by rain splash, sheet flow, rill flow, and gully flow, {For detail see chapter no.—3.9.} we may effectively utilise this basic strategy to restrict the process. It also improves the stability of soil aggregate to increase the surface roughness that will reduce the velocity of water and wind.

Soil conservation measures generally include (1) biological and agronomical measures, (2) soil management measures and (3) mechanical measures or combination of these measures.

Agronomic and biological measures utilize the tilling process and role of crops or vegetation to minimizes soil erosion.

Soil management is concerned with ways of preparing the soil to promote better growth of crop or vegetation and to improve its structure and roughness of land so that it become more resistant to erosion. Mechanical measures try to manipulate the surface topography to control erosion.

Mechanical methods are tried to control the erosion when the process has advanced, soil management methods when the process has already started, and the biological as well as agronomical methods at the initiation of the process.

The study conducted by Voetberg and Morgan shows (Table – 4.2) the effects of biological, mechanical and soil management measures on soil erosion (detachment and transportation) caused by rain, run-off and wind separately. The final outcome of the study clearly indicates that the biological measures, as a whole, have a clear edge over the other measures particularly for the erosions caused by rain and run-off water. Out of the different biological measures **covering of soil surface** is the best technique. The second most effective measure is **increasing surface depression storage**. Combination of these two can reduce soil erosion to the maximum possible extent. The **manuring and fertilisation** under the soil management measure have moderate to good effect on all types of soil erosion ; actually this measure is an indirect one that finally works as the covering of soil surface through vegetative growth of crop. The mechanical measures, particularly **contouring and ridging** as well as **terracing**, have a reasonable effect on soil erosion. The other few specific methods have very good effect on some specific type of erosions.

In dryland farming, which is generally practiced in the cultivable field where there are no established erosion hazards like rilling and gulling, the agronomic and biological measures should always be the first choice to conserve the farm soil. These measures reduce the raindrop impact, increase the infiltration rate and produce more output. {The biological and agronomical measures and soil management are discussed in the Chapter no. – 5.2. & 5.3.}.

Mechanical measures are generally adapted where the farmland is affected by established soil erosion process or the process is likely to get established within a short period. The judicial application of mechanical measure in combination with proper biological method can effectively control or stop the soil erosion in farmland.

Table 4.2. The effect of various soil conservation practices on the detachment (D) and transportation (T) of soil particles.

Practice	Rain splash		Effect of Runoff		Effect of Wind	
	D	T	D	T	D	T
(A) **Biological and agronomical Measures**						
(1) Covering soil surface-	strong	strong	strong	strong	strong	strong
(2) Increase surface Roughness-	nil	nil	strong	strong	strong	strong
(3) Increasing surface depression storage-	medium	medium	strong	strong	nil	nil
(4) Increasing infiltration-	nil	nil	medium	strong	nil	nil
(B) **Soil Management**						
(1) Manuring and Fertilization-	medium	medium	medium	strong	medium	strong
(2) Sub-soiling, Drainage-	nil	nil	medium	strong	nil	nil
(C) **Mechanical Measure**						
(1) Contouring & ridging-	nil	medium	medium	strong	medium	strong
(2) Terracing-	nil	medium	medium	strong	nil	nil
(3) Shelter belts-	nil	nil	nil	nil	strong	strong
(4) Waterways-	nil	nil	nil	strong	nil	nil

Voetberg, 1970 and Morgan, 1980

Soil Conservation Measures for Arable Land

Water Erosion Control Measures

1. Mulching and crop residue management;
2. Contour farming;
3. Strip cropping (contour strip, field strip, buffer strip);
4. Inter or mixed cropping with erosion resistant crops, and creepers or climbers;
5. Tree farming (alley farming, farm forestry, agroforestry);
6. Ridges and furrows /dead furrows methods;
7. Grass waterways;
8. Contour/graded bunds (including vegetative bunds) or trenches;
9. Gully plugging;

Wind Erosion Control Measures

1. Vegetative cover and stubble mulching;
2. Strip cropping in association with tree farming;
3. Wind breaks and shelter belts;

4.2.3. Gully Control and Restoration of Gullied Farmlands

Aforestation and/or improper soil and water conservation measures lead to sheet and rill erosion that sometime result in gullies. Sheet and rill erosion, if not controlled in the initial years through vegetative and cultural measures, gullies will be the ultimate result (within five years). These are often 1 to 5 meters deep depending on the rainfall, topography and soil. For sustainable restoration of gullied land vegetative methods are far more effective than mechanical methods. However, in critical situation combination of these two may be adapted.

Mechanical structure like small dams, usually 0.4 to 2.0 m in height are, made across the gully from locally available materials such as earth, wooden planks, brushwood or loose rock (Table 4.3) etc. This structure is used to trap sediment thereby reducing channel depth and slope. These structures have a high risk of failure but they provide temporary stability and are therefore used in association with vegetative treatment of gullied and surrounding lands. Even though they are temporary, the dams have to be carefully designed. They must be provided with a spillway to deal with overtopping during peak-runoff flow. They must be constructed at spacing appropriate to the slope of the gullied channel {see 4.4 3. - Of this chapter for more information}.

Like any natural channel, every gully has its own head and specific catchment area. The runoff water of a heavy rain moves towards the head of the gully from its catchment area and finally comes down to flow along the gully with a great speed. In that situation the gully control structures/dams are at risk of getting destroyed. This flowing water frequently carries huge sediment load that provides the flow more power of destruction.

To avoid such a situation, that ultimately gives the check dam a longer life and restrict the gullying process, the entire catchment area should be covered by dense vegetation, particularly around the head of the gully.

Revegetation is used in gully erosion control as most sustainable method that may restore the gullied land for human use. The area around the gullies should be treated with grasses, legumes, shrubs and trees or combinations thereof, aided in the early stages by mulching if necessary. The research findings of the 'Huang He Conservancy Commission' shows that trees (afforestation) can reduce runoff in the gullied areas by 65 to 80 percent and soil loss by 75 to 90 percent. Growing grass will reduce runoff by 50 to 60 percent and soil loss by 60 to 80 percent (Gong and Jiang, 1977).

By planting trees and herbs on the steep slope of the gully sides and raising crops on bench terraces on the gentler slopes of the divides, the land can be stabilised and sediment prevented from entering the gullies. By supporting these measures with mechanical control techniques of check dams and reservoirs along the gully beds, the sediments outflow from gullied area can be reduced significantly (Jiang, Qi and Tan, 1981). The best vegetative treatment for tunnelled areas is to establish a dense, uniform grass cover. Where conditions are too marginal for grass to grow, shrubs and trees will have to be used but species with a good system of lateral rather than vertical roots should be chosen.

Table 4.3. Gully Plugs of Various Materials and Location of Gully Plug

Slope of nala/gully Bed (%)	Width of nala/ gully bed (m.)	Type of check dam	Vertical interval (m.)
0 – 5	(a) up to 4.5	Brush wood	up to 3
	(b) 4.5 to 10.5	Earthen	2.5 to 3
	(c) 7.5 to 15.5	Sand bag/gabion	-
	(d) 7.5 to 15.5	Masonry	-
5 – 10	(a) up to 4.5	Brushwood	up to 3
	(b) 4.5 to 6.0	Earthen	1.5 to 3

4.3. ON-FARM SOIL & WATER CONSERVATION MEASURES AND STRUCTURES

4.3.1. Tillage

Tillage is an essential management technique of any farmland. Tillage, one of the major part of land preparation, provides a suitable seed bed for plant growth and helps to control weeds. It also influences intake rate of water and reduction of soil loss by creating obstruction to surface flow. The roughness created by tillage operations is reduced over time by raindrop impact and water and wind erosion. However, the soil loss (SL) by water erosion decreases with increasing roughness (R) according to the relationship –

$$SL \sim e^{-0.5R}$$. (Cogo, Moldenhauer and Foster, 1984).

The effect of tillage on a soil depends upon the shear strength of the implements used, the nature of the confining stresses and the direction in which the force is applied. The tillage tools are designed to apply an upward force to cut and loosen the compacted soil, sometimes to invert it and mix it, and to smooth and shape its surface. A distinct shear plane is formed from the base of the tool, which is concentric in shape in case of tines but is modified where the tool turns and inverts the soil as with a ploughshare.

In case of tillage work done by the tractor, the tractor itself apply force from above and compact the soil. The pattern of compaction depends upon tyre pressure, the width of the wheels and the speed of the tractor. This may result in an increase in shear strength through an increase in bulk density and in the number of clods larger than 30 mm in diameter. However, these effects are often more than offset by decreased infiltration and increased runoff due to wheeling that creates zones of concentrated erosion. This compaction generally extends to the depth of the previous tillage, up to 300mm for deep ploughing, 180mm for normal ploughing and 60mm with zero tillage (Pidgeon and Sone, 1978). {Different tillage practices used for soil and water conservation are discussed in Chapter 5.2.}

4.3.2. Mulching

It is the covering of the soil surface. It helps to conserve the soil moisture and maintain the moisture regime in root zone area by reducing the evaporation loss. Reduced evaporation potential under mulch reduces the loss of moisture from sub-soil as well as from underground water by inhibiting the capillary rise. It also helps to conserve the soil by protecting them from erosive agents like water and wind; otherwise they will detach and drive out the soil particles from their original position.

Mulching is practiced in different ways depending on the area, crops and agronomical practices of a farm. Mulching by organic material / residue is found to be the best particularly in dryland areas. The best result can be obtained by covering 70% to 75% of the cropped area. For any kind of commercial cultivation in drylands mulching should be done throughout the year. However, mulching in a scientific manner is not at all practiced in dryland areas though mulching is the best tool for conserving soil moisture present in the root zone. {Details of mulching is discussed in Chapter- 5.5.}

4.3.3. Contouring

Any operation along the contour of an area or farm is termed as contouring. The effectiveness of different contouring methods depends on the steepness and length of the slope. The type and nature of surface soil is also another factor that helps to determine the effectiveness of contouring methods. Depending upon the suitability there are many types of contouring methods like *contour cultivation, contour bund, contour ridge, contour trench, contour strip cropping, terracing*, etc. Almost all these methods are effective in dry land farming {For details of different contour methods see Chapter-5.2.}.

To implement any contour operation the suitable contour lines are first to be determined. For large areas it can be determined by engineering survey {described in Chapter- 3.10.}, which ultimately creates a **contour map** of the area. On the other hand, for small to medium sized area, a few simple methods are available to determine the contour lines. A brief description of these methods are narrated hereunder.

Simple Way to Setting Out Contour and Graded Contour Lines and Slopes in a Small to Medium Sized Farm Land.

The vertical distances, horizontal distances, slopes, contour lines etc are determined basically by the surveying works of qualified or skilled surveyors using sophisticated equipments. But there are several home-made equipments that are being used or can be used by farmers to determine these measurements. Few common tools like *measuring tape,* 2 to 5 metres long *measuring rod, plumb bob, carpenter level, poles* & *pegs* is required along with these equipments, which are easily available in any market. The home-made equipments and their functions are given below:

- **Boning Rod. To Determine Horizontal Lines and Slopes.** It is a simple 'T'-shaped item made up of wood. Its height is normally 100cm and the length of cross piece is 50cm. The width of the frame is 10cm. Usually 3 to 4 boning rods are required for work. While working, the sun should be kept behind the observer. With the help of eye-sight and by raising or cutting the ground surface, the top of all the bony rods (minimum 3) are placed in a line to create a horizontal line or slope.

- **The N-Frame Level. To determine Slopes and Contour Lines.** This instrument is a goal-post like wooden/iron structure of 0.9m height and consists of 2 to 3 m long main lath with two legs. On the middle of the main lath a carpenter level is firmly fixed. Through this equipment one can find two points of equal elevation. Now placing one leg on the previously created point a third point of equal elevation may be created. Repeating this process a line of same elevation i.e. the contour line can be created. In the same process a second contour line of desired height difference can be created with this equipment.

  To create desired slope (of a ditch or water ways) a partial modification of the equipment has to be made i.e. one leg has to be shortened. Now, placing the short leg in the starting point and by turning the N-frame around that leg the desired point of required slope can be found.

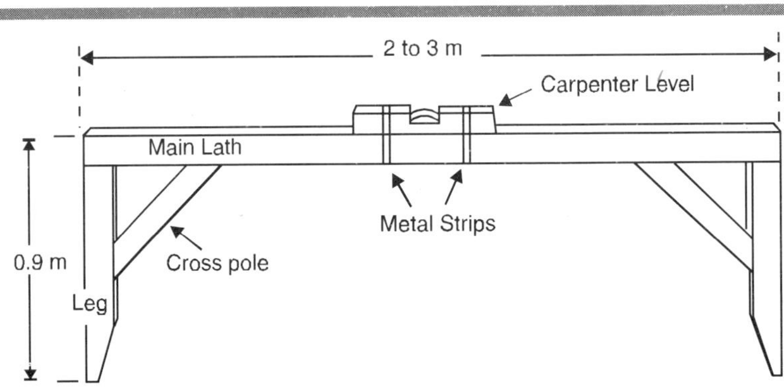

Fig. 4.3. N-frame Level

- **The Flexible Tube Water level. To determine contour lines and elevation differences.** This equipment consists of two portable wooden staffs of 2m heights and a 14m long transparent flexible tube of small (1.5 to 2cm) diameter. The tube is fixed along the length of the staffs from top to bottom, and is graduated as per scale. The tube is then filled with coloured or muddy water up to the 1m marks of the staffs placed in same plane. No air bubble is allowed in the water column of the tube.

  Now, wherever the two staffs are placed vertically straight on two points of same elevation, the free water surface in the tubes will maintain the same mark of the staffs. With this principal one can create contour lines and different elevations of a land surface.

Fig. 4.4. Flexible Tube Water Level.

4.3.4. Trenching or Pitting

In this method the trenches or pits are dug out across the slope and along the contour, suitable to the vegetation/crop present in that area. These structures store runoff water as well as eroded soil by reducing the runoff velocity and providing the runoff water more time for percolation. As a general practice, in all cases of trenching and pitting, the down slope side is bounded by ridge made-up of the dugout soils and the ridges are planted by suitable plants/crops fit in that situation. {Different types of trenches and pits are described in Chapter- 5.2.1.2.}

4.3.5. Terracing

Terraces are the large (in width) plain surfaced earth embankments constructed across the slope to intercept surface runoff and convey it to suitable outlet at a non-erosive velocity and to shorten slope length. They thus perform similar function to contour bunds whose size is smaller and the design specifications are not very stringent. They differ from one another in respect of spacing and length, the location of terrace out-lets, the gradient and dimension of terrace channel and in the lay out of the system. This difference basically depends on soil, its slope, and rainfall, its type and obviously on the cropping pattern. On the basis of that they are broadly divided in three types: Level terrace, Graded terrace and Bench terrace. {Details of the same is explained in Chapter- 5.2.1.2.}

4.3.6. Waterways

The safe passes of excess water collected from surface run-off is generally provided through channels called waterways. It is one of the important aspects of soil and water conservation measures because it indirectly helps to reduce the erosive power of overland flow and frequently guided the runoff water into the water harvesting tanks or farm ponds in the down slope for future use. Actually the main purpose of waterways is to dispose off excess runoff water safely from a specific area.

The waterways should be carefully designed and its dimensions must provide sufficient capacity to carry the peak-runoff from a storm with a ten years return period. Now the estimation of peak runoff is a vital part to design a waterway particularly for small areas. It also helps the dryland farming in many ways particularly in respect of implementation of different water harvesting structures. However, a complete surface water disposal system generally has three different types of waterways. They are *diversion channel, terrace channel* and *grass waterway* or main waterway.

Diversion channels are placed upslope of a farm or plot to intercept the water running off the slope above and divert it across the slope to a grass waterways. These channels are provided with slight grade and a suitable outlet for its proper function and life.

Terrace channels are generally placed upslope of terrace bank to collect the runoff water from inter terraced area. It is constructed across the slope with slight grade so as to convey the water with non-erosive velocity to a suitable outlet i.e. to main waterways placed along the slope.

Grass waterway or main waterway is the main outlet of a field, which runs down the slope with same grade as the slope, to carry the runoff water supplied by the diversion channels or terrace channels to any suitable outlet. The suitable outlet may be a stream or its smallest branch, a reservoir/tank that are generally designed on natural depression in the down slope. These main channels are generally lined with grass up to10^0 slope. On steeper slope the channels should be lined with stone or like materials and that should be up to 15^0 slopes. On the hill sides with alternating or combination of gentle and steep sections, use of different drop structures in the steeper portions in the grass waterways are recommendable. The successful vegetated waterway is dependent on good conservation treatment of its watershed, which reduces the peak rate of runoff and the volume of runoff carried by the waterway. {For details of Grass Waterways see Chapter- 5.3.8.3.}

4.3.7 Estimation of Volume of Peak Runoff of a Small Catchment Area

Several methods have been developed for estimating the volume of peak runoff from small areas where no measured data exist. These include (1) *Rational formula,* (2) *Cooks method,* (3) *Hydrologic soil cover complex number method,* and (4) *Table method*. The basic assumptions made for all the methods are: uniform rainfall intensity spread over the entire watershed for a duration at least equal to the time of concentration of that watershed. The first three methods are widely used and suitable in different situations but require a few meterological or other information, which may not be readily available. The fourth one involves minimum information thus suitable for unknown remote places.

Rational method: We will discuss the 'rational method' in brief to have an idea about this method of estimation, which is simple and widely accepted. In this method the peak rate of runoff is estimated by the following equation.

$$Q = \frac{1}{380} C\,I\,A \text{ (in metric units)}$$

Where, Q = peak runoff (m^3/sec).

I = intensity of rainfall (mm/hr), for a duration equal to the time of concentration and for a given frequency.

A = area of the catchment.

C = runoff coefficient (ratio of peak runoff to rainfall intensity, from curve)

The Hydrologic Soil Cover Complex Method: This method calculates runoff on the basis of recharge capacity of the area. Here $\mathbf{I_a}$ is the initial quantity of interception, storage in depression and infiltration of rain water in the area that must be satisfied before runoff occurs from a rain. It is assumed that the ratio of the actual runoff (Q) and the rainfall (P) minus the initial utilization (I_a), i.e. maximum possible runoff (P-I_a) is equals to the ratio of actual retention (P-Q- I_a) to the seepage capacity (S). On an average it is assumed that I_a = 0.2 S.

Hence the formula is: $$\frac{Q}{P - I_a} = \frac{P - Q - I_a}{S}$$

Therefore, $$Q = \frac{(P - 0.2\,S)^2}{(P + 0.8\,S)} \text{ mm}$$

Here the seepage capacity (S) is required to be calculated by using two tables on rainfall and moisture conditions.

Table method : This method was developed by Hudson (1981) for use in tropical Africa. In this method a table was created considering the area (A) and Catchments Characteristics (CC) and the runoff, in cubic meter per second, for a 10 year frequency. An area-weighted score based on the vegetation, soil and slope conditions determines the value of catchment characteristics (CC). The typical scores for each of these factors are shown below:

Table 4.4. Catchment Characteristics

Cover		Soil type		Slope	
Forest/heavy grass-	10	Deep, well drained soils (sands)-	10	Flat to gentle (0–3°)-	05
Scrub/med. Grass-	15	Deep, moderately pervious soil (slit)-	20	Moderates (3–6°)-	10
Cultivated land-	20	Fairly deep & permeable soils (loams)-	25	Rolling (6-9°)-	15
Bear or eroded-	25	Shallow, impeded drainage-	30	Hilly or steep-	20
		Clay or rocky soil surface for Mountains	25		
		Impervious or waterlogged surface-	50		

Now using the total area of the catchment (A) and the value of the catchment characteristics (CC) an estimated runoff can be obtained from the table given as follows:

Table 4.5. Estimation of Peak Runoff

Peak runoff as a function of catchment characteristics and area

A \ CC	25	30	35	40	45	50	55	60	65	70	75	80
5	0.2	0.3	0.4	0.5	0.7	0.9	1.1	1.3	1.5	1.7	1.9	2.1
10	0.3	0.5	0.7	0.9	1.1	1.4	1.7	2.0	2.4	2.8	3.2	3.7
15	0.5	0.8	1.1	1.4	1.7	2.0	2.4	2.9	3.4	4.0	4.6	5.2
20	0.6	1.0	1.4	1.8	2.3	2.9	3.6	4.4	5.3	6.3	7.3	8.4
30	0.8	1.3	1.8	2.3	2.9	3.6	4.4	5.3	6.3	7.3	8.4	9.5
40	1.1	1.5	2.1	2.0	3.5	4.5	5.5	6.6	7.8	9.1	10.5	12.3
50	1.2	1.8	2.5	3.5	4.6	5.8	7.1	8.5	10.0	11.6	13.3	15.1
75	1.6	2.4	3.6	4.9	6.3	8.0	9.9	11.9	14.0	16.4	18.9	21.7
100	1.8	3.2	4.7	6.4	8.3	11.6	14.7	18.2	21.8	25.6	29.9	35.0
150	2.1	4.1	6.3	8.8	11.6	14.7	18.2	21.8	25.6	29.9	35.0	40.6
200	2.8	5.5	8.4	11.7	15.3	19.1	23.3	28.0	33.1	38.5	45.0	52.5
250	3.5	6.5	9.7	13.2	17.2	21.7	27.0	32.9	39.6	46.9	55.0	63.7
300	4.2	7.0	10.5	14.7	19.6	25.2	31.5	38.5	46.2	54..6	63.7	73.5
350	4.9	8.4	12.6	17.2	23.2	30.2	37.8	46.3	53.8	62.5	71.5	81.0
400	5.6	10.0	14.4	19.4	25.6	33.6	42.2	51.0	60.0	69.3	79.5	90.0
450	6.3	10.5	15.5	21.5	28.5	36.5	45.5	55.5	65.5	76.0	86.5	97.5
500	7.0	11.0	17.0	23.5	31.0	40.5	51.0	62.0	73.0	84.0	95.0	106.5

A is the area of the catchment in hectares, CC is the catchment characteristics from previous table, and the runoff (in cubic metres per second) is for a 10-year frequency.

The adjustment factors to runoff values are placed in the following table :

Table 4.6. Adjustment Factor to run off

Main factors	Sub-factors	Adjustment
Rainfall intensity:	High intensity-	multiplies by- 1.0
	Low intensity-	multiplies by- 0.75
Catchment shape:	Long, narrow-	multiplies by-0.8
	Square/circular-	multiplies by- 1.0
	Broad, short-	multiplies by- 1.25
Return period:	2 years-	multiplies by-0.9
	5 years-	multiplies by- 0.95
	10 years-	multiplies by- 1.0
	25 years-	multiplies by- 1.25
	50 years-	multiplies by- 1.5

Now the estimated peak runoff may be adjusted by multiplying with the adjustment factor stated above.

4.3.8. Measurement of flowing water

Measurement of water (flowing water) is an essential aspect in dryland farming. It is very much important to measure water for runoff, irrigation, well/storage discharge etc, not only for a dryland farming situation but also for any kind of water use.

It is easy to measure the water at rest i.e. water in tank, pond, reservoir or as standing water in the field. It is actually the volume of water calculated from the size and shape of the tank, pond, plot etc. It is expressed as cubic meter or liter or hectare-centimeter or acre-feet.

The measurement of flowing water is a somewhat complicated job and commonly expressed as cubic meter per second (cumsec) or cubic feet per second (cusec). From this flowing rate and the time of flowing one can estimate the quantity of water. For example, water flowing at the rate one cubic-foot per second for 24 hours equals to 1.9835 (or 2) acre-feet of water.

This measurement is of two types, **(a) measurement of flowing water in open channel** and **(b) measurement of flowing water in pipes**. Now, we will discuss in brief about few of these methods for necessary knowledge.

Float method: It is a cheap and simple method to measure flowing water in open channel, but not very accurate. The formula expressed in this method is Q = A V where, Q is discharge, V is mean velocity of the flow, and A is cross section area of the channel (uniform for 20 to 25 m long).

The velocity will be measured by determining the velocity of a floating object, allowed to flow with the current for at least a distance of 20 to 25 meters. To get the mean velocity, the measured velocity of the float (consider as the velocity of upper surface of water with minimum resistance) is to be multiplied by 0.8.

Example: if the average depth of a 1m width semicircular channel is 0.5m and the considered length is 25m then the cross-section area (A) ($[\pi \times r^2]/2 = 3.14 \times (0.5)^2 / 2 =$) is $0.39m^2$. Now if a float takes 40 seconds (average of several reading) to cover the length (L) of 25m then the velocity (V) of the flowing water of the channel is ($[L/T]0.8=[25/40] \times 0.8 =$) 0.5M/SEC. Thus the flow (Q) is (0.39/0.5=) $0.2m^3$ or 200 liters per second.

Current meter: The velocity of water moving in a channel can be measured more accurately by current meter. It is an instrument, having cups or propeller fitted in a rotating shaft, which is attached to a speed counter. For each current meter a rating curve is provided, by which the velocity of water can be calculated.

Direct discharge method through weirs: It is a reliable and practical method to measure flowing water in open channel. Weir is a notch of a regular form through which water flows. The weir can be made up of wooden plank, metal sheet or concrete with an opening cut (notch) on the top edge. It may be (1) sharp crested weirs or (2) broad crested weirs.

Sharp crested weirs are of three types, e.g. (i) rectangular weir, (ii) trapezoidal (1:4 side slope) weir, and iii) triangular (90° 'V' notch) weir.

For measuring the rate of flow in a channel, a weir of known dimension is placed across the channel and the head of flow (H) over the crest of the weir will be measured. Now if the length of the crest (L) is equal to the width of the channel then the corresponding equations for three types of weirs, to estimate the discharge (Q), are as follows (in metric system):

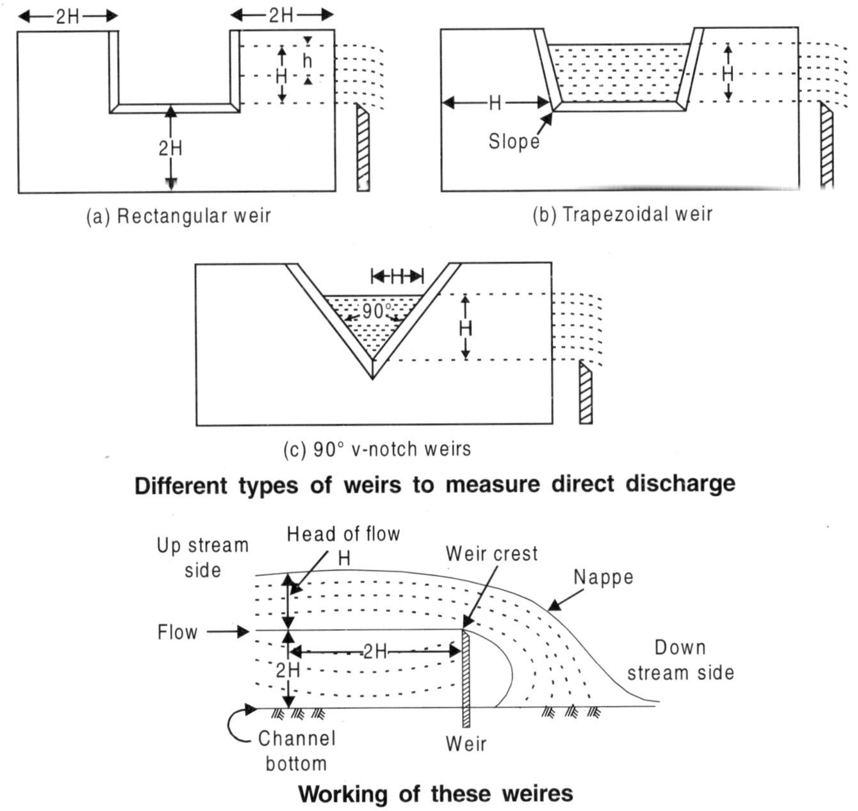

Fig. 4.5. Different types of weirs for measurement of discharge

(*i*) $Q = 0.0184\ L\ H^{3/2}$

(*ii*) $Q = 0.0186\ L\ H^{3/2}$

(*iii*) $Q = 0.0138\ H^{2.48}$

Where, Q = Discharge in liter/second, L = length of crest in c.m. and H = Head over the crest in c.m.

Now if the width of the channel is bigger than the length of the crest, which is very common, then the water approaches the notch in a converging manner. In such a situation the effective length (L) of the crest will be 0.2H less (L – 0.2H) and this modified L will be used in the above formulae.

There are many other methods to measure the flowing water in open channel e.g. 'orifices', 'broad crested weirs', 'flumes', which are not discussed here.

Measurement of flowing water in pipes: The simplest method of measuring the flowing water in pipes, at apparently in a uniform rate, is to fill a container of known volume by the pipe and record the time required. Thus the rate of flow can be calculated by dividing the volume by the time. The apparent flow rate can be judged by placing the end of the pipe horizontally above the ground surface, if possible, and measure the distance of fall of water from the pipe to the ground. For uniform flow rate, the distance will be constant.

But, when the flow rate is not uniform, average rate may be obtained by frequent measurements to get accurate data. For example, if the valve of a pipe is opened to maximum for three different times and takes 1.75, 1.50, and 2.00 minutes to fill a 20 liter container then the average is 1.75 minutes. Hence, if one wants to irrigate a piece of land with 286 liters of water by that pipe, the pipe should be kept open to its maximum for (286×1.75/20) 25 minutes.

The more accurate methods to measure the flow rate in pipes are *'venturi'*, *'pitot tube'*, *'elbow meters'* etc. However, we will not discuss about these methods here.

4.3.9. Drop Structures

It is a weir structure generally used as spillway to dispose runoff or any kind of flowing water safely towards its natural course. The drop structure is efficient for controlling relatively low heads, normally up to 3 meters. There are three principal types of drop structures, viz. *Drop spillway* (Figure: 4.4), *Drop inlet spillway*, and *Chute* used in check dams, low earthen dams and in waterways. All these structures are used to protect soil from erosion caused by fall of moving runoff water from heights. Out

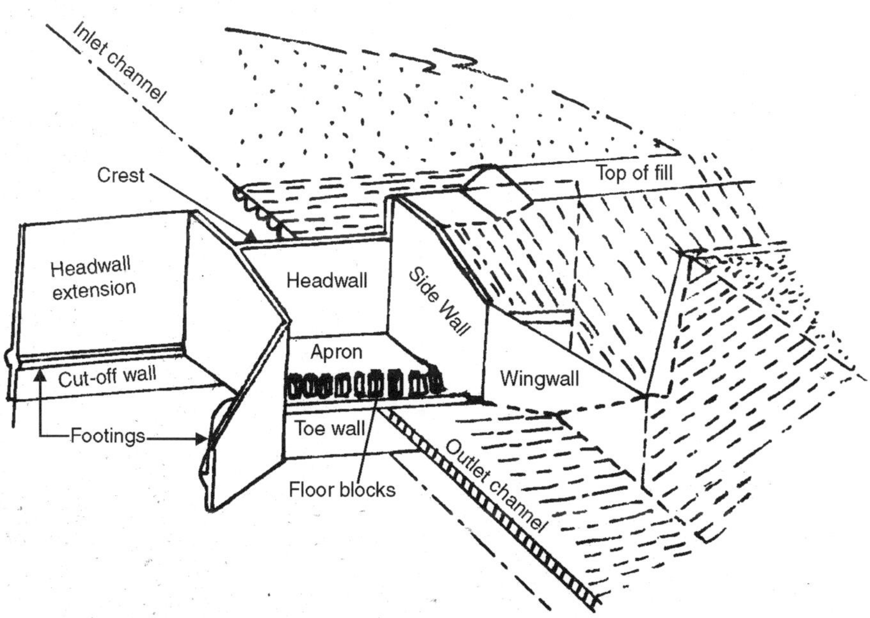

Fig. 4.6. Common Drop Structure

of these structures Chute is used in high slope (30° to 40°) whereas drop inlet and common drop structures are used in moderate and low slopes respectively.

Most of the commonly used structures consist of four major components, such as embankment, inlet, conduit and outlet. However, the selection of drop structure and its design is dependent on rainfall pattern, soil and obviously on slope of the land. Except earthen embankment the structures like inlets, conduit and outlet can be constructed by reinforced concrete, plain concrete, rock masonry, stone or concrete blocks, and gabions.

In common drop structures, strong and wide embankments are used. In chute, the open channel type conduit happens to be narrow and long and the outlet may be an apron provided with any type of energy dissipater to minimise the erosive effect of dropping water. The conduit receives water from the inlet and conducts it through the structure up to the outlet. In drop spillway and chute conduit it is of open channel type and in case of drop inlet structures it is of close type in the form of a box or pipe. Water flow passes through the wire opening present at inlet, drops through conduit to an approximately level surface, called 'apron', and then passes into the down slope channel. Sometimes instead of apron, a small water body called 'stilling basin' is used to reduce the erosive power of falling water effectively.

The role of different drop structure is as follows:

- Grade stabilisation in waterways, channels, gullies and different outlets.
- Reduce soil erosion and sediment load of streams and rivers.
- Safe disposal of overflow of different water storage structures or reservoirs.
- Control flowing of irrigation water.

4.3.10. Farm Ponds

In relatively flat terrain, where the topography does not offer itself to construct an embankment to create small or medium reservoir, ponds can be constructed to store rainwater. This structure is very much useful in farming purposes particularly where no other source of water is present. This excavated or dugout structures are usually constructed in the lower most part or natural depression of a farm in specific geometrical shapes like square, rectangular or circular, considering the type & size of catchment area.

However, in undulating terrain farm pond can be constructed and can be fed by diverted surface runoff (from terrace etc.) or by seepage of ground water or by diverted stream/spring flow. Sometimes it may be a combined structure made up by excavation of soil together with embankment by soil.

Farm pond can be used for irrigation, fishery, water for cattle etc. However, the selection of site of pond or ponds in a farm is a vital aspect to be considered initially. Proper selection of site takes necessary care for the reduction of construction cost and best utilization of the pond. Multiple farm ponds are always recommendable to increase the efficiency of storage and utilization of rain water, which also reduce the construction cost. Proper seepage control measures should be taken while constructing the farm pond {see chapter- 5.1.2.5.}. For dry season necessary evaporation control methods may be adopted {see chapter- 4.1.3.1.}.

A convenient depth is chosen for the pond, which usually varies between 3 to 5 mt. depending upon the situation. Greater depth requires high excavation cost disproportionate to the additional capacity it provides and may increase the seepage loss due to porous sub-surface and increased hydraulic pressure. But greater depth reduces the evaporation loss on volume basis as it expose a minimum water surface area in proportion to volume. In some situation, particularly in very dry areas, the greater depth with adequate measure of seepage control will be very much effective.

The farm ponds should be provided with controlled inlets so that the erosion does not take place at the inlet portion. Outlets are also provided in these ponds to discharge excess water safely in rainy season. These structures are generally effective in farms of dry areas with a slope up to 4%. The watershed of farm pond is to be determined properly and if necessary can be designed in a way that the annual runoff

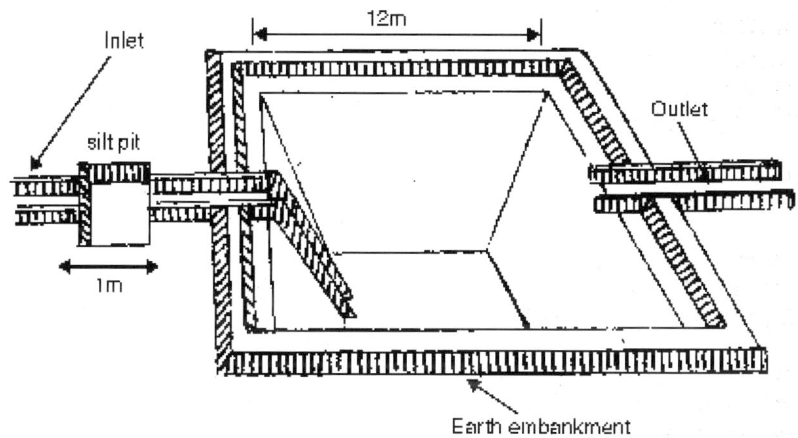

A—Sketch of a Excavated/dugout farm pond (small)

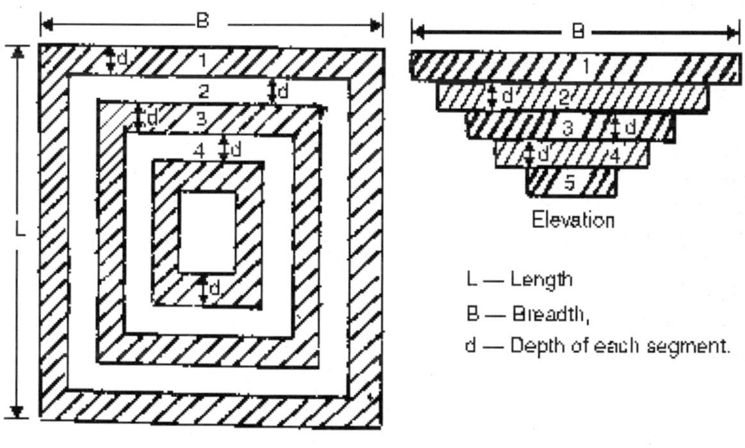

B—Cross section of Excavated/dug out farm pond (large)

Fig. 4.7. Excavated / Dug out Farm pond.

is sufficient to fill up the pond. The watershed area should be maintained carefully to minimise the soil erosion to minimise the silting possibility. The dugout soils should be disposed out of the watershed area otherwise rainfall will wash the material back into the pond.

One specimen dimensions of a small farm pond is given below:

Bottom – 6mt.× 6mt.; Top – 12mt. × 12mt.; Depth – 3mt.; Side slope – 1:1;

Silt pit size at inlet – 1mt. × 1mt. × 0.5mt;

Inlet width – 2mt.; Capacity – 250mt.^{3};

Operational area – 1 hectare (approximate).

In case of somewhat bigger and multipurpose farm pond, a step-like cross section for excavation is usually adapted in dryland areas particularly in light textured soils. This method of construction is practiced since long past. However the estimated volume of this type of ponds are calculated by the Prismoidal formula, stated as follows:

$$V = (A + 4B + C) / 6 \times D$$

Where: V = volume of excavation (m^3);

A = area of excavation at ground surface (m^2);

B = area of excavation at mid-depth point (m^2);

C = area of excavation at bottom (m^2);

D = average depth of the pond (m).

4.3.11. Check Dam

Check dam is a small permanent type of obstruction that checks the flow of water running through any watercourse and intercept the eroded soils carried by it. These structures are placed across the nala or gully with necessary passage for excess water to flow. There are different types of check dams (Figure:4.6) depending upon the slope, soil, rainfall and available raw materials.

(a) **Brush Wood Check Dam.** These dams are constructed by driving wooden pegs into the floor of nalas and gullies along the cross section. Then the brushwoods are intertwined along the driven pegs. These dams are adaptable in all areas irrespective of soil and rainfall.

(b) **Loose Boulder Check Dam .** These are porous checks dam constructed by using loose boulders available nearby. These dams are also adaptable in all areas irrespective of soil and rainfall.

(c) **Gabion Check Dam.** These are the dams made of wire woven baskets filled with stones. These are more effective in the areas of high slope and high rainfall. In gullied areas it is a very effective measure.

(d) **Permanent Check Dam.** These masonary structures are generally used in big nalas and gullies. In case of dryland farming it is of very limited use except for creating reservoir in suitable situation to store water to be used afterwards. However, the economy of this use should be calculated properly. However,

Water Harvesting (WH) and use of water for farming in Semi Arid/Arid Areas

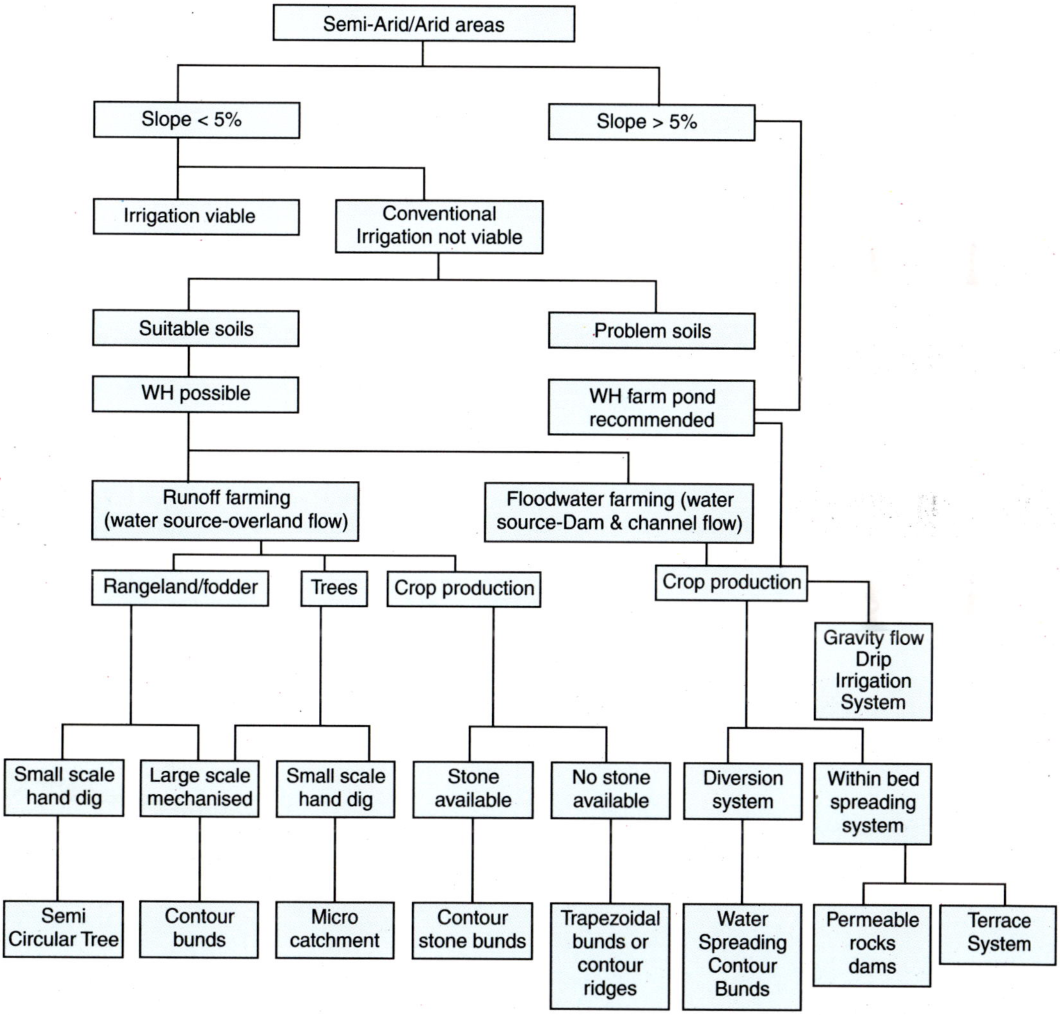

PLATE-I

Cement Nala Bund

Low earthen (in bag) dam

Staggered trenches

Loose boulder check dam

Embankment pond

Farm pond

Terrace cultivation

Low cost drip irrigation

PLATE-II

Fig. 4.8. Different Types of Check Dams.

water storing check dam can be constructed on community basis and generally constructed by stones or bricks joined by cement mortar. This type of structure should be provided with proper drop structure to pass the excess water safely.

Gully can be effectively plugged by using different kind of check dams placing at different intervals. {Details of gully plug by check dams are narrated in 4.2.3.}

4.4. COMMUNITY SOIL & WATER CONSERVATION STRUCTURES

4.4.1. Nala (Natural Watercourse) Bund

These are the embankments or bunds or check dams placed across the nala or natural watercourse to halt and store the flowing water and regulate the overflow through wire or channel (Figure : 4.9). This can be adopted in low to medium rainfall areas. These embankments are constructed by using general engineering principles. In steeper slope series of nala bunds are more effective. The nala bunds should be always designed in a way that there must be some arrangement for safer disposal of overflowing water to down stream and thus enhancing its life span. Considering average peak flow of the nala the design of bund and its spillways are to be determined.

This structure helps to conserve both the soil and water in dry areas. It provides the flowing water more time to percolate into the soil, which increase the moisture status of underground strata and thus helps the ground water recharge process. It induces an underground flow of water towards the down slope and in process reduces the immediate runoff loss of rainwater. At the same time by reduction of the erosive power of the flowing water the gulling process of erosion has been reduced. These nala bunds, as a barrier to the water flow, reduce the sediment carrying capacity of the water and as a result the eroded fertile soils are deposited in the up slope of the bund.

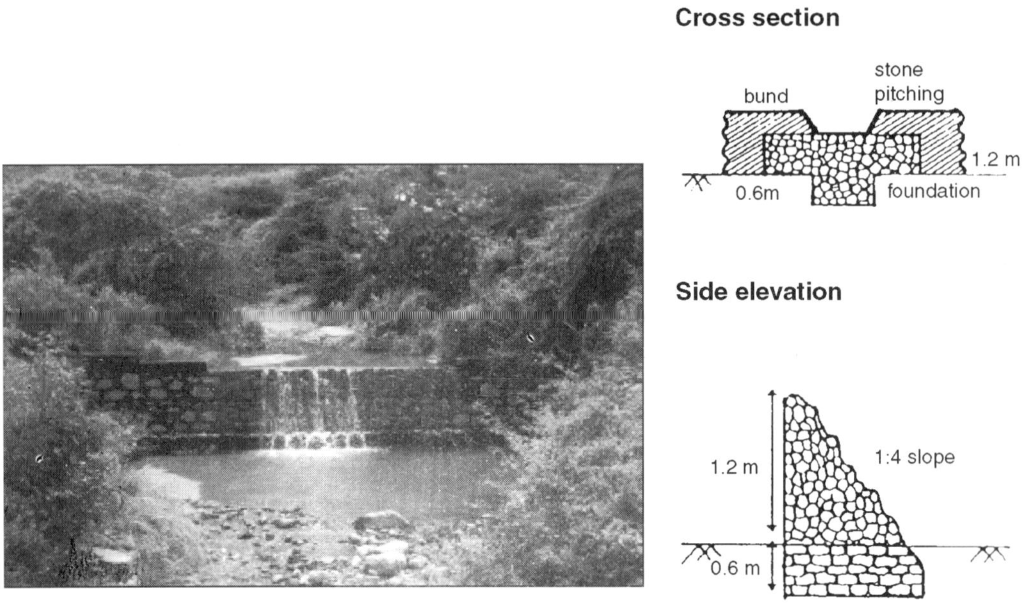

Fig. 4.9. Use of Bunds and Checks in Nala (Waterway).

4.4.2. Vegetative Filter Strips

The strips of suitable (erosion resistant) plants species are planted across the permanent watercourse or nala (Figure : 4.10) to check the velocity of the flow and thus arresting the eroded soil in the up side. This structure may be adopted in all areas irrespective of soil and rainfall. In steeper slope it may be combined with nala bund to extract maximum response.

Fig. 4.10. Vegetative Filter Strip

4.4.3. Surface Water Bodies/ Small Reservoirs

Structures set up to store the runoff water always requires the estimate or information or idea about (1) the watershed or catchment-area of that structure, (2) its runoff producing characteristics and (3) its soil and land use topography. With these information the structure of tank/reservoir may be designed. At the time of designing the size of water body, treatment for seepage loss, purpose of use of that water should be considered properly. With this information one can assess the economy of these structures. The different structures used in dry land farming are narrated below.

4.4.3.1. Percolation Tanks

Percolation tanks are constructed in the bed of natural paths of the run-off water or are constructed in the outlet of diversion channels or waterways where the land topography were flattened or depressed. It is generally constructed in the up streams when the terrain is slopy (4% to 5%) and annual rainfall is poorly distributed over the season that provides the water little scope to percolate. In such areas nalas or natural flow paths get flooded for short periods during the monsoon and remain dry for rest of the year. It is therefore necessary to arrange for the rainwater to soak into the ground and as a result an underground flow of water towards downslope will be created. This underground flow can augment the moisture regime to the crops in the down slope as well as increase the sub-surface recharge of wells and other water bodies present in the down slope.

The excavation of this tank with an earthen embankment having necessary drop structure would be very much effective. The name of this structure itself suggests that no seepage control measures are required, on the contrary regular de-siltation is recommendable. The bed of this tank may be utilised in the 'rabi' season for cultivation of any suitable crops using the residual moisture present in the gathered silt. In the areas of small land holdings this structure is generally effective as community structure.

However, percolation tank, in favourable situations, may be constructed for individual use where the seepage water is used for crop production in plots just below a small percolation tank. (Figure : 4.11) The plot is so designed that the seepage water moistens its root-zone area automatically and the excess water is diverted through the sides of the plot. It is basically a model, which is narrated with the help of the following figures.

4.4.3.2. Embankment/Surface Pond

It is the type of reservoir where earthen embankments store the water on ground surface utilising topographical advantages. The advantage of natural barriers in the sides of the reservoir is generally used for construction of this type of structure. The excavation from the base of the reservoir is required only to build the embankments. This structure is very much effective up to 4% slope. On higher slope, particularly in terraced area, this type of pond (relatively small in size) may be useful to store the harvested water from the upslope terraces (Figure: 4.12).

A sufficient top width in the embankment is provided to serve as a road, connecting both the banks of the water body. On the constructional part, the side slopes of up-stream and down-stream embankments are generally kept 3:1 and 2:1 respectively. In case of embankments with porous soils, a *core wall* of impervious soil is provided to cut down the excessive seepage losses. A *key trench* with the foundation is provided to secure the embankment to, overcome sliding action. Other supportive structure like *rock*

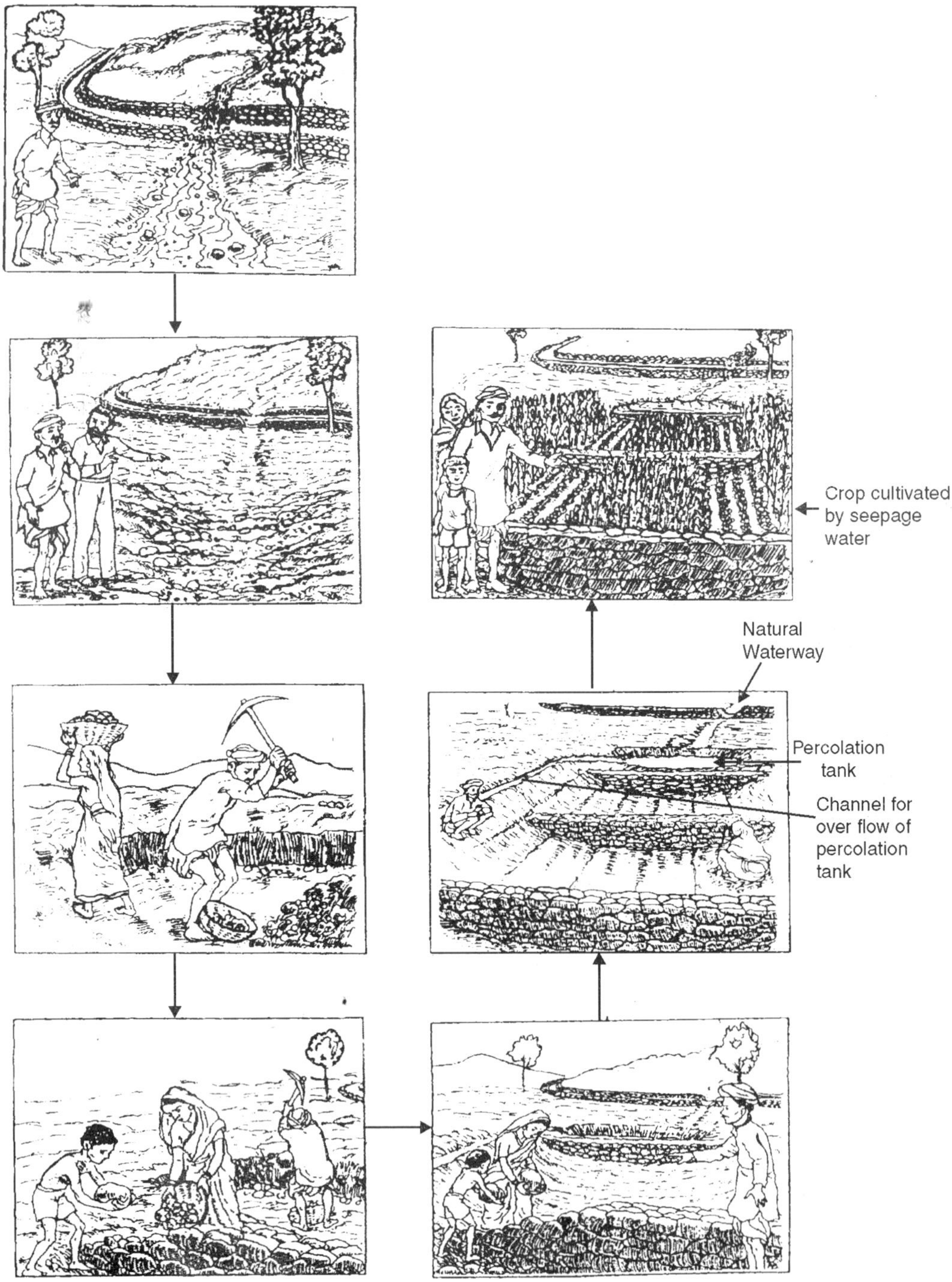

Fig. 4.11. A Model of Utilisation of Percolation tank for Individual Purpose (taken from the book :) "Resources management in rainfed dryland," published by 'Myrada')

Fig. 4.12. Surface Type Water Harvesting Structure in Slopy Land.

toe is provided to the embankment to check the grade of phreatic line. There should be an outlet structure for safe disposal of excess water or to avoid overtopping.

This type of structure is most common throughout India, particularly in dry land areas. It may be constructed on individual basis or on community basis.

4.4.3.3. Low Earthen Dam

This type of structure is generally constructed on community basis or constructed by the pubic authority. However, in some cases this structure may be implemented by any individual farmer whose farm is situated in and around the place where this type of structure is most suited. Its economic potentiality is very high in dry areas in respect of fishery, fish nursery, and selling of water apart from own use.

However, this low earthen dam/mini irrigation dam may be constructed across any suitable waterway to create a small water reservoir for providing irrigation water to crops or for other needs. The following criteria are to be considered for selecting the site and design of this type of dams.

1. There should be no possibility of the dam being breached by floodwater.
2. The seepage line should be well within the toe at the down stream face.
3. The up-stream and down-stream faces should be very much stable.
4. The foundation sheer stress of core wall should be within safe limits.
5. The free flow of water should be controlled.
6. The dam and foundation should be safe against piping and underground mining.

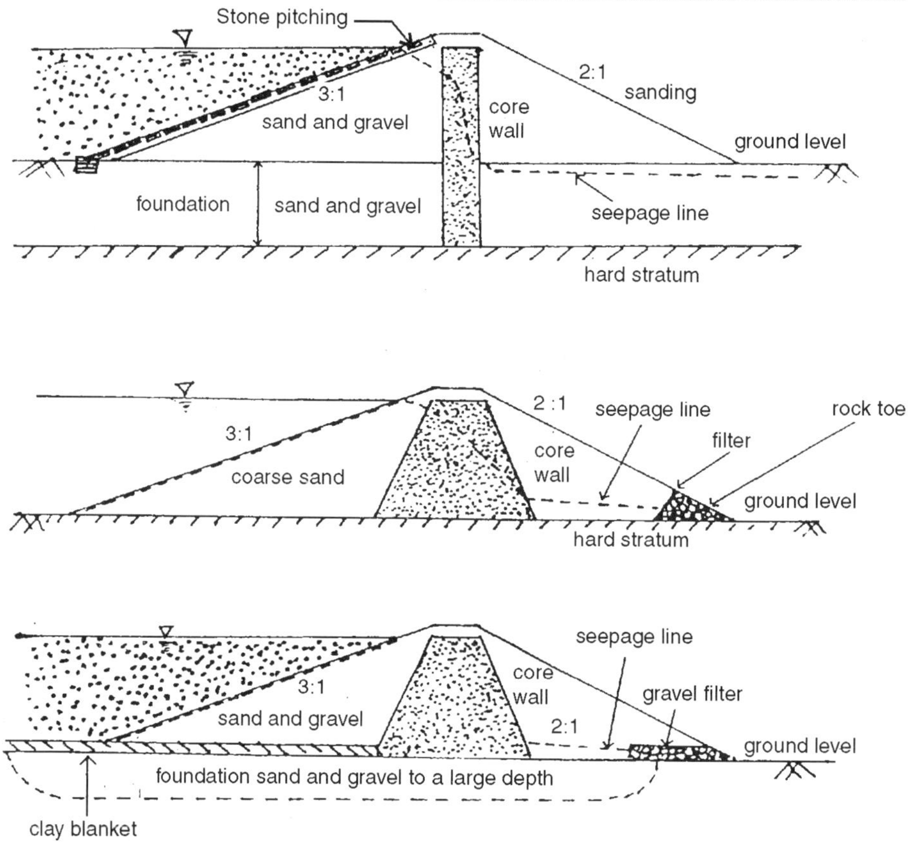

Fig. 4.13. Cross Sections of Different Types of Low Earthen Dams.

7. The up-stream face of the embankment should be properly protected.
8. The well-defined catchment area of the dam should be managed properly in respect of soil and water conservation measures to enhance the life and efficiency.

Cross sections of earthen dam with different type of foundations –

A comparative statement of water harvesting structures in respect of different aspects is narrated below in tabular form.

Table 4.7. Comparative statement of water harvesting structures

Structures	Soil Type	Slope	Utilisation	Beneficiary
1. Farm pond	Red and black	Up to 4%	For irrigation purposes and for fishery.	Individual
2. Percolation tank	Soils with light textured sub-soil	4% to 8%	Ground water recharge and subsurface moistening of down slopy plots.	Community/ individual
3. Embankment/ surface pond	Red and black soils	Up to 4%	For irrigation, fishery and for ground water recharge.	Individual/ community

Structures	Soil Type	Slope	Utilisation	Beneficiary
4. 'Nala' bund	Red, black and lateritic soils	Upto 4%	Gully reclamation, irrigation and ground water recharge.	Community
5. Low earthen dam	Red, black and lateritic soils	4% to 8%	All shorts of community purposes including ground water recharge	Community

4.5. BIOLOGICAL TECHNOLOGY

4.5.1. Seed Technology

Seed is basically a dormant or arrested state of plant with storage of food covered by a seed coat. Farmers all over the world have produced, stored and used their own seed. Modern cultivation methods with high quality seeds gradually changed this inherent practice, but, apart from rich and prospered cultivators all other farmers are bound to use their own seeds. In dryland areas the farmers have no other option as they have little money to purchase the high cost seeds and at the same time the companies are not interested to supply different kinds of seeds in the dry areas.

In this context it is always better for them to know and practice the scientific seed technology according to their capacity to maintain the age-old practice of growing, storing and using their own seeds. Instead of purchasing the high cost hybrid or genetically modified seeds, every year the indigenous or locally adapted high yielding seeds of all necessary crops or its varieties should be grown and stored scientifically. These high yielding varieties have demonstrated greater resistance against many problems of the farm and are less expensive i.e. more economic for the farmer. Farmer can adapt established on-farm hybridisation methods to produce hybrid seeds. At the same time, proper selection of quality parent material for cross-polinated crop can help to produce better quality seeds.

There are many factors that influence a farmer's choice of seed. However, apart from the seeds, which is engineered to produce higher yields, if any other variety is chosen by a farmer, a few simple activities can be practiced by improving seed harvest, processing and storing.

Procedure of Seed Collection

1. Choose and mark the healthy plants just prior to harvest, where the produce appears to be of good size, shape and colour.
2. Collect seeds when the crop is fully matured. When the seeds are in the pulp (e.g. Gourds, Brinjal etc.), pick the selected and marked fruit in fully ripe stage. Where the seed is synonymous with the crop like cereals, pulses, oil-seeds etc. dry the seeds on the plant itself (except groundnut) and pick just before it is dropped by the plant. The size and specific weight of the seed should be considered while collecting the same.
3. Cleaning of Seeds. Cleaning of dry plant waste would be done first in case of grain seeds. Then it may be dipped into water for further cleaning and

eliminating the seeds whose specific gravity is low. Wash seeds that have juice or flesh adhered to them (like Tomato etc.) with cold water. Rub seeds surrounded with mucilage (Brinjal etc.) gently with coarse sand and then wash off using a sieve.

4. Drying of Seeds. Seeds should be collected in a bright day after 10 am and they should not be moist by dew or rain. Then dry the collected seeds properly before storing it for future, except the recalcitrant seeds, which by definition loses viability after drying. The seeds which have dried well snap and crack easily when broken. By every one percent (1%) decrease in seed moisture, longevity will be doubled. Proper sun drying is sufficient in most cases. Under Intense sunlight (11 am to 2pm) seeds dry rapidly and this may lower the germination capability. Uniform drying by spreading them in thin layer with occasional turning is the best method of drying.
5. Treatment of Seeds. After proper drying the seeds may be treated to protect them from seed-borne diseases, storage insects and soil insects if prevalent in that locality. Bio/Organo Chemicals or additive available locally or in market may be used for the treatment. However, locally available Bio/Organo additives (like neem leaf etc.) should get preference over chemical pesticides or fungicides.
6. Labelling. In any possible way the labelling of the seed, specifying crop and its variety, date and place of collection should be done.
7. Storing of Seed. The seeds should be stored in a place safe from pests, rodents, mould, micro-organisms and moisture. At the same time it should be safe from any natural adversity like cyclone, flood etc. Various indigenous and improved or modern seed storage systems are available for use and ultimately the storage structure should be placed in a cool and dark place. Interestingly every 5 % decrease in storage temperature doubled the longevity of the seed.

As the dryland farming require different type of crops in different time to utilise the available moisture, the availability of seed is a very important aspect to look into. So it is preferable that the seeds should be always at farmer's disposal.

Seed Testing. Testing of seed should be done before planting. In dryland farming sowing time is very much restricted and specific, so the confirmation of germination of the seeds are very much vital. Now the in-situ methods of seed testing are as follows–

Apart from few special seeds (like hard coated seeds etc.) the other seeds are placed in 100 numbers on a well-soaked news paper/blotting paper and are floored by a polythene sheet. Then roll or fold the paper along with the polythene sheet in such a way that no opening is left and leave for five days in a cool and dark place. After 4 or 5 days unroll the paper and count the germinated seeds. In case of below 80 percent germination it is better to cancel the lot of the seeds or increase the seed rate proportionately.

4.5.2. Drought Hardening of Plants/Crops

Many plants that recover from a period of water stress or subjected for several days to moderate dehydration has increased drought resistance. When hardened plants are subjected to drought, their protoplasm shows a lower viscosity and higher permeability to water than that of similar but non-hardened plants. Thus the adapted protoplasm does not become rigid and brittle as soon as that of upgraded plants, and is better capable to hold water against dehydrating force. Higher rates of photosynthesis, lower rates of respiration, and a higher root:shoot ratio also characterize hardened plants.

Drought tolerance can be acquired in plants through physiological means, and the thus acquired tolerance is heritable. Seeds are soaked in water and then air dried in shade; this process is repeated three or four times before sowing in the field. Alternate moistening and drying of seeds increases protoplasmic viscosity, temperature threshold of seedlings to survive dehydration. Such increase in drought tolerance of tomato plants, which was retained up to two subsequent generations, was noted by Marlyanova and Zhurikhin (1971). However, all crops or varieties do not respond in same way by this method.

Not only seeds but also the seedlings were alternatively moistened and dried (Henekel-1966). In consequence their capacity to withstand drought increases throughout the life of the resulting plants and also in next two to three generations in cereals, two in tomatoes, upto six in sunflower. The drought-resistant wheat Novyj (new) was produced in this way from the high yielding wheat variety Washington. In Israel, hardening pre-treatment of seeds increased the resistance of leaves of sorghum to water loss, but had no effect on yield under conditions of sever drought (Jacoby and Oppenheimer 1962).

Other methods of pre-sowing hardening of seeds have also been investigated; solutions containing various inorganic salts, organic acids, menthol, (2-chloroethyl) tri-menthyl-ammonium-chloride, auxins, and extracts from different plant tissues were used. Seed treatment with 0.09 percent solution of succinic acid for three hours before sowing improves germination, promotes plant growth and increases crop survival.

Biotic Stress. Apart from the abiotic stress caused by water, temperature, soil and any unanticipated aberrant weather conditions, the biotic stress caused by the insect pest, diseases, poor nutritional management and finally the genetical incapability may create serious problem to the growth of crops/plants of dryland area. These biotic problems should be combated with chemical or biological or cultural or genetical measures or with the combination of them. To avoid this biotic stress the following biological technology may be followed in dryland farming.

4.5.3. Application of Bio-Chemicals

Why Bio-chemical. Till recent past the chemical control of pest and disease was the only accepted and used method in farming practices. But recently we realise and confirm that these chemicals, which are mostly poisonous, are doing tremendous harm to the mankind from different directions. All these chemicals, irrespective of its poisoning capacity, are slowly and gradually poisoning our flora and fauna including us in different ways. It is also slowly toxicating the soil and water of our environment. These chemicals are indirectly creating more dangerous pests which becomes resistant to any of that chemical and forces us for searching new brands of poisons. In high rainfall areas where the environment receive and release good amount of water, the adverse effect of these chemicals including fertilisers are less due to the dilution effect. But in dryland areas, due to low rainfall, there is almost no dilution effect of these chemicals including fertilisers. Naturally the adverse effects are more severe in comparison.

As we are bound to protect our crops from the pests and diseases to feed ourselves, we should apply these chemicals judicially and in combination with other methods of control like biological control and cultural control.

Application. So in dryland farming application of chemicals should be reduced to a minimal level. It should always be applied alternately with bio-chemical products like Neem-products (cake, oil, solution), Nafatia-products (solution), Tobacco-products etc. either in crude form or in industrial form. Most importantly, in dryland areas, the application of said chemicals should be followed next to the cultural methods, from hand picking to application of different traps, and newly introduced biological measures. {For detail see chapter No. 5.3.6.2.}

4.5.4. Breeding and Genetic Engineering

The progress of conventional breeding for increased drought resistance without reducing productivity or quality of the produce has been slow. This is obviously because of drought resistance character of plant, which is governed by several factors e.g. root number and extension, resistance to water flow, stomatal size and frequency, leaf water potential, abscisic acid (ABA) and proline accumulation, cuticular resistance and waxiness.

So far, success achieved in developing drought-resistant productive varieties has been more a matter of chance than of directed breeding. Actually, it is very difficult to understand and assess quantitatively the interaction between growth stages because resistance to moisture stress or drought is the result of many frequently independent morphological and physiological traits. The present trend is to breed varieties or lines for superiority. In general, no satisfactory physiological, anatomical, or morphological characteristics have yet been defined which could serve as reliable criteria in selecting drought resistance. Most of the traits conferring moisture stress resistance are polygenic. Nevertheless, many attempts have been made to use xerophytic characters as indices of drought resistance for breeding purposes, such as the osmotic pressure of germinating seeds, viscosity, etc.

Breeding for high yield under dryland conditions can be achieved either by eliminating the traits that limit the productivity and cause fluctuation from season to season or by combining traits that increase the productivity potential of the plant. Breeding for early maturity is one of the examples of this idea.

With the advent of *genetic engineering* methods, the opportunities for nutritional upgradation of coarse cereals have improved and it need efforts. All coarse cereals can be used in formulating and developing a range of ready-to-eat foods, snacks, extruded products, grits, popping, composite flours etc. Quality protein maize, waxy maize, baby corn, pop sorghum, special sorghums for different culinary purposes, white bajra, the nutritious ragi and several of the millets have a role in developing diversified food preparations and products. Now, if this development of coarse cereals, basically dryland crop, is continued then it can be possible to adapt these improved crops in dryland areas in future.

Participatory Plant Breeding. Poor farmers can select and store the best seed for sowing in the following year. Now, the adapation of farmer participatory approaches for selecting and breeding have a good potentiality to improve the existing varieties. The scientist /breeders should collaborate with farmers, who should be consulted at every stage to set the breeding objective. Participatory plant/crop breeding may be done with locally adopted and selected parental materials or it may be done between locally selected cultivars and high yielding varieties in farmers field. The main advantage of this practice is that whatever the development occurs, through this process, farmer will adapt immediately. Steps for participatory plant breeding:

(a) Set the breeding objectives.

(b) Identify the parent material.

(c) Decide the model.

(d) Crosses to be made.

(e) Monitor the progress - Follow the participatory variteal selection approach of new lines identified from participatory plant breeding.

(f) Facilitate entry of the said lines in formal trial system.

(g) Prepare trial system proposal, if success is achieved.

4.6. AGRICULTURAL METEOROLOGICAL INSTRUMENTS

Dryland farming is very much sensitive towards change of any climatic factor, particularly the rainfall- intermediate short-span drought spell in between two rains in rainy season and uncertain occurrence of pre & post rainy season rainfall. In dry season prevalence of high temperature, high wind speed, and high evaporation reduces availability of water and makes dryland farming very much difficult.

In view of the above factors, knowledge about local climatic elements and the inter-relationship of crop growth and productivity with various meteorological factors become extremely important. Hence, it is essential to know practical system and procedure of field level measurement of important meteorological elements like rainfall, temperature, humidity, solar radiation, evaporation etc., which have significant influence on crop production.

4.6.1 Measurement of precipitation

Rain gauge : This simple and standard instrument consists of (1) a funnel with a rim of diameter127 mm, (2) a collecting jar/can or receiver with narrow mouth and handle, (3) a cylindrical body to accommodate the funnel and jar, and (4) a splayed base. These items are assembled together and fixed firmly on a masonry or concrete foundation of 60x60x60cm size sunk into the ground. It should be done in such away that the rim of the gauge will be exactly 30 cm above the ground level and perfectly level.

Now rain water will be collected through the rim into the jar or receiver, which may be graduated in milliliter (ml) or a separate measuring cylinder, is used to measure the amount (ml) of collected rainfall. According to the diameter size (127mm) of the rim, 126.7 ml collected rainwater means 10 mm of rainfall. Sometimes measuring cylinder, directly graduated in mm considering this conversion ratio, is available for direct measurement of rain.

For measuring snowfall, a special gauge, similar to rain gauge but different in few aspects, is used and is called snow gauge. Its constituents are same but it is mounted on a iron stand at a height above the average snow fall level. Sometimes 200mm rim diameter is used instead of 127 mm. However, for smooth operation, a known quantity of warm water is poured into the receiver that helps the collected snow to melt and accumulate properly.

To know the intensity (mm/hr) of a particular rain or a specific period of a rain, the measurement of rainwater should be taken for that specific period. There are automatic rain gauges or recorder which records the amount of rainfall continuously.

Simple method: Rainfall can also be measured, quite easily and accurately, without a rain gauge. A metal or plastic bucket is placed in an open place, just like the rain gauge, to collect the rain water. After a specific rainfall, the collected volume of water is measured by a measuring cup/cylinder. Then the volume of water (m^3) is divided by the area (m^2) of the mouth of that bucket to obtain the amount of rainfall in m or mm.

For example, say, a bucket with diameter of 26cm is placed to collect rainwater. After a rain it collected 192.4ml of water. Now the area (A) of the bucket is (ð r^2 = 3.14 x 13^2) 530.7cm^2 and the volume of rainwater is measured as 192.4cm^2. Hence the amount of that specific rainfall is (192.4 / 530.7) 0.36cm or 36mm.

Duv Devani dew gauge: It is a simple instrument that helps to measure dew. It consist of a rectangular wooden block (32x5x2.5) coated with red paint which favours retention of dew deposited on it. The gauge, placed on the ground in an open place, is observed just at the time of sunrise, and the dew deposited on the surface is compared with a set of photographs of a 'standard dew album'. The album represents different dew scales (1 to 8 represents 0.02 to 0.35 mm of dew respectively) and thereby the actual dew can be measured. There are also facilities or patterns in the dew album for occurrence of dew and rain and for the occurrence of rain only.

In order to measure the dew formation in different heights (5, 25, 50, & 100 cm above the ground), four similar wooden blocks can be placed in those heights on a fixed structure. For measurement the same procedure shall be followed.

4.6.2. Measurement of Temperature & Humidity of Air

Stevenson screen: In general we are aware that thermometer is the measuring instrument of temperature. However, to measure air temperature the thermometer should be protected from direct solar radiation, neighboring object including ground surface, precipitation, and from heat losing, particularly at night. Hence, for measurement of air temperature, a specially designed box that allows free circulation of air inside is used to house the thermometers, which is called **"Stevenson screen"**.

Stevenson screen is a well accepted slandered size box with double layer louvered in all the four sides and a double layer roof having a gap of 2-3 inches and a plain floor. It is generally made up of teak or similar wood. The box is painted white and mounted on wooden posts in such a way that the floor of the Stevenson screen is fixed 4 feet (1.22 m) above the ground surface and the door is facing north to avoid entry of sun ray.

This specially designed box or Stevenson screen houses four types of thermometer: two for temperature measurement and the other two for humidity measurement. For measurement of temperature, 'maximum' and 'minimum' thermometers are placed at the back side of the box which records the maximum and minimum temperatures of a day, and can provide the mean temperature and fluctuation of temperature for the day.

For measurement of humidity there are wet-bulb thermometer and dry bulb thermometer fixed in the left and right walls of the Stevenson screen. Dry bulb thermometer is a plain/normal thermometer that will show the normal/dry air temperature. The wet-bulb thermometer is also a normal thermometer whose bulb is covered by a piece of muslin or any hygroscopic cloth, which is continuously kept moist by a cotton thread attached with a container having distilled/rain water.

Now when water evaporates from the muslin the latent heat requirement is drawn from the bulb of the thermometer indicating the reduction of temperature than that of dry –bulb thermometer. This reduction will continue till the air, just surrounding the wet bulb, becomes saturated. This is known as wet-bulb temperature. If the air contains more water vapour, the difference of temperature between dry-bulb and wet-bulb thermometers will be less and vice-versa. Now by using 'Hygrometric table' (of 1000mb for altitude below 1500 feet and 900mb for more than 1500 feet) the relative humidity as well as dew point can be measured for any specific temperature.

4.6.3. Measurement of Direction & Speed of Air Flow

Wind vane & Anemometer: Wind vane is a freely rotating arrow like structure installed 10 feet (3.05m) above the ground surface. The head of the arrow will denote the direction of wind and the geographical directions are prefixed with the **wind vane**. There should be no obstruction, and if any, that shall be away from the wind vane at a distance of at least 10 times of its height.

Wind speed is measured normally by **'cup counter anemometer'**. It is also a free rotating device consisting of three or four hemispherical cups. These cups are in motion when pressure difference developed between the two faces of the cups due to slightest flow of wind. The number of rotations made by the cups in a specific time can be counted by a counter and is directly calibrated into kilometers. The installation of the anemometer is similar to that of wind vane.

4.6.4. Measurement of Evaporation

Pan Evaporimeter: It is an instrument for measurement of evaporation, which consists of a large cylindrical pan made up of 20 gauge copper sheet tinned inside and painted white outside. Its diameter is 120 cm and depth is 25 cm. A small diameter (5cm) cylindrical tube like structure called still-well is provided inside the pan with three circular holes at the bottom to obtain undisturbed water surface inside that well avoiding ripples of that pan. The reference point is provided by a brass rod fixed at the centre of the still well. The top of the rod tapers to a point exactly 190 mm above the base of the pan. The pan rests on a white painted wooden grill. The pan is provided with standard wire mesh to prevent loss of water by any external agency like animal, bird etc. A thermometer is provided just below the water of the pan to measure temperature.

For measuring daily evaporation, first the water is filled up to the tip of the rod at 08.30 hrs IST. Next day at the same time water is poured (in case rainfall is less than evaporation) or taken out (if rainfall is excess of evaporation) with the help of measuring cylinder up to the level equal to the tip of the rod. Now first the actual height of pan water (mm) is required to be obtained by adding/ subtracting the water poured/ taken out by measuring cylinder (in mm). Then subtracting the actual level of pan water from the original level of pan water in mm (at the tip of the rod) will give the daily evaporation figure in mm/day.

4.6.5. Measurement of Evapotranspiration

LYSIMETER: It is a field level instrument to measure evapotranspiration (ET) of a particular crop/plant. It can also be used for other measurements related to crop/plant growth.

It is a large tank, filled with soil (transported or in-situ) to grow or maintain any crop/plant, and is supported by necessary weighing mechanism. The depth of the container should well cover the root-zone considering the capillary rise. It may be provided with drainage facility at the bottom to measure deep percolation loss occurring from the soil block of the lysimeter. It is to be located in the field where same type of crops/plants is grown as in the lysimeter. Some important aspects are to be considered while constructing a lysimeter. These are: heat advection through wall, drainage facility similar to the adjacent soil, filling of soil etc. That will ultimately provide a similar layer-wise status of adjacent soils.

ET rates are determined from the weight changes in lysimeter by using water balance equation:

$$ET = \text{weight change} + \text{water added} - \text{percolation.}$$

This weighing type lysimeters allow an evaluation of each direct and indirect element of the water balance equation. Weighing is accomplished either manometrically or by a suitable scale system and can be linked with a computer programme.

CHAPTER 5

MANAGEMENT OF DRYLAND FARMING

The success of any improved technology depends on the meticulous management of its various components. This is true in case of dry land farming also. Most of such elements of dry land farming technologies are vital constituent of nature e.g. water, soil, crop/plant, animal including birds, and man. As no life, from the minute unicellular microbe to the most improved form manifested in man, can survive without water, this natural element is the most important component to manage in a dryland farm.

Frequently, dryland farming is performed in extreme conditions in respect of water status of soil and socio-economic situation of the people living in dry areas. Therefore, for growing any crop in such areas, water, in whatever quantity it is available, must be conserved to the fullest extent. For that matter, *organic matter* and *mulching* are two important aspects that can profusely influence conservation of soil-water for crop growth. These two can change the perspective of dryland farming to a great extent. So, to give appropriate thrust on this, the management of organic matter and mulching is discussed separately besides water and soil management.

On the other hand without active participation of man, who will finally implement the technologies of dryland farming on individual or community basis, the successful execution of dryland farming will not

be possible. Here comes the social aspect of dryland farming thus the *social managements* is incorporated in this chapter to provide necessary information.

We have already identified a few specific technologies in the previous chapter and most of them are directly or indirectly related with all the components mentioned above. Now, this chapter narrates the management of water, soil, crop/plant, animal (including fish), organic matter, mulching, and social aspects independently and one by one.

5.1 WATER MANAGEMENT

Water is the most important and essential input for crop production in dryland areas. It profoundly influences photosynthesis, respiration, absorption, translocation and utilisation of mineral nutrition and other processes. Naturally, shortage of water affects the growth and development of a plant directly and thus quantity and quality of yield is affected immensely.

Water management is generally divided into two sections, one is to manage excess water and the other is to manage shortage of water. In dryland, the first part of the water management is generally irrelevant except providing safe waterways for storm rainwater. So here we concentrate mainly on the other part i.e. the management of available water for an efficient and economic utilisation in respect of crop growth. It also deals with the management of life-saving moisture supply in a specific period to bridge between two conducive growth periods.

After giving all possible efforts to conserve or harvest and store the most precious water, it is now the necessity and duty for us to utilise it properly and economically in field condition. The said utilisation is always considered in respect of growth of crops/ plants. Proper management of water to increase the farm production automatically boosts up the availability of drinking water for human and farm animals.

Now the subject 'water management in dryland' will deal with four main aspects that are **Rainwater management, Residual water management, ground water management** and **Irrigation water management**.

However, before going into the details of the management aspects we should go through the idea of " water use efficiency". The clear conception of 'water use efficiency' helps a lot to understand the events and procedure of water management in a dryland farm.

5.1.1. Water Use Efficiency

The yield of crop or the crop growth is achieved by the use of a given amount of water. This is a very important information for production of any kind of crop especially in areas where moisture is scarce.

Depending upon the consumptive use and transpiration, two types of water use efficiency, 'field water use efficiency' and 'crop water use efficiency', are generally expressed.

Field water use efficiency is the ratio of crop yield (Y) to the amount (total) of water used in the field that can be represented by water requirement (WR) of the crop.

So **field water use efficiency = Y / WR.**

Crop water use efficiency is the ratio of crop yield (Y) to the amount of water depleted by that crop that can be represented by evapotranspiration of crop (CropET).

So **crop water use efficiency = Y / CropET.**

The term 'crop yield' used here itself can be defined in terms of total growth (i.e. dry matter production) or in terms of marketable product.

Out of these two efficiencies, the concept of 'crop water use efficiency' is of fundamental interest while 'field water use efficiency' is of greater practical importance. However, in both the cases if we can increase the crop yield (Y) the water use efficiency will increase. But this option has a great limitation in the form of genetic capacity. Drastic reduction of transpiration is also very difficult. However, reduction of evaporation (through mulching etc.) and somewhat reduction of transpiration in extreme hot period can increase the water use efficiency significantly. Besides crop-ET, if the water loss through run-off and deep percolation is reduced then the total water requirement (WR) will also be reduced and this will increase the field water use efficiency significantly. In dryland areas this can be effective through technological development of crop production practices.

In dryland farming it is mandatory to achieve the higher water use efficiency and it should be done by application of both the options, i.e. increasing yield and reducing water requirement, simultaneously. Proper soil management and proper crop management can manipulate the water requirement in favour of the crop and can also increase the crop yield. So the management of soil, water and crop are designed and implemented very meticulously because they are very much inter-linked and interdependent.

Few package of practices of soil and crop management are listed below :

Soil Management. (1) different Tillage practices, (2) Mulching, (3) Application of organic matter, (4) Reduction of evaporation and deep percolation loss, (5) Reduction of soil loss, and (6) increase the fertility status of the soil.

Crop Management. (1) Proper selection of crop or plant species and their suitable variety, (2) Identify proper crop rotation, (3) Appropriate method and time of seeding or planting, (4) Proper supply of nutrition, (5) Reduction of evapotranspiration loss in extreme dry periods, (6) Control over weed, pest and disease infestation, and (7) irrigation.

The details of these points are discussed elaborately in the respective sections and whenever necessary.

Planning of Water Management For Dryland Farm

The water management system of a dryland farm includes structures required at water source, rainwater collecting and outlet structures, erecting of water-control structures like bunds, trenches, terraces etc, grading or levelling of land on contours, and laying out of irrigation system.

Water management planning requires a thorough assessment of soil (in respect of texture, structure, topography, ground water situation, infiltration and permeability, depth and slope etc.) and crop (root-system, life-span, consumptive water need, critical periods, market value, position in crop rotation, yield response in relation to moisture supply). After a proper assessment of these factors including topographic survey, a dryland farm is divided into a number of blocks and plots primarily in relation to available water and its supply. In dryland farms rainwater is the primary and sometime only source of water, so the discussion on management of rainwater is very important.

5.1.2. Rain Water Management

It is the most important and initial part of water management but it is the toughest part to manage. High intensity is the basic feature of rainfall in dryland areas, which provides very little scope of percolation and seepage to the total amount of rainfall received by the land surface. This rainfall produces high amount of run-off that frequently instigates erosion of soil. The rainwater has to be managed where it falls and that has to be followed along its flow. Thus ***catchment becomes the logical unit for the management of rainwater***.

On slopy lands (above 1%) rainwater flows down creating rills, gullies and ravines. On flat lands it causes problems of water stagnation, water logging, etc.

Principally rain water management involves (1) reduction of impacts of rain drops, (2) increase the retention of rain water at and above ground surface, (3) creating barriers to obstruct the run-off, which increase the 'time of concentration', (4) creation of suitable structure for storing of a portion of run-off water, and finally, (5) reduction of percolation loss, and (6) safe disposal of excess rain water through proper laid down water ways.

5.1.2.1. Reduction of Impact of Rain Water

Impact of rainwater on bare soil detaches soil particles, destroys granulation and also helps in the transportation of soil causing water erosion. It is very important to reduce the great force of the dashing raindrops while hitting the soil surface. If the soil surface is covered, completely or partially, then the impact of great velocity of raindrops can be reduced to a considerable extent. The covering of the soil surface may be done by growing vegetation or applying materials (mulching) on the soil surface. Erosion resistant crops such as cowpea, groundnut, greengram, blackgram etc may be grown to act as a cover crop. The sowing of these crops may be done in such a way that the soil surface is covered with dense canopy of the crop during peak period of rainfall. {Cover crops are discussed in Chapter- 5.3.3.} For reduction of impact of raindrops, trees and grasses may also be grown.

Spreading of mulching materials on the soil surface not only reduce the impact of raindrop, but also effectively control soil erosion. It also improves physical, chemical and biological atmosphere of the soil. {Detailed discussion on mulching has been made in Chapter 5.5.}

5.1.2.2. Increase the Period of Retention of Rain Water

It is the opportunity time for infiltration of rainwater. The measures found promising for improving infiltration and reducing runoff are discussed under :

(a) land shaping,

(b) tillage, and

(c) organic matter management.

(a) Land Shaping

1. **Raised Sunken Bed (RSB).** This system of 6m wide and 30 to 35cm high bed provides drainage to the upland crops, keeps the part of runoff water stored in sunken beds, facilitates the profile recharge and tends to stabilize root zone moisture during periods of prolonged dry spell. The performance of soyabean and sesamum planted on raised beds improved significantly; the 'kharif' crops can also be successfully grown on relatively flat topography or slow permeable deep Vertisols of high rainfall areas. A reasonably good performance of soyabean and paddy during rainy season and gram and wheat in winter season demonstrated the potential of this technology.

2. **Broad Bed and Furrows (BBF).** The BBF system consists of broad beds separated by sunken furrows (Fig. 5.1). The seedbed is made into wide beds of 1.05m alternated with furrows of width 0.45m. The width of the broad bed depends on the number of rows of crops to be grown and their spacing adapted. The bed width and crop geometry can be varied to suit the cultivation and planting equipment. The slope of the furrows may be between 0.4 -0.8 per cent on Vertisols. This method encourages moisture storage in the soil profile and works best on deep Vertisols where annual rainfall is 750mm or more. It is also quite effective when combined with dry sowing. It has not been very successful in areas of less dependable rainfall or on Alfisols or shallow black soils. This system with improved crop management practices gives better control on runoff, due to good crop cover during the rainy season, periodic land smoothing, slope adjustment, controlled

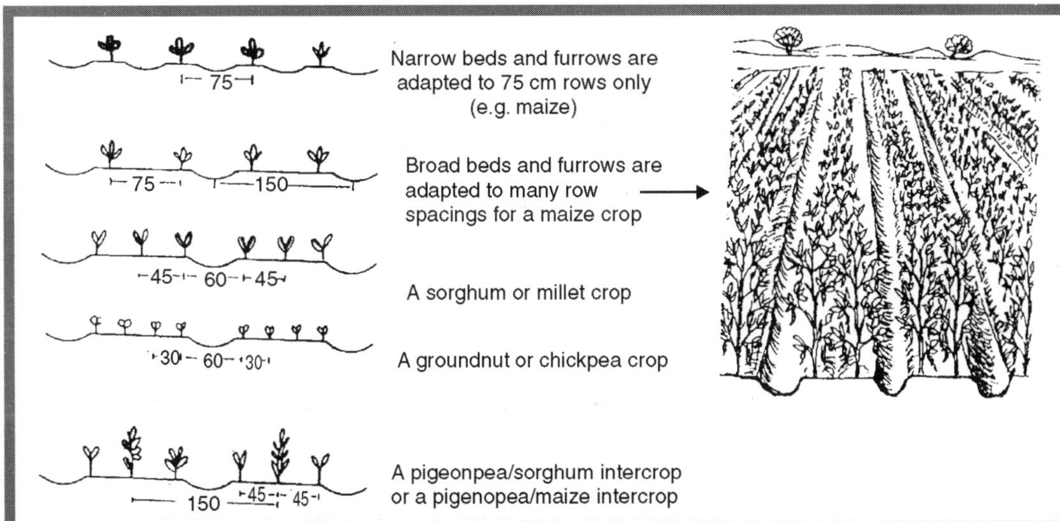

Fig. 5.1. Different kinds of bed and furrow system, including Broad bed & furrow (BBF).

runoff velocity as it flow through many furrows and disposes through grass waterways. This system performs better in the disposal of excess water than the flat system because of lower values of sediment load in the runoff. It is beneficial for the Vertisols, which often have waterlogging conditions when disposal of runoff is essential with least soil loss.

3. **Land Levelling/Development.** Among the various land shaping approaches land levelling is very important. The levelling process avoids ponding of water in troughs and provides uniformity in soil moisture. In case of tree crops few land development works can store rainwater, which can be utilised by that tree (Figure 5.2.).

Fig. 5.2. Land Development for Tree Crops in Slopy Land.

4. **Scoop.** Scoop is another method of improving in-situ soil and water conservation in dryland areas. It is somewhat similar to the range-pitting method practiced in grassland. It provides more time for water to infiltrate into the soil that would otherwise be lost as runoff. A significant increase in pearlmillet grain yield in the 20mm storage capacity scoops (2.42t/ha) was observed over the flat seedbed (1.79t/ha).

(b) Tillage

Tillage alters the physical characters of soil like porosity, bulk density, surface roughness and hardness of pans. It helps to control weeds, influence intake of water and reduction of soil loss by creating obstruction to surface flow of water. Rough and cloddy soil surface resists the force of wind from causing erosion. Tillage is generally performed during land preparation for sowing or planting of seed or seedling. Tillage done during non-crop situation is known as 'off-season tillage'. It helps in increasing rainwater infiltration, decreasing weed problem and minimising water evaporation by a mulching effect. The tillage used to get this mulching effect is called 'mulch tillage'. On Alfisols, it allows earlier sowing and has increased crop yield by 7 to 9 percent over control plot. {For details of different types of tillage see Chapter- 5.2.1.3.}

(c) Organic Matter

Incorporation of organic matter in soil brings about a marked improvement in physical, chemical and biological properties of soil besides addition of nutrients. It improves soil aggregates, infiltration capacity of soil, moisture-holding capacity of soil, and reduces evaporation loss from soil. {Details of organic matter and its management is described in Chapter- 5.4}

5.1.2.3. Creating Barriers to Obstruct the Runoff Water

Unobstructed runoff water with its erosive velocity increases the loss of rainwater along with soil. The velocity as well as the momentum of runoff water can be decreased to a considerable extent by land shaping, construction of suitable structure and some agronomic measures.

1. **Land Shaping.** Discussed earlier {5.1.2.2.}.
2. **Construction of Structures.** There are many structures, which actually lay across the slope to act as a barrier to the flow of runoff. These structures reduce the length of slope that ultimately decreases the velocity and momentum of runoff. These structures are commonly placed in series. They obstruct the flow repeatedly providing more time for percolation of water and settle down the sediment load of the flow. Some time this obstructed water, accumulated behind the structure, is used as supplement irrigation water in intermittent dry spells. However, selection of structures to be constructed in a field is dependent on slope of the soil, type of the soil (sandy, clayey etc), crops to be grown, implements to be used, nature of rainfall of the area and finally peak runoff generated in that area. The common such structures are *contour bund, graded bund, contour ridge, tied ridge,* different types of *terraces, trenches,* etc. {The details of these structures are described in Chapter- 5.2.1.2.}.
3. **Agronomical Measures.** Properly implemented different agronomic measures like *contour cultivation, strip cropping* etc can effectively reduce water loss from the field. On the other hand poorly designed agronomic measures enhanced the soil and water loss. {The details of these measures are discussed in Chapter- 5.2.1.3. & 5.2.3.1.}

5.1.2.4. Water Harvesting Structures

Water harvesting is a method to induce, collect, store and conserve surface runoff of the locality. This collected water can be used for agriculture and other purposes in arid and semiarid regions (Fig. 5.3). Each and every structure has a collecting or catchment area and a command or utilization area. In dryland area the catchment area is usually in higher proportion compared to command area. This harvested water may be stored for future use or can be allowed to percolate for recharging the ground water.

Fig. 5.3. Different Use of Harvested Water Stored in a Farm Pond.

From the view of operational management the structures are divided into two groups: Individual structures and community structures. *Farm pond, check dam, nala bund, vegetative filter strip, percolation tank, embankment bund, surface pond, low earthen dam* are the water harvesting structures, which may be utilised in dryland areas. {The details of these structures are discussed in Chapter- 4.3. & 4.4}. The main causes of loss of harvested water are evaporation from water surface {For details of reduction of evaporation loss see Chapter- 4.1.3} and seepage through the structures. Seepage losses occur due to the high permeability of the beds and walls of the structures. High permeability denotes the lighter texture of the soil and the porous nature of substrata. So suitable seepage control methods can be employed to overcome this problem.

5.1.2.5. Seepage Control

Various seepage control methods have been tested in the field and research projects. Interestingly, more effectiveness involves more cost, which may not prove economical under any circumstances. However, depending upon farming activity and economic condition of farmer, the following methods may be tried.

1. **Puddling.** Seepage rate of newly constructed ponds or tanks can be reduced considerably by puddling the floor of structures by man or animals or both.
2. **Compaction.** In well-graded soil, compaction of the bed of the structure results in sizable reduction of seepage rates. The top 20 to 25 cm. depth of soil is scarified, pulverized, and then compacted at optimum moisture content. The thickness of

compacted seal depends upon the head of water to be stored. But a 15 cm. thick layer is generally maintained.

3. **Clay Layer.** Sometimes a clay layer is used to cover the porous bottom of a pond or tank or other such structures to make them leak proof. For 3 mt. depth of stored water, a 25 cm. thick clay blanket may be useful.
4. **Bentonite Treatment.** Bentonite is a very fine textured colloidal clay, which is mixed with surface soil of the pond bed and compacted as described in the above item no. 2.
5. **Addition of Sodium Salt.** Addition of deflocculated chemicals like sodium carbonate, sodium chloride and tetra sodium polyphosphate to the surface soil of the reservoir's bed deflocculates soil of the bed. Thus the seepage rates can be considerably reduced. For these substances to be effective the clay content of the soil should be at least 15% and the cation exchange capacity (CEC) should be greater than 15 milliequivalent/gram. Sodium carbonate, which is more durable, is generally applied at the rate of 0.48-1.45 kg/m^2. After every 2 to 3 years an extra amount of sodium carbonate is to be added to continue the effectiveness.
6. **Soil-Cement Lining.** Well graded sandy soils mixed with cement in the ratio of about 10-12:1 by volume are found to reduce the seepage rates from ponds or tanks considerably. The higher is the cement content the greater will be the effectiveness. It requires proper watering and curing for good setting and long life.
7. **Asphalt Lining.** This lining consists of the mixture of sand and gravel bond together with asphalted cement. This watertight layer is not costly but encourage growth of vegetation, which easily penetrate the lining making them defunct rapidly.
8. **Stone slab and Brick Lining.** Where stones are available abundantly, they are laid in cement mortar (1:3) on the bed and sides of the storage structure, which are quite effective in reducing seepage loss. Bricks can be used instead of stone but a lining with rich cement mortar (1 cm thick) on the surface of that lining is required to obtain desirable effectiveness.
9. **Cement Concrete Lining.** It is the most effective method of seepage control, but the initial cost is very high and therefore, is uneconomical for common farming purposes. A cement concrete mixture (1: 3 : 6) is applied at thickness of 3 to 5 cm. Measures should be taken against the cracking of this layer through proper curing and maintenance in due course.
10. **Polyethylene, Plastic, Rubber Films.** The films or sheet of these materials can be used as open or buried layer in the bottom and side surfaces of reservoirs and waterways to cut off the seepage through them. It shows varying degree of success, but require special precautions against punching or cutting, rodent burrowing and weathering.

5.1.2.6. Safe Disposal of Excess Rainwater

Graded bund, Graded trench, terrace, waterways, and in some cases drop structures are the methods and tools that have in-built arrangement of safe disposal of excess rainwater. {All these items are discussed in chapter 4 and in 5.2.}

5.1.3. Residual Water Management

Residual water is the water retained by soil after completion of overland flow, deep percolation and crop uptake. The accumulation of this water may increase through resultant effect of proper rainwater management. The residual water is present in soil strata and available to the plants only. Loss of this water from the soil occurs only through evapotranspiration {For details of reduction of evapotranspiration loss see Chapter- 4.1.3.}. So crop management is the best means to manage the residual water along with some other methods like mulching, tillage etc. The proper manipulations of cropping {see 'crop management' in Chapter- 5.3.} ensure the maximum utilisation of residual moisture, whereas the other methods like mulching, tillage etc. {described in chapter 4} reduces the loss of residual water that can be used for crop production.

5.1.4. Ground Water Management

To control or influence the behaviour of ground water, i.e. fluctuation of water table, is almost an impossible task for an individual. However, on community basis, the availability, seasonality and fluctuation of ground water can be managed to a great extent. Proper and all-round implementation of the water conservation measures, like contouring, bunding, trenching, pitting, terracing, check dams, percolation tanks, etc can improve the ground water situation. But to get proper effect, these measures have to be implemented by each and everyone of the locality .

Interestingly many of the said measures can be implemented in an individual farm. And adding up these individual efforts will create the community feelings, which finally implement the rest community works like construction of percolation tanks, check or low earthen dams etc.

Apart from construction of these structures, maintenance of the same is also an important job for management of ground water. This can be done individually as well as on community basis. This jobs are as follows:

1. De-siltation should be done regularly in all the structures.
2. Regular maintenance of dams etc should be done.
3. Restricted use of underground water in dry spell should be maintained.
4. Spacing of wells should be properly maintained.
5. Restriction on diameter of dug-well should be imposed.
6. Used water should not be allowed to flow above the ground.

Undertaking the above measures wide fluctuation of ground water may be controlled. To know the seasonal or intra seasonal ground water level the data of ground water of a particular farm for few consecutive years is required.

Measurement of Ground Water Table. It is a simple method which requires an observation well (in absence dug-well may be used) and a measuring instrument like graduated metal rod of about 5mm diameter, or an electric sounder with a graduated cable or a measuring tape with 10cm long hollow metal pipe attached to its one end. By inserting the measuring material into the well, the distance of ground water level in the well from ground surface may be calculated. It may be measured weekly and may be continued for few years and tabulated in the following format:

Sl.No – Date – Time interval – Depth of Water table – Rise or fall of water table– Remarks

In case of open well, measurement just after a heavy rain should be avoided.

{For storing of underground water see chapter- 4.1.4.}

5.1.5. Irrigation Water Management

After utilising the soil moisture properly the crop requires supplement water for avoiding a kind of water stress to obtain maximum yield as per the potentiality. Supply of this supplement water to any crop is generally termed as irrigation.

In dryland areas the term "irrigation" does not carry the same sense as it does in the areas where water is plenty. Here irrigable water is very much limited. Irrigation in dryland areas primarily mean application of supplement water to save maximum cropped area from the effect of water stress in their active growth period. Now adaptation of proper water conservation methods {described in chapter 5.1} can generate more irrigable water for economic farm production. So in dryland area, irrigation requires close observation in regard to the availability of water, its periodicity, crop and its water need and water use efficiency. This should be followed by a proper irrigation planning.

In dryland farming storing of protective or supplement irrigation water in different water harvesting structures is very much important. Water stored in 'farm ponds' and application of the same at the critical juncture helps to boost the yield of cereal crops up to 40 to 60 percent (Randhawa and Venkateswarlu, 1980). But high cost of supplement irrigation and poor farm-gate price of field crops frequently established it as an uneconomic venture. Vegetables like chilli, brinjal, tomato, bottle gourd, etc. yield more with supplement irrigation and is economically better option than cereals (Vijayalaxmi, 1987). Protective or supplement irrigation to different perennial fruit crops suitable in dry areas, like pomegranate and *ber* proved highly rewarding.

Any irrigation process requires three types of basic information that are to be worked out before starting the process. They are (1) **when to irrigate**, (2) **how much to irrigate** and (3) **how to irrigate**. Depending upon the crop, which are selected in relation to the availability of irrigation water in a particular dryland, the above information are to be collected and on that basis the irrigation water should be managed.

The ultimate objective of irrigation water management is to achieve highest *water use efficiency*. However, in dryland farming, apart from augmentation of the yield of the selected crop, increase of the WUE through minimisation of wastage of irrigation water, which ultimately expands the cultivation area of that crop is the focal point of scientific irrigation management in drylands.

Plenty of research work has already been performed on when, what amount, and how to irrigate a particular crop. The findings of these research works provide a basic idea to work out a farm plan in relation to these three factors. This crop-wise research works were done by different scientists in different research stations; still the in-situ experience with the basic knowledge always provide best result as they are highly variable in respect of soil, localised climate, varieties of crop and infrastructure available in that farm. Now we can discuss how one can be able to manage these three aspects efficiently in dryland.

5.1.5.1. When to Irrigate / Time of Irrigation

This particular aspect of irrigation is generally called 'scheduling'. An ideal irrigation schedule must indicate when irrigation water is to be applied and the quantity of water to be applied. The irrigation needs of crops are decided by the evaporative demand of the ambient atmosphere, soil-water status and plant characteristics. Thus the criteria for scheduling may be grouped into the categories like (1) soil-water regime of the rootzone, (2) characteristics of crop, and (3) meteorological criteria.

(1) Water regime of Soil. Irrigation scheduling based on soil-water content is the most accurate and dependable method. Determination of available soil water in rootzone of the crop before the time of irrigation is most important. {For determination of soil moisture of rootzone see Chapter- 3.7.1}. The optimum soil water regime means the range of available soil water in which crops do not suffer from water stress. So when the soil moisture of rootzone in a specified crop is depleted to a particular level it is to be replenished through irrigation.

To determine the time of irrigation one should start right from the condition of a soil following a heavy rain or an application of irrigation. Immediately after watering the surface soil will be filled with water. This soil is said to be saturated soil and is at its maximum retentive capacity. If no further water is added to the soil, after a few days the surface soil will attain field capacity. After attaining field capacity, the soil water decreases through evapotranspiration of crops, which remove moisture from the soil rapidly until wilting point is approached. Capillary rise of water from root zone area and subsequent evaporation from soil surface also influence this depletion of soil moisture towards wilting point. This is the water regime of rootzone when irrigation is very much required. The time of attaining wilting point varies according to soil type and standing crop. Hence the time of next irrigation also vary.

Rooting depth of crops are usually recommended for different soil conditions including status of soil water. However, potential rooting depth is not as important as depth of maximum water absorption by roots in normal condition. For example: (1) shallow rooted crops like onion and spinach absorb maximum water only from 10-15cm (4-6 inch)depth whereas its potential root depth is 60cm (2ft); (2) crops with medium rooting depth like chillis, beans, cucucumber etc draw most of their water from 25-40cm (10-16inch) below the surface, whereas its potential rooting depth is 1.2m (4ft); (3) similarly the deep rooted crops, grapes, watermelon etc, collect water 1.2-1.4m (4-6ft) depth though its potential root depth is about 2m (7ft). Hence application of water into a crop as a single irrigation requires necessary calculation considering the root depth of that crop and it can be determined by the following formula:

$$W = p_r \times AW \times \frac{d}{E_a}$$

Where, W = amount of water required to bring the soil to field capacity (FC) (lit/m^2)

P_r = the proportion of available soil water and water that the plant can actually use without experiencing water deficit and is estimated as 0.5.

AW = available water of plant (Field Capacity – permanent wilting point) (mm / m)

d = root depth of crop (m).

E_a = water application efficiency (water in root zone-water applied)

Example : (1) In a sandy soil the available water storage capacity of the root-zone is 120mm/m and 50% of which has been used up. The area is laid out in 100 number of 1.5m × 1.5m beds with a plan to grow annual vegetables having adequate rooting depth of 0.45m (1.5ft). Now to irrigate the beds to field capacity with a irrigation system of 80% application efficiency, the amount of water needed for irrigation would be :

$$W = 0.5 \times 120 \times \frac{0.45}{0.8} = 34 \text{ mm of irrigation water, i.e. 34 liters/m}^2$$

Now, for each bed ($1.5 \times 1.5 \times 100 = 225m^2$) the water requirement is

$$= 34 \times 225 = 7650 \text{ liters of water.}$$

(2) In the same situation if the soil is of clay loam type and the plant available water is 200mm/m then the W = 0.5 × 200 × 0.45 × 0.8 = 56mm = 56 liter/m2; and for the bed 56 × 225 = 12656 liters of water.

In crops like maize, wheat, etc application of irrigation at 25 per cent depletion of available soil moisture (lower level of optimum soil water regime) is adequate. For drought resistant crops like sorghum, pearl millet, cotton etc it is sufficient to irrigate at 50 percent depletion of said water. Every crop has a characteristic optimum water regime and the same for different crops are tabulated as follows :

Table 5.1. Optimum Soil Water Regime For Different Field Crops in India.

	Optimum Soil Moisture Regime (in %)
Sorghum, Maize, Barley, wheat, green gram	100 to 50
Ground nut, Tomato	100 to 60
Pearl millet, Green fodder of sorghum, Lucerne, cluster bean	100 to 75
Cotton	100 to 20

Source : Dakhinamurti et. al. (1971).

Available water of the soil for plant use depends on the type of soil and rootzone depth of the plant. In general, fine textured soils have a wide range of water between field capacity and permanent wilting point than coarse-textured soils.

Texturewise range of available soil water is tabulated below :

Table 5.2. Range of Available Water in Soils.

Soil Type	Available soil water (in cm / meter depth of soil)
Fine sand	2 – 4
Sandy loam	4 – 11
Silt loam	6 – 13
Clay loam	10 – 18
Clay	16 – 30

Source: Michael (1990)

(2) Characteristics of Crops. Based on the physiological characteristics, different criteria of crop are considered to schedule irrigation. Among different criteria, critical growth stage and potential evapotranspiration are two vital aspects of scheduling irrigation.

Critical Growth Stage of Crop. The crop water demand is not the same as the crop water requirement. Even though, for successful cropping, regular supply of water is necessary. But withholding of water during certain physiological phases adversely affects the economic production whereas during certain other phases it does not. These physiological phases are called 'critical stages'. As the amount of irrigation water is generally scarce in dryland and is sufficient only for a few irrigations, it is essential to time the watering at the most critical stages.

So the information about critical stages of different crops helps to manage the irrigation water in such a way that the yield is not affected by water stress. Many scientists are doing a number of studies and research works, to identify the critical stages of different crops. The cropwise list of critical stages is given below:

(a) Alfalfa grass (1) Just after cutting;

(b) Wheat (1) Crown root initiation stage (20 days after sowing);
(2) Booting stage (end of vegetative stage to emergence of ear);
(3) Heading stage (end of booting stage to start of flowering);

(c) Maize (1) Tasselling stage (45 to 55 days after sowing);
(2) Silking stage (56 to 62 days after sowing);
(3) Grain formation stage;

(d) Sorghum (1) Booting stage;
(2) Blooming stage / Flowering stage;
(3) Milky stage / Grain formation stage;

(e) Cotton (1) Flowering stage;
(2) Boll formation stage;

(f) Sun flower (1) Seedling stage (if possible);
(2) Flowering stage;
(3) Grain / Seed development stage;

(g) Small grains (1) Booting stage;
(2) Heading stage;

(h) Sugar cane (1) Maximum vegetative growth stage;

(i) Ground nut (1) Pegging / Flowering stage;
(2) Pod formation stage;

(j) Soya bean (1) Maximum vegetative stage (if possible);
(2) Flowering stage;
(3) Grain filling stage;

(k) Tobacco (1) Knee height stage;
(2) 3 weeks before topping;

(l) Potato (1) Continuous, in tuber formation stage;

(m) Tomato (1) Flowering stage;
(2) Fruit development stage;

(n) Cabbage (1) Head formation stage;
(2) Head enlargement stage;

(o) Fruit plants - It depends on the type of fruit plants. However, the moisture level in soil should be very good at-
(1) The time of new branch/leaf initiation;
(2) Flowering stage;
(3) Fruit development stage.

The manuring and fertilizing should be preceded by irrigation.

Potential Evapotranspiration. Several approaches have been studied and on that basis attempt have been made to create crop-wise irrigation schedule. For wider adaptability the climetological approach to assess potential -ET seems to be easier. Using the climatic factors like sunshine, temperature, relative humidity and wind velocity the empirical formulas were developed by the researchers, which can be used to estimate the potential-ET or crop-ET. The Food and Agricultural Organisation (FAO) has suggested Blaney-Criddle and Penman methods, in modified form, for estimation of crop-ET.

These empirical formulae needs some fine tuning in respect of condition of crop growth and the local conditions. Here crop growth generally refers the size of canopy cover of the crop. Upto 80% of crop canopy cover, the evaporation from soil is higher than the transpiration by crop. But with full canopy formation the water need is at its peak and transpiration takes the major share of the ET. Now knowing the open pan evaporation (Eo), pan factor (Kp) and coefficient of crop (Kc), the evapo-transpiration of crop field (ETcrop) may be expressed as –

$$\text{ET crop} = \text{Eo} \times \text{Kp} \times \text{Kc}$$

Table 5.3. Crop Coefficient (Kc)

I	II	III	IV	V	VI
Rice	1.1	1.15	1.3	0.95	0.6
Maize	0.5	0.85	1.05	0.8	0.6
Banana	0.5	0.85	1.0	0.9	0.75
Cotton	0.5	0.7	1.05	0.8	—
Groundnut	0.4	0.7	0.95	0.75	0.55
Sorghum	0.4	0.7	0.95	0.75	0.55
Sugarcane	0.5	0.8	1.0	0.75	—
Sunflower	0.4	0.8	1.0	0.8	0.45
Tobacco	0.4	0.7	1.0	0.9	0.7
Onion	0.6	0.75	0.9	0.95	0.9

I – Crop, **II** – Initial growing stage, **III** – Root development stage, **IV** – Maximum vegetative stage, **V** – Before maturity stage, **VI** – At harvest stage.

In this backdrop, we may understand that during water shortage (not below the permanent wilting point) in non-critical stages, the crop may reduce transpiration loss. It will have only a small effect on yield thus increasing the water use efficiency. This knowledge of cropET and critical stages of moisture stress provides a clear-cut idea about the application of supplement water in the form of irrigation to a particular crop.

Apart from crop evapotranspiration and the critical stages of crops, the soil moisture status (in the root zone) is also considered while planning the schedule of life saving irrigation. These informations are necessary to estimate the amount of water required for each application.

While the cropET is generally considered for irrigation schedule, for scheduling of the seeding/planting the soil moisture content should be considered. The soil moisture content is commonly measured by –

- The amount of total water in a given amount of soil, which may be expressed as –

$$\underset{\text{(\% by weight)}}{\text{Soil moisture content}} = \frac{\text{(Wt. of moist sample)} - \text{(Wt. of oven dry soil)}}{\text{(Wt. of oven-dry soil)}} \times 100$$

- The stress of tension under which the water is held by soil in different moisture content stages may be measured by the instruments like 'Tensiometer', Electrical resistance etc.

Irrigation frequency

It is the schedule of irrigation that brings the soil again to its field capacity. This is important information to maintain maximum yield and can be calculated as: $i = W / ET_m$, where i = irrigation interval in days; W = water requirement to bring root-zone soil into FC when 50% of available water is depleted; and ET_m = maximum evapotranspiration per day of the crop or crops of the field growing under optimal condition ($ET_m = K_c \times ET_0$).

If in sandy loam soil W is 34mm (consider the earlier example) and ET_m is 4mm/day then the watering frequency to bring the root-zone into FC is (34mm / 4mm/day) 8.5 days i.e. 8-9 days. In case of clay loam (W = 56mm) this figure will be (56/4=) 14 days.

(3) Meteorological Criteria. Attempts have been made to develop empirical formulae using different meteorological parameters for estimating the evapotranspiration and consumptive use of water by crops for scheduling irrigation by water budget method. The daily evapotranspiration loss is deducted from the soil water reserve in root zone after irrigation and a balance is worked out. When the soil water balance shows that the soil water is depleted to predetermined level (the lower level of optimum soil water regime), irrigation is applied to replenish the water lost through evapotranspiration. However, this approach of scheduling irrigation to crops is complicated for an ordinary farmer.

5.1.5.2. How Much to Irrigate / Water Requirement For Irrigation

Knowledge of gross water requirement (WR) is one of the basic informations of crop planning in relation to the available water. The WR is generally expressed as total quantity of water, regardless of its source, required by the crop or diversified patterns of crops in a given period of time for its normal span of growth under field condition at a place.

This total requirement of water is generally computed by aggregating the loss of moisture due to– (1) evapotranspiration (ET), (2) application loss, (3) water required for land preparation, planting etc. This may be expressed mathematically as follows –

$$WR = ET + \text{Application loss} + \text{Special needs.}$$

Water requirement is, therefore, the demand of crop. However; the supply would consist of contribution from any of the sources of water. The major source being the rainfall (ER), soil profile/ sub-surface contribution (S) and supplement water i.e. irrigation water (IR). So numerically the quantity of irrigation water may be stated as – $IR = WR - (ER + S)$.

The farm irrigation requirements is generally expressed as height of standing water. It depends on the irrigation need of individual crop. From various experiments the cropwise water requirements has been estimated as follows:

Rice – 120 cm	Wheat – 40 cm	Jower – 50 cm
Maize – 62.5 cm	Bajra – 50 cm	Sorghum – 50 cm
Groundnut – 45 cm	Pulses – 30 cm	Cotton – 60 cm
Chillies – 60 cm	Sugar Cane – 250 cm	Banana – 200 cm
Tomato – 95 cm	Onion – 100 cm	Turnip – 240 cm

Irrigation Requirement

It is the *depth of irrigation* water, exclusive of (1) precipitation, (2) carry over soil moisture, (3) ground water contribution, and (4) any other gains in soil moisture in root zone, that is required for consumptive use of a crop/plant to yield at par with its potential. In other way, it is the amount of irrigation water required to bring the soil moisture level, in the effective root zone, to field capacity. Thus it is the difference between the field capacity and the soil moisture content of the soil. It may be termed as net irrigation requirement. However depth of irrigation water depend on the extent of soil water depletion at the time of irrigation and are calculated by the following formula:

$$d = \sum_{i=1}^{n} \frac{Fci - Mbi}{100} \times Asi \times Di$$

Where,

d = Net depth of water to be applied or net irrigation, in cm.

Fci = Field capacity of the ith layer of soil, in per cent.

Mbi = Water content of the ith layer of soil just before irrigation, in per cent.

Asi = Apparent specific gravity of ith layer of soil, in gm/cm^3

Di = Depth of ith layer f soil in the root zone D.

n = Number of soil layers in the root zone.

Now to get the gross irrigation requirement the field efficiency of the irrigation system / methods is generally considered. It may be expressed as follows.

$$\text{Field irrigation requirement (gross)} = \frac{\text{Net irrigation requirement}}{\text{Field efficiency of irrigation method}}$$

Naturally the irrigation methods or systems with higher efficiency, which is represented as percentage, require less amount of water. It is an important aspect of dryland farming, where availability of water is very limited.

Irrigation Efficiency

Irrigation efficiency indicates how effectively the supplement water is being utilised for crop development. That means how the irrigation water is utilised by crops to augment the production. To increase efficiency of precious irrigation water, reduction of water loss in irrigation process is the first job.

To increase *irrigation water use efficiency* one have to manipulate the conveyance, application, distribution, and storing of irrigation water. For that matter four distinct components of irrigation efficiency such as conveyance efficiency (Ec), application efficiency (Ea), distribution efficiency (Ed), and storage efficiency (Es) are to be understood and managed properly. The expression of these efficiencies are stated below:

$$Ec = \frac{\text{Water available at delivery point}}{\text{Water delivered at source}} \times 100$$

$$Ea = \frac{\text{Water available at root zone}}{\text{Water available at delivery point}} \times 100$$

$$Ed = \left(1 - \frac{y}{d}\right) \times 100,$$

Where d = Average depth of water stored in the crop root zone after irrigation.

y = Av. numerical deviation between water required in root zone and d.

$$Es = \frac{\text{Water stored in root zone after irrigation}}{\text{Water required in root zone}} \times 100$$

$$\text{Economic efficiency of irrigation} = \frac{\text{Actual total production}}{\text{Potential total production}} \times 100$$

5.1.5.3. How to Irrigate / Methods of Irrigation

Water may be applied to crops as irrigation in different ways. The selection of method for a farm depends on the amount of available water, types of crop, type of land and soil, climate and obviously on economics. Some common methods are generally practiced in farms. Out of which flood irrigation method is most common and furrow irrigation method is the next. These methods are applied in rectangular check basins or borders in plain lands and contour basins or borders in undulating / hilly areas. Furrow irrigation methods are commonly used in crops planted in rows. In tree crops ring type check basin method is generally used.

The said methods are also used in dryland areas with lots of uncertainty. However, for proper dryland farming few vital points should be taken into consideration at the time of selecting the methods of irrigation. The points to be considered are: (1) availability of irrigation water, (2) structure and profile of soil, (3) rate of evapotranspiration, (4) schedule and pattern of rainfall, (5) topography, (6) status of ground water.

Considering all these points and facts the common methods of irrigation in dryland farming are prescribed as under.

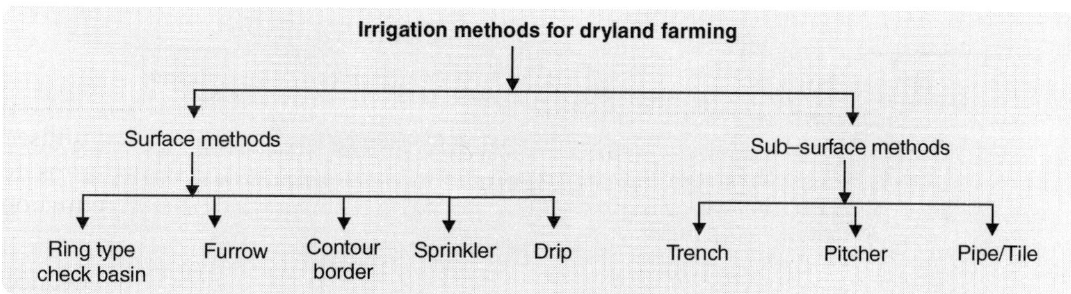

Surface Methods

Most of the common surface methods generally used for irrigation purpose are of no use in dryland situation due to very low efficiency. Still few of them may be practised in dryland situation considering the status of available irrigable water and in a more technical way. For these methods proper water lifting devices and scientific transportation of irrigation water is must. A brief discussion on these issues and practices are given below :

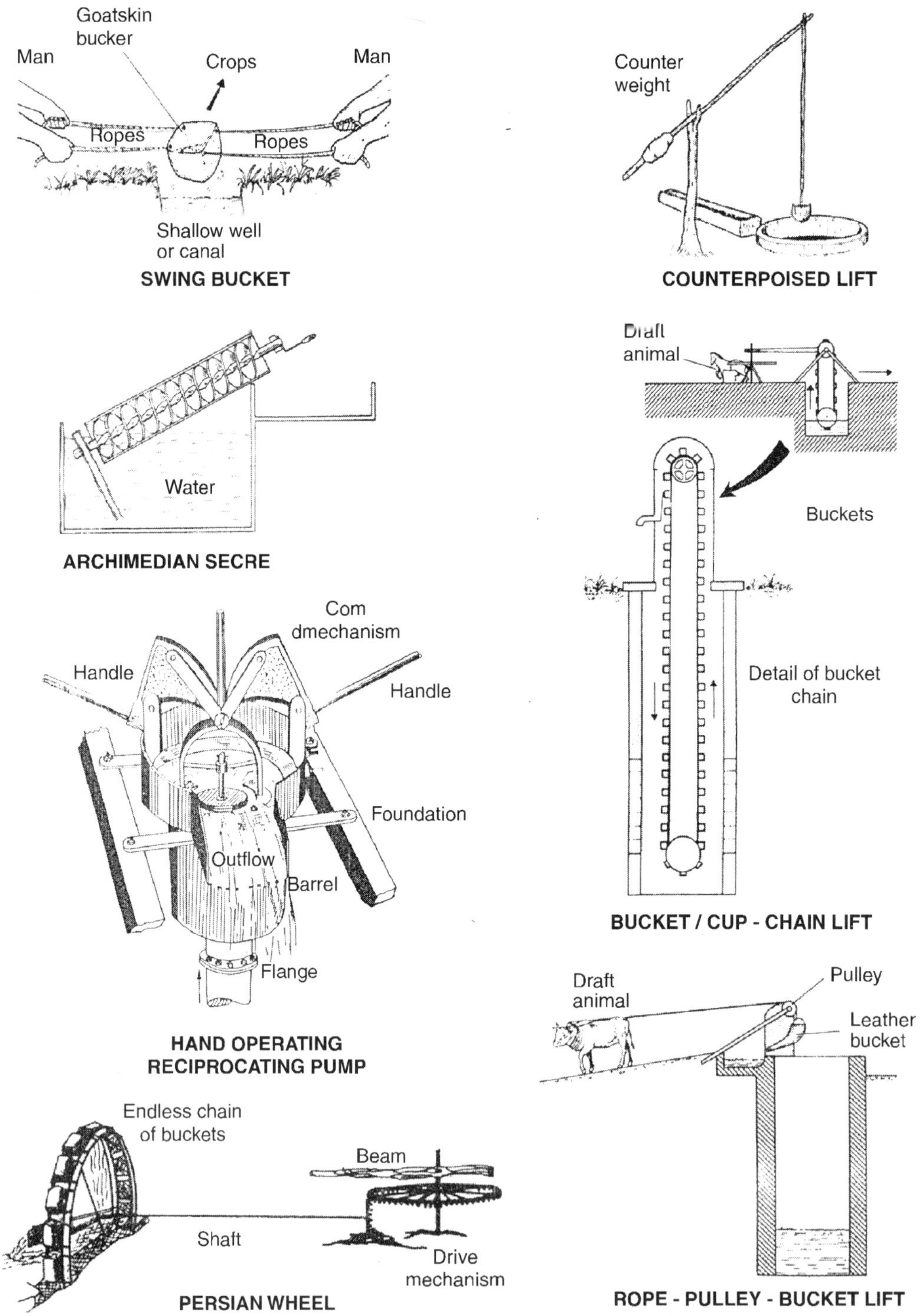

Fig. 5.4. Sketch of some hand/animal operated water lifting devices

Water lifting devices for irrigation

In dryland, where availability of water is less and most of the resources are surface water or dug-well, i.e. the water level is available within 10m of ground surface; simple human or animal driven lifting devices can be efficiently used for farming purposes. Some of these devices can be greatly improved. However, as the power delivered by humans and animals is not constant and the fluctuation of water level in dryland is more, these devices have some limitation on its efficiency, but it will certainly deliver something in every occasion. On the other hand, use of pumps can be costly to install and run and are prone to failure. Frequently in dryland areas the rate of supply of water from the source does not match with the rate of discharge of the pump.

Swing bucket, Counterpoised lift, Archemedian screw, Bucket/cup-chain lift, Hand operating reciprocating pump, Persian wheel, Rope-Pulley-bucket lift etc. are some of the common devices used in different parts of the world from ancient times. Some illustrations of these devices are given in the figures and the associated information is given in the table (5.4) to have a readymade idea.

Table 5.4. Some information about the water lifting devices

Devices	Used source	Lift	Power	Discharge	Irrigated area
Swing bucket	Surface water	1 -2	2 man	0.15	0.3 -0.5
Counterpoised lift	Both	2-5	1 man	< 0.1	0.1-0.3
Archimedian screw	Surface water	1 – 1.5	1 man *	0.1	0.3 -0.5
Bucket/cup-chain lift	Dug well	6 - >10	1-2 man/ animal*	1-1.5	2-5
H O Reciprocating pump	Both	4	1-2 man	3-6	4-10
Persian wheel	Surface water	1-2	1 animal	2 (assume)	5 (assume)
Rope-pulley-bucket lift	Dug well	10	1 man & 1 animal	1 (assume)	2-4 (assume)

Pumps: Pumps are the mechanical devices generally powered by electricity or by any powerful means. They can provide much higher discharge rates and some of them can lift water from much more depth. Correct choice of pump appropriate for a particular situation, is a critical factor predominantly in dryland areas.

There are two main types of pumps viz. '*turbine pumps*', which use a rotating impeller to move the water, and '*reciprocating pumps*', which have a piston with valve and a check or foot-valve assembly. When both these types of pumps are housed above the ground and draw water by suction from below (within 8m) are generally called *suction pump*. If the groundwater lies deeper the impeller or piston assembly can be installed down the well inside the water to lift out water from the well and is therefore called *lift pump*. However, electrically powered turbine pumps are most suitable for installation down deep wells.

Transportation and application of water for irrigation

In dryland, planning and designing for transportation of irrigation water is an important aspect to increase water use efficiency as high as possible. In case of transportation of irrigation water from public/community sources one have little scope to participate In its minute and important technical parts.

However, community or personal, whatever may be the conveyance system, some basic principal is required to be followed for transportation of irrigation water from that source to the point of application in the field. These principals are: 1) reduction of seepage loss by proper lining of the channels, 2) Management of diversion points of the channels, and 3) speed of the water flow in the channel.

Now considering all sorts of recommendations, derived from numerous experiments, it shows that plastic pipe or plastic lined channel is best for dryland farming. It will zero the seepage loss and substantially reduce the frictional loss to increase the speed of the flow. However, frequent diversion of the flow sometime stands as a limitation.

On design front, for a plastic lined channel, there should be no excavation in the channel bed and should be constructed by making two parallel bunds alongside the

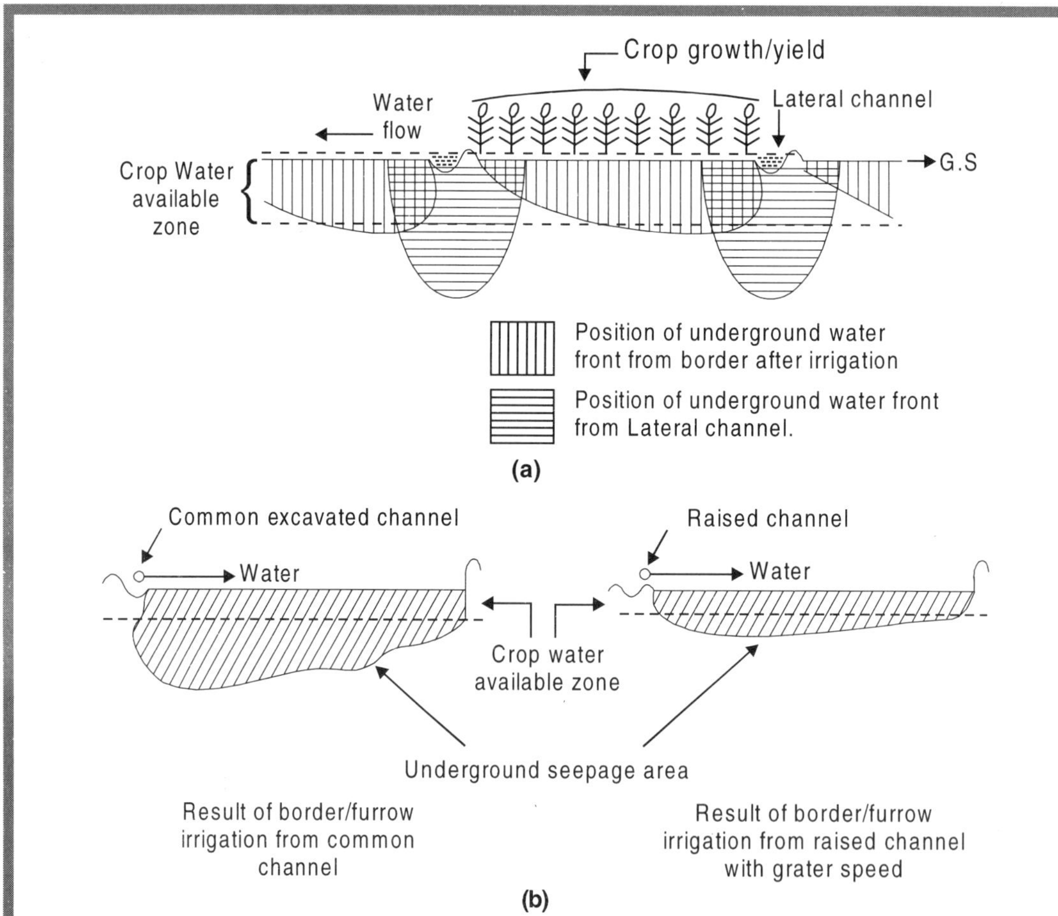

Fig. 5.5. (a) Result of well planned border method of irrigation. (b) Effect of raised lateral channels to irrigate basin, border & furrow methods of irrigation.

channel bed. This will ensure quick delivery, easy irrigation planning, utilizing the last drop of water. Integrating all these advantages will reduce the seepage loss from the field bed (fig. 5.5).

Border method

It is the flooding of small rectangular beds one by one to irrigate close growing crops like cereals etc. The beds are either sunken or surrounded by small ridges. Depending on the soil type the width and length of the plot is determined. In dryland farming, depending on the soil, width may be 2 to 3 meter and length may be 6 to 10 meter. For sandy loam the size may be reduced further and in fact shortening of width and length of border to certain limit increases efficiency. However, in heavy type of soil it works better.

Depending upon land situation, particularly the slope, and crop considered the borders may be constructed differently and are termed as *graded border* or *level border* or *contour border*. In slopes more than 2-3 per cent contour border is recommended.

Finally in dryland situation this method of irrigation is not generally recommended and if require this method of irrigation may be designed in consideration with the situation of available water resource, type of soil and crop.

Furrow method

It is the gradual making of the land or bed into ridge and furrow of definite size and shape. Basically it was a network of furrows. Sometimes the bed of the border, as said earlier, is divided into ridge and furrow. However, the height of ridge and depth of furrow is commonly not built simultaneously but step by step from seedling stage to flowering stage. This system is applicable to row crops like potato, maize, sugarcane etc.

Leveling of the land is an important condition to adapt this method of irrigation. Depending upon the type of crop the width of ridge and furrow is determined, but the length of ridge and furrow is dependent on type of soils and position of channel.

Skip furrow method is a modified furrow method suitable for dryland and for the crops requiring relatively less amount of water. Here the width of the ridge is double to that of the furrow and two rows of crop is accommodated in that ridge.

Ring Type Check Basin. This is a suitable method in dryland farming particularly for tree crops and other commercial herbs and shrubs. To increase its field efficiency pipes (made of bamboo, plastic etc.) or specialised structures may be used instead of field channels that reduces the transportation loss. To reduce the evaporation loss the basin may be mulched properly.

Sprinkler. In this method water is sprayed / sprinkled into the air and allowed to fall on the crop that somewhat resembles rainfall. Power is required to sprinkle the water.

In many dryland areas hand or manual sprinkling of harvested or collected water is a common practice. This is not only used for saving the life of crop but also for maximising the return of that crop in that situation. In that practice a container is used to swing it with water with a disturbance in the opening of that container by

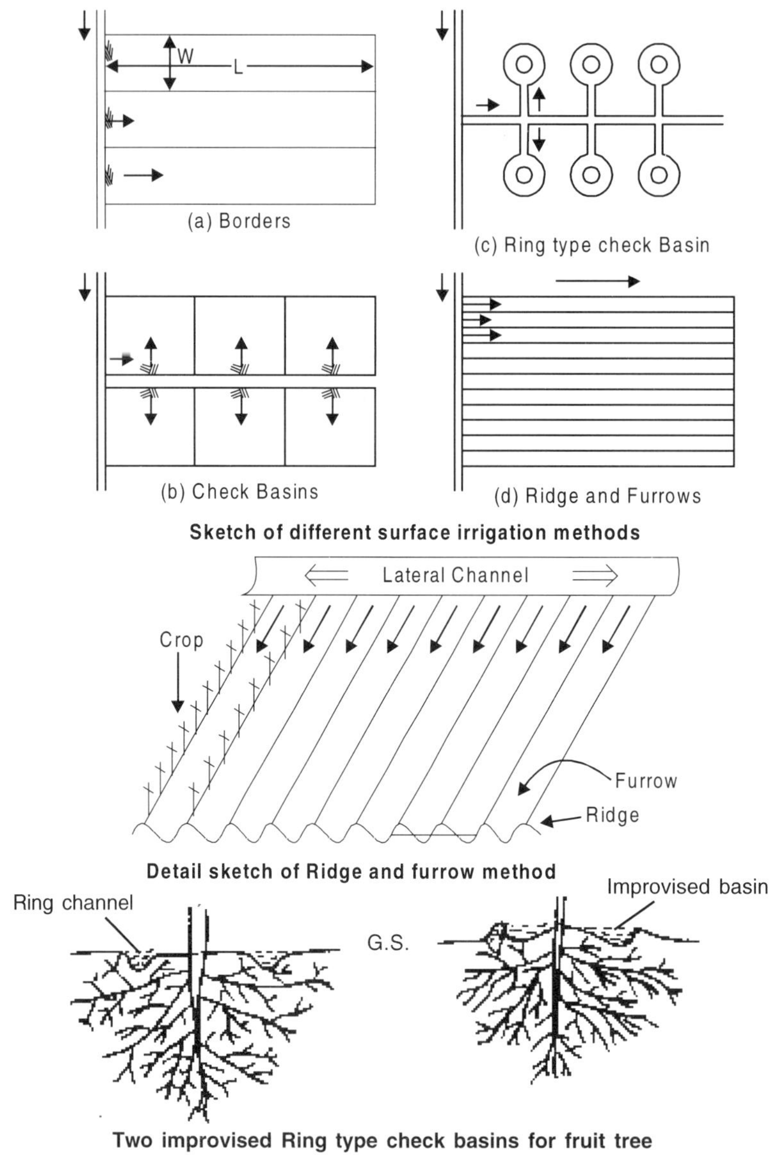

Fig. 5.6. Different types of surface irrigation method.

hand. Naturally this method has low distribution efficiency. The advantage of this method is that it requires almost no fixed cost and the only expenditure is manual labour which is abundant and cheaper in our country.

In mechanical sprinkler systems a pump or a pressure head is used to sprinkle the water. In general, all these systems of irrigation give rise to higher water use efficiency. Through this system about 30 to 50% of water can be saved and hence by introducing this method for closed spaced crops, more than 50% additional area can be brought under irrigation. In dryland areas, depending upon the available water, this system may be designed by careful selection of spraying equipment and operating pressure. Through this system the amount of irrigation water required to refill the crop root zone can be applied nearly uniformly at a rate to suit the infiltration rate of the soil, thereby obtaining efficient irrigation. This method is suitable for almost all the crops (expect jute and rice). However, this method is very much useful for cultivation in sandy soils and in areas with undulating topography.

The main problem of using this system in dryland is the high temperature and hot wind that decreases the application efficiency by increasing evaporation loss while spraying. To avoid this problem, sprinkler may be used after sunset. Besides, this system requires high initial investment in comparison, which is not economically viable with the existing farming practices in dryland. By increasing gross as well as net cropped area with cultivation of high value crops, this system of irrigation has been proved to be a viable option for dryland farming.

Two basic types of sprinkler systems, depending upon the spraying mechanism, are generally used in the field. One is *rotating nozzle type* and the other is *perforated pipe type*. The first type sprays water through one or two small sized nozzles fixed in an equipment called head. While spraying, the head rotates on its own axis and sprays water in a specific circular area as evenly as possible. This spraying is also termed as 'rain'. In the second type, the end pipe of a irrigation pipeline is perforated with small holes in a specially designed pattern or in the form of rotating head to distribute water fairly uniformly. The main advantage of perforated type is that it requires less pressure (0.5 to 2.5 kg/cm^2). According to the design of this system and placement of the perforation it may spray water on both sides of the pipe and can cover a strip of land from 5mt to 15mt while travelling through the strip mounted on a wheel. On the other hand, rotating type small perforated pipes are able to spray water in circular area smaller then the rotating nozzle type heads. The perforated pipe type sprinkler irrigation system can be operated with the help of overhead tank only.

Both these two types may be designed in different ways like (1) portable, (2) semi-portable, (3) semi-permanent, (4) permanent and (5) solid state. Out of these, number one is cheapest in respect of initial investment and the cost rises gradually as the number increases. For this reason, apart from tea, the portable and semi-portable sprinklers are becoming popular for cultivation of various crops in India.

In dryland farming the water available in farm pond or well or other water harvesting structures are generally used for cultivation through any surface irrigation methods or through hand spraying. Now if they use the small sprinkler system, particularly the perforated pipe systems, they can increase the command area of the water source, thus increasing the yield.

5.1.5.4. Drip Irrigation

It is the system of irrigation where water is applied on the ground surface above the root zone area drop by drop or by trickling. In India since ancient times people practiced this principle for individual plants for different interests. For example, they placed water filled pot, with cover, over the plant like 'Tulsi' with a hole in the bottom, plugged by cotton or any other permeable material through which water fell drop by drop in the dry season. The drip irrigation is the most useful system of water application to the crops in water scarce areas and in the soils having high infiltration rate. It minimises the conventional losses of irrigation water like deep percolation, seepage and evaporation at the time of conveyance and distribution. About 40 to 70% water can be saved by this method and the yield can be increased by 10 to 100% for various crops.

The modern or mechanical method of drip irrigation is accomplished by using small diameter plastic pipelines having perforations at selected spacing, which are connected with main and sub-main lines and the water comes out from these perforations by a small pump or by any pressure head. The perforations are fitted with a device called 'emitters' or 'drippers' to deliver water drop by drop with almost zero velocity. This method is very much suitable for individual plants. The initial cost for the equipment is considered to be the prime limitation for large-scale adaptation, particularly in dry areas where coarse cereals predominantly exists. However, in wide spaced crops the initial cost is reduced considerably and is also proven to be much economical. It is also viable for less widely spaced high-value crop in dryland farms.

Advantage of Drip Irrigation System

1. Spot application of water reduces evaporation and other losses of irrigation water.
2. Only target plants receive water, reducing weed problem.
3. Saves water up to 70% in comparison to other irrigation methods.
4. Low rate (1-5 lit/hr) of application of water moisten only the root zone thus avoiding any kind of percolation or seepage loss.
5. This system of watering maintains an ideal soil-water-air relationship optimally favourable to root system. This can increase the growth of plant or yield by 10 to 100% while compared to other irrigation methods.
6. This sustainably moistens the root zone but creates a dry micro-climate around the canopy thus reducing incidence of pests and disease.
7. This system of irrigation provides a unique scope to apply fertilizer through irrigation water, called **fertigation**; this increases the efficiency of fertilizer use and reduces the loss of nutrients supplied through fertilizers up to 40 to 60 percent.

Disadvantages of Drip Irrigation System

1. High initial investment particularly for close spaced crops.
2. The tricky technicality and inadequate technical input.
3. Maintenance, particularly for clogging of drippers.
4. Not suitable for plant population based field crops.
5. Planning and layout do not provide sufficient scope to increase the area in due course.

Components of Drip Irrigation System

A drip irrigation system consists of six distinct components viz. (1) Power generating segment, (2) Filtration system, (3) Fertilizer/agro-chemical injection segment, (3) Water distribution or pipeline segment (4) Water emitting segment and (5) control head segments.

1. Power generating arrangements. This arrangement is the starting point of the drip irrigation system. Pressure is required to send water from its source to the ejection or emission point. This shall overcome the head loss occurred in between due to filtration unit and frictions along the distribution pipelines and its bends. Generally requirement of pressure is not very high but for a large system the total pressure requirement is higher and it is provided with pumps of required power. For smaller units **gravitational head** can be used to create necessary pressure. The job of this pressure head is to supply water up to the delivery point where it will be released drop by drop almost with zero velocity. However, the drip irrigation system operates under pressure as low as 0.15 to 0.2 kg/cm^2 and as high as 1 to 2 kg/cm^2.

2. Filtration System. Proper water filtration is essential for smooth functioning of drip irrigation system. Physical clogging of emitters or drippers can be kept under control through this filtration system. **Effect of clogging-** Even a small percentage of emitter clogging can drastically affect the uniformity of water application. With, as low as, 5 percent emitter clogging, the statistical uniformity can be reduced to approximately 75 percent.

When physical factors (sand, silt, and other suspended materials) is the main reason of clogging, good quality filters are used in a drip system, operated at 1000 lit/hr/m^2. However, three basic types of filters are used in drip irrigation system either individually or in combination.

- **(a) Sand filter:** It can effectively remove sand particles and hence used as pre treatment. Suspended other solids may be removed by sand filters (Fig- 36).
- **(b) Media filter:** This type of filter consists of fine gravel and sand, placed in a pressurized tank, followed by ring-way valve. It can reduce both sand and suspended materials efficiently.
- **(c) Screen filter:** It is the simplest design, made up mostly of plastic for removing suspended materials effectively. Depending upon the quality of water the filtration system may include a single or combination of screens.

3. Fertilizer and other agro-chemical injection segment. A separate air tight fertilizer mixing tank is installed, which acts as a pressurized vessel with inflow and outflow pipes/tubes. It is connected with main water line having a pressure regulation valve in between two connecting points (fig:- 36). Actually a portion of irrigation water collects fertilizer mixture through sucking (ventury) or through pressure difference (mixing tank) or by the help of external force (injector) and return to the main irrigation flow.

Other agro-chemicals like pesticide, fungicide, nematicide etc, which are systemic in action, may be applied to the tank, in proper dose, and is finally applied to the crop through irrigation water.

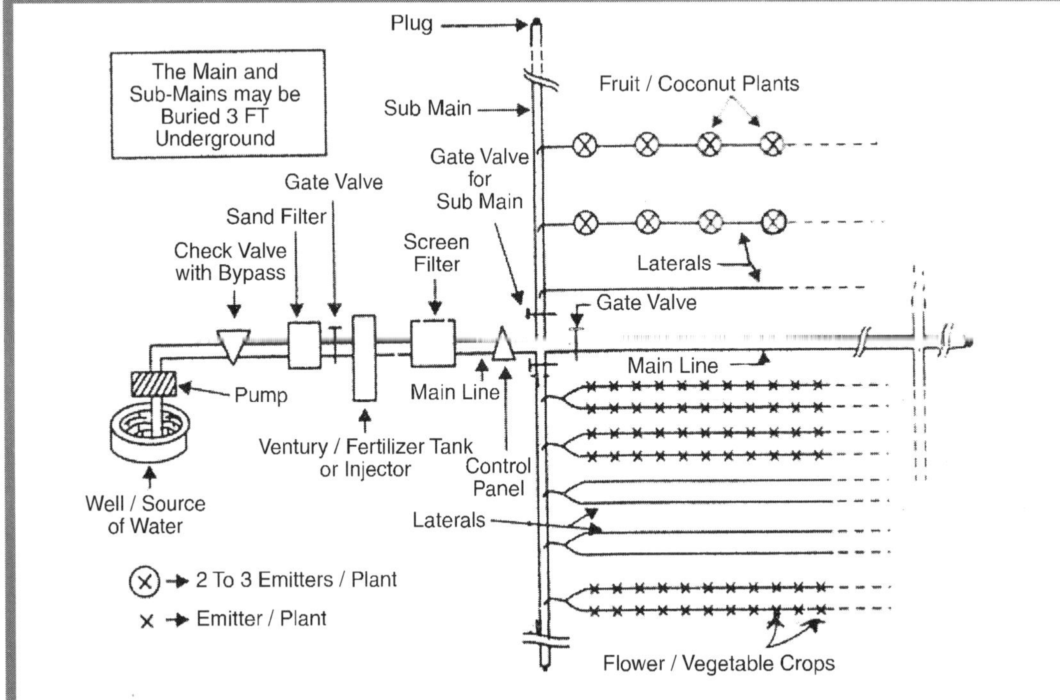

Fig. 5.7. High pressure drip irrigation system and its components.

A few 'organo-phosphates' and 'carbamade' insecticides for systemic control of sucking/piercing insects like thrips have been moderately successful. Many other systemic insecticides have been found effective through drip system against respective pests.

Phytopthora diseases are the main reason for which fungicides are applied through drip irrigation system. Contact fungicide like 'ethazol', 'benomyl' and systemic fungicides like 'ridomil' may be used through drip system to control Phytopthora diseases like 'root rot'.

Applying contact nematicides through drip irrigation system is rapidly gaining popularity, as it is required in lower doses.

4. Water distribution or piping system. The distribution line consists of a network of a main line, sub-mains, and/or laterals, generally made up of either 'high density polyethylene' (HDPE) or 'low density polyethylene' (LDPE) 'poly-vinyl chloride' (PVC) pipes. The main line is larger in size with a range of 25 to 50 mm diameter, depending upon the size of land and number of emitters. The sub-mains, having diameter of 15 to 35 mm, are generally used for larger plots with different blocks. The buried main and sub-mains are commonly made up of HDPE PVC pipes. However, the laterals are normally made up with LDPE PVC pipes of smaller diameters e.g. 10 to 20 mm.

The appropriate fittings like elbow, tee, cross, plug etc, made up of HDPE PVC or mild steel, are available and used to connect the pipes of the distribution system of a drip irrigation system.

5. Water emitting segment. It is the device placed inside or outside of a small hole of 1mm diameter in the lateral lines, which are called as emitter of dripper. These drippers are spaced along the laterals according to the requirement of crop and emit water almost at zero velocity with very low discharge rates i.e. 2 to 4 liters/hour.

Different types of drippers have been designed and are used nowadays. Constant discharge with specific rate, capacity to avoid clogging, adaptability to force flashing, suitability of placement, and cost are the points considered during designing of drippers.

6. Controlling system. The function of drip irrigation control system is to measure, monitor, and regulate the flow of water running through the drip irrigation system. It may be done through simple valves to most sophisticated computerized control system. However, a drip irrigation system shall at least consist of different kinds of valves, meters, and switches, which work in isolation or in combination in a pre-planned mode.

Clogging of Drip irrigation system

Proper monitoring during installation and running of a drip irrigation system is necessary to avoid clogging, which imposes one of the major problems. For this matter the following aspects should be taken care of.

(a) Quality of irrigation-water, the basic reason of clogging of the drippers. Analysis of water not only provides the information about its applicability according to crop requirement but also gives an idea about the possible clogging problem in a drip-irrigation system. So it is essential to estimate water quality before making a plan for installation of a drip-irrigation system.

There are four factors associated with irrigation water that causes 90 percent of clogging in a drip irrigation system. Physical, chemical and biological aspects of irrigation water have been rated as 35, 22, and 37 percent responsible for clogging, respectively.

Physical factors of water means the presence of inorganic particles, like sand, silt, and clay, organic materials and microscopic aquatic plants (phyto-plankton) and animals (zoo-planktons), which frequently passes through ordinary filters and reach the emitting parts leading to emitter clogging.

Chemical factors, i.e. presence of chemicals/salts that leads to deposition of Calcium Carbonate ($CaCo_3$), Iron-oxides etc in the narrow passages of the emitters and starts clogging.

Biological slimes, filaments, bacteria, algae and other microbial deposition are the most serious factors that frequently clog the drippers. Generally this occurs in presence of iron and hydrogen sulphide.

(b) Solutions against clogging: Prevention and reclamation of emitter clogging is the main aim to fight against clogging. Out of prevention methods (1) selection of proper emitter, (2) proper filtration of irrigation-water, (3) regular field inspection, and (4) scheduled pipe line flushing are the basic four methods, which reduce the clogging of emitters. These are done individually or in combination as per necessity.

Reclamation of partial or fully clogged system may be done by the following methods.

1. **Chlorination**: Addition of chlorine is the primary means to control clogging occurred due to the activity of microbial organisms (algae, fungi, bacteria) present in the irrigation water. Calcium-hypo-chlorite, sodium-hypo-chlorite (1-2 mg/lit on continuous basis) and chlorine gas are used to reduce or remove clogging of emitters.
2. **Acid treatment:** The chemical salt deposition, which causes clogging, can be cleaned by acid treatment. A solution of dilute hydrochloric acid (36 %) of 0.5 to 2 % by volume of water, if introduced in laterals for 10 minutes, can effectively remove $CaCo_3$ deposition thus reducing clogging.
3. **Pipe line flushing:** Flushing the main, sub-mains and laterals with pressurized water can remove the potential clogging materials like sand, silt, organic/ biological matters present in the system. The flushing valves shall always be provided at the end of the main streams and manual flushing or some type of lateral flushing device is required occasionally (may be once in every six month) to augment the drain valves. In case of severe clogging, the pipeline flushing may be done by air compressor with a pressure of about 70 meter head of water.

Drip irrigation system for small farmers

This system utilizes the gravitational force and does not require any external power to operate. It works with low pressure of about 1.5 to 2 meters water height. It can be planned in the following way.

1. A 200 to 500 liter or more capacity water tank is mounted on a platform of 2 meter height to store water for irrigation through drippers at the root-zones of the targeted crop. The water from source can be drawn manually or otherwise and stored in the tank after completion of one irrigation cycle for a specific area designed according to the capacity of the tank.
2. An outlet pipe of 25 to 40 mm diameter made up of black alkathen (BAP) or PVC pipe is fitted as main line at the base of the tank. The diameter of pipe is dependent on the number of emitters i.e. requirement of discharge.
3. A gate valve is provided in the mainline to regulate the flow from the tank. A filter, hand made with fine brass or copper wire of 200 mesh sieve or simple water filter used for domestic purposes, may be fixed just after the gate valve (fig: 5.6). If sub-mains are required, a separate gate valve for each sub-main is to be provided to maintain the schedule of irrigation. The end points of the mains and sub-mains are to be closed with plugs. Joints are to be fitted and, if needed, clamped properly.
4. As per laid out beds and according to the spacing of crop the laterals are placed on the ground before planting, one end of which shall be connected firmly with main or sub-main pipe line and other end plugged by simply bending the end point and tying it with a thread. Laterals are generally made up of 8 to 12 mm diameter LDPE pipes having emitters along the length, with a discharge capacity of 0.6 to 2 liters/hr, spaced according to the spacing of the crop.

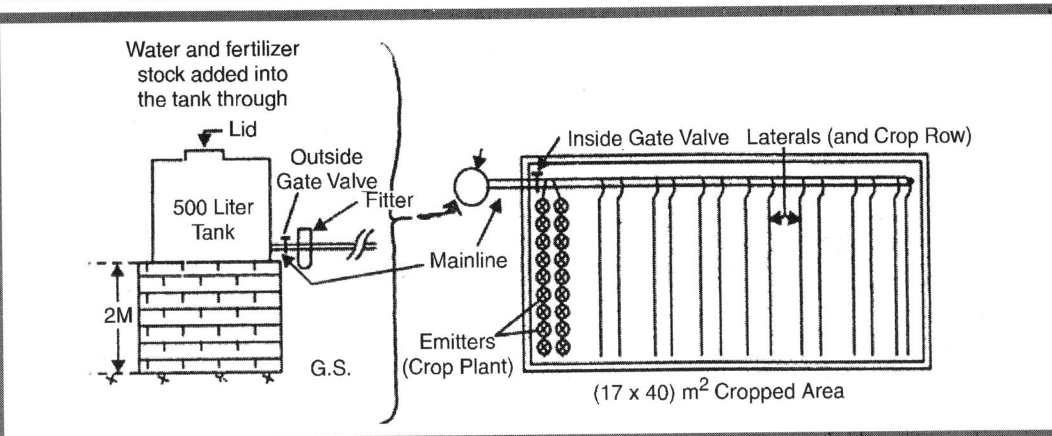

Fig. 5.8. Low pressure drip irrigation system for vegetable or flower.

Planning of this drip irrigation system: The main obstacle of this system is the length or the laterals, which has to be restricted to a certain limit, depending upon the number of emitters. However the following information is essential to plan such a low-pressure drip irrigation system.

- Water requirement of a single plant in liter/day.
- Size of storage tank in liters.
- The number of drippers/plant.
- Time gap for refilling the storage tank in hours.
- The number of plants that can be irrigated by the storage capacity of the tank in one cycle and the size of the land in respect of length of laterals.
- Plan of fertilizer application through this system.

Example for low pressure drip irrigation system

1. **For vegetable cultivation.** Tank size – 500 liters, Crops – Tomato, egg plant, cucumber, gerbera, rose, etc. Distribution – only mainline (25mm ID) and laterals (8-12mm ID) providing one emitter (1 lit./hr) per plant, Area – approximately 700 sq. m. Maximum length of lateral – 17 to 18 m. Fertigation– nitrogen and potash is applied in the tank to feed the crop through irrigation water.
2. **For tree crops.** Tank size - 200 to 500 liters, Crops – Mango, Guava, ber, lemon, pomegranate etc., Distribution - mains – (38mm ID), sub-mains (25mm ID), and laterals (12mm ID) providing 3 emitters (2lit/hr/emitter) per full grown plant. Area – 0.2 ha.

This system of irrigation can provide highest degree of water use efficiency. It can increase net-cropped area by way of saving water. It also improves the quality and quantity of the crops. It also reduces the cost of land preparation and labour cost for application of fertiliser etc.

So, for dryland farming it would be the best possible way to utilise precious water for irrigation purpose.

5.1.5.5. Sub Surface Methods

Though theoretically sub surface method seems to be an effective irrigation method in dry land farming as it protects irrigation water from evaporation, still except pitcher method no other methods are economically or technically viable in dryland areas.

However in established plantation, depending on land situation, the trench method of irrigation is proved to be a viable method. Here the trenches are dug in between rows and water, manure and fertilizers are generally applied in that trench. It also helps to catch hold of runoff water in the trench.

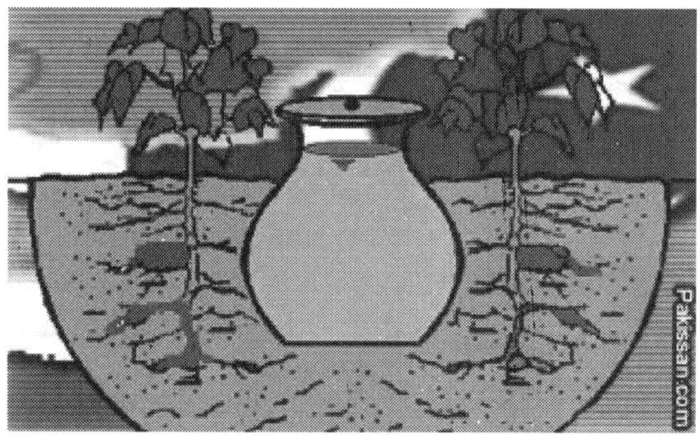

Fig. 5.9. Pitcher Method.

Pitcher Method is a very useful means of subsurface irrigation method in dryland areas especially for young tree crops. An earthen pitcher/pot of 10 to 15 liter capacity, with a hole at the bottom, is buried at the rootzone very near to the seedling (Figure 5.9). It is buried in a way that its neck remains just at the ground level and is provided with a lid on its mouth. After pouring water into the pot the lid is placed tightly to minimise the evaporation loss. The hole at the bottom is plugged with a cotton or jute twig to allow the water of the container to trickle out. Fertilizer and other agro-chemicals may be applied through this method. This pitcher method will be stationary in a same place minimum for three years if the young plant survived properly, and then, if not broken, the pot can be removed and reused at another site.

Pitcher irrigation in comparison to irrigation with simple watering-can is more labour saving or economical. This practice has spread in a surprisingly slow pace though its potential for small land-holders is considerable. Tilman (1982) estimated that one person of average strength irrigates effectively 3500m^2 through pitcher method of irrigation in lieu of 500m^2 by watering-can.

Sub-irrigation. In some natural or artificial situation it is possible to irrigate crops by applying water below the ground surface, hence it may be possible to manipulate the position of underground soil water below the root zone. By utilizing the natural impervious layer/sub-soil or construct a barrier below the root-zone soil to cause soil water to back up and rise close enough to the surface to water the crop is a good and viable option to increase the efficiency of sub-irrigation (fig-5.10).

With a separate approach the perforate pipes buried just below the root-zone can also be used for sub-surface irrigation. Fertilizer and pesticides may be applied through this irrigation system.

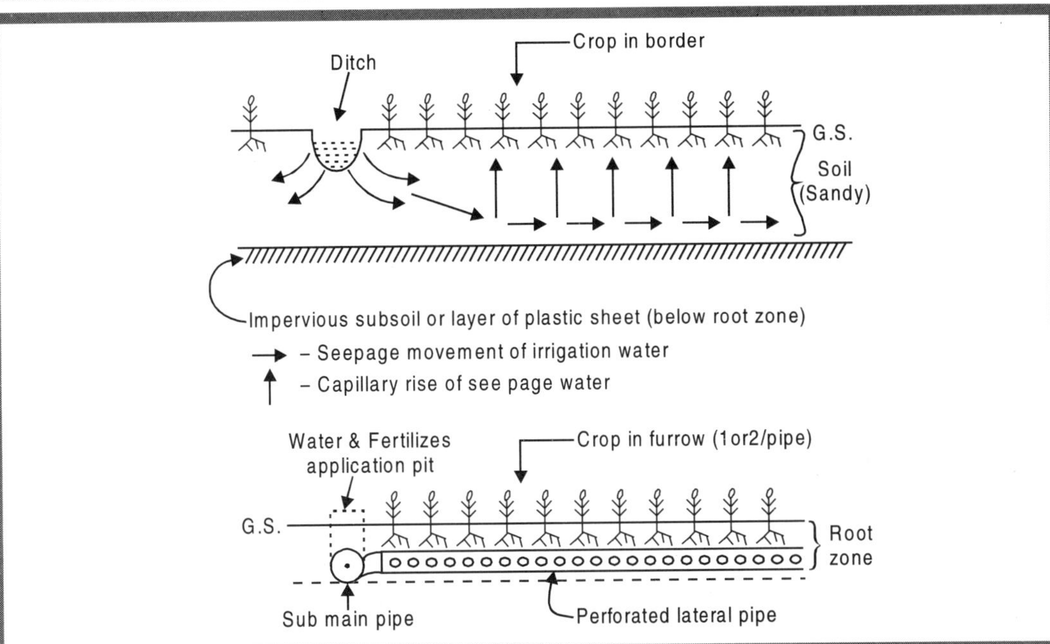

Fig. 5.10. Two methods of sub-irrigation.

Care is needed for all types of sub-surface irrigation in respect of rising of water table too close to the surface which may cause salinization and/or destroy beneficial organisms. This system of irrigation seems to perform best when the soil is fine textured and the land is level.

Water Absorbent Polymer

A sample composition of the polymer chain :

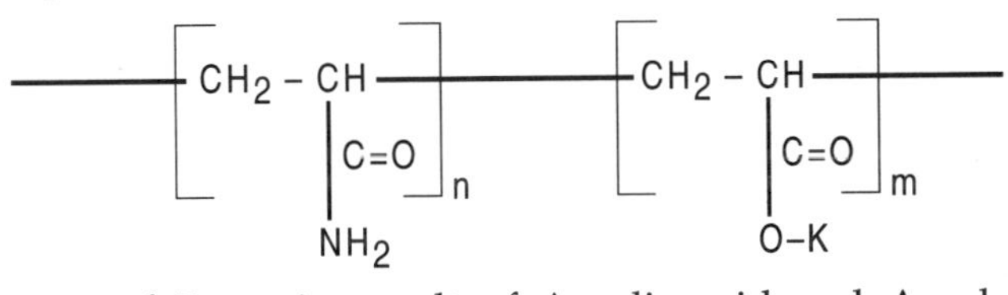

It is a copolymer of Potassium salt of Acrylic acid and Acrylamide. It degrades naturally in the soil, up to 10% per year, to CO_2 (carbon dioxide), H_2O (water), and N_2 (nitrogen) compounds. These degradation products are water-soluble metabolites and are non-toxic.

Therefore, this polymer is fully compatible and safe to the environment or can be termed as ecofriendly material. It is basically an inert polymer and does not contaminate soil and thus, does not contaminate the plants, other organisms of soil and ground water.

Functions : It is an absorbent and retainer of water. When water molecule comes in to contact with this polymer chain it is drawn into the molecule by osmosis and become gel (Hydrogel),. Its water retaining capacity is 200 to 400 times of its own weight, depending upon the soil. When soil mixed with this sand like polymer at an appropriate rate (4-10 kg/ acre or 10-25 grams / plant or pot), depending on the crop, the water holding capacity of the soil increases many fold and reduce the percolation loss.

The absorbed water release in to the soil for plant/crop on demand (drying) as function of absorption-release cycle in phases to the extent of 95% of absorbed water. This absorption-release cycle can occur up to hundred times and this polymer is active in the soil effectively maximum up to 6 years and takes ten year to degrade completely.

Advantage of water absorbent polymer

- Increase water-holding capacity of the soil up to 100%
- Reduce water requirement of crop up to 50%.
- Provide buffering effect of fertilizer availability by preventing leaching thus reduce fertilizer consumption.
- Cut down loss of irrigation / rain water from soil through percolation and evaporation.
- Improve porosity of soil.
- Reduce mortality of seedling and transplant materials significantly, particularly in rain-fed condition.
- Increase crop production or yield up to 40% with same input.

5.2 MANAGEMENT OF SOIL

Soil management is a vast subject particularly in relation to its wide variability. Its requirement also differs from field to field, crop to crop and farmer to farmer.

Soil is an important component of dryland ecosystems. The drylands have a cafeteria of soil situations. Alfisols, Vertisols, and aridisols are the principal soil orders in dry areas, although Entisols and Inceptisols also occurrs in few places, especially in toposeences. The major characteristics of dryland soils are – a generally low level of organic matter, alkaline to slightly acidic reaction in the surface, Calcium carbonate ($CaCo_3$) accumulation somewhere in upper layers of soil (150cm.), coarse to medium texture, low microbiological activity and high proneness to erosion. These soils suffer from a variety of constraints that limit successful crop production. The constraints develop due to the nature and properties of soils in association with very low moisture status. These are accentuated by erosion caused by water and wind, nutrient depletion by cropping, salinisation by high rate of evaporation and compaction or hard setting. However, the dominant soil related constraints in drylands are:

1. **Tillage and Seeding Problems.** In Vertisols, optimum moisture range for tillage is very narrow. In Alfisols and Entisols, surface soil layer is sandy and dries out rapidly. Vertisols are hard when dry and have very plastic consistency when wet.
2. **Seedling Establishment Problem.** In Vertisols, the aeration of the seedling root zone is very poor. In Alfisols and Entisols the surface crusts develop on drying and limit seedling establishment.

3. **Poor Soil Fertility.** Available major nutrients i.e. N, P, K, is deficient in drylands. Micronutrients such as Zn, S and often Fe are also lacking in many dryland soils. Poor organic matter content magnifies the magnitude of this constraint.
4. **Uneven Soil Wetting.** It happens as a result of non-uniformity of soil profile.
5. **Salinity and Shallow Water Tables.** Vertisols in regions of granite-gneiss geology having clay loam subsoil have perched water tables during the rainy season. Soil salinity commonly occurs in the soils where E T rate is very high. Irrigated condition, in these areas, increases the severity of the problem.

Sustainable crop production in drylands calls for scientific understanding of different property or aspects of soil and its management. The dryland soils should be managed in a way that the constraints of the soil are either eliminated or reduced to a considerable extent for successful crop production. To manage a soil in a sustainable way for better crop production the physical, chemical and microbial aspects are to be managed properly and concurrently. Now we may discuss the physical, chemical, and microbial aspects of soil separately.

5.2.1. Physical Management of Soil

It is the manipulation of physical properties of soil i.e. texture and structure by tilling, land developing or shaping works or otherwise to change its various aspects in favour of plant growth. This can be influenced by (1) land shaping, (2) creation of different structures, (3) tillage and other intercultural operations, (4) application of silt or imported soil and other soil stabiliser. Before discussing these four items in details we can discuss the management of soil structure in different textural classes of soils.

The coarse-textured soils have good aeration, drainage and are easy for tillage operation but lack the capacity to absorb and hold sufficient moisture and nutrients. They need granulation and it is only possible through addition of organic matter. The long chains of organic molecule, the polysaccharide gums of soil bacteria and growth of fungal and actinomycetes hyphae act as binding agent for the coarse soil particles and as a result increases the water and nutrient holding capacity.

In fine-textured soils the potential plasticity and cohesion are always high because of the presence of large amounts of colloidal clay and structural management of this kind of soil is a hard task. The tillage of this type of soils must be carefully timed. Like coarse-textured soil, incorporation of organic matter in clay soils induces better granulation. Here the sod crops should be utilized properly and the crop rotation should be planned to attain maximum benefits.

5.2.1.1. Land Shaping

Land shaping is a fundamental practice and is essential for agricultural purposes. Physical alteration of soil is the main act required here and land levelling is a commonly practiced land shaping activity.

There are a few land shaping systems that are practiced for better soil and water management; they are **raised sunken bed** (RSB), **broad bed and furrow** (BBF), and **scoop**. However, land shaping plays a great role in preparation of seedbed, providing equal facility to each and every plant of a crop population, but management of water is the ultimate subject linked with land shaping. For that matter land shaping is discussed in detail in 'rainwater management under water management. {See Chapter-5.1.2.2.}

Land leveling and grading: According to the present idea, it is the process of preparing or modifying or reshaping the land surface to obtain a required and modified land surface suitable for agriculture. Basically the method of this development is cutting of high areas and rising of low spots of a specific land area in order to remove the surface irregularities to make it a plane surface.

The primary idea behind this process is to cultivate field (grain) crops in such a land and the secondary purpose is to conserve soil and water. It any one thinks to utilize this type of land for any other purpose like fruit orchard etc. then process of land leveling and grading will be a separate proposition and will be more simplified one. Principally, for sustainable development, it is always wise to take the soil and water conservation measures and cultivation of crops with minimum change of land surface.

However, there are several factors related to existing process of land leveling and grading, which are mentioned below.

1. **Cropping pattern.** The crop rotation and the nature of individual crops in that rotation is an important determining factor to plan the land leveling and grading criteria.
2. **Method of irrigation.** Different methods of irrigation have their own requirement of ground/land condition like area of plot, slope of that plot, etc.
3. **Rainfall characteristics.** Rainfall characteristics, like intensity and duration of the average single rain, are important factor to determine the maximum and minimum allowable limit of grade of the field.
4. **Topography.** The plan of land leveling and grading is very much dependent on topography of the field. Actually the amount of cut and fill, thus the cost of leveling, has a strong bearing with topography.
5. **Soil characteristics.** Soil characteristics like soil profile, soil texture, etc has a strong relation with land levelling and grading.

5.2.1.2. Creation of Different on Farm Structures

Different field structures are frequently used to control the movement as well as status of the soil. The selection of available technology like **'Ridging'** or **'Bunding'**, **'Terracing'**, **'Trenching'** or **'Pitting'** etc. {mentioned in Chapter – 4.3} depends on the objective of soil management and are normally employed in conjunction with agronomic measures.

Bunding/Ridging

(a) Contour Bund. This is the earthen bank, 1.5 to 2 m. wide, laid across the slope and along the contour to act as a barrier to runoff. This barrier or series of barriers help to settle down the soil particles and hold the water in the up-slope side (Fig. 5.11). It also breaks up the slope into segments shorter in length than is required to generate

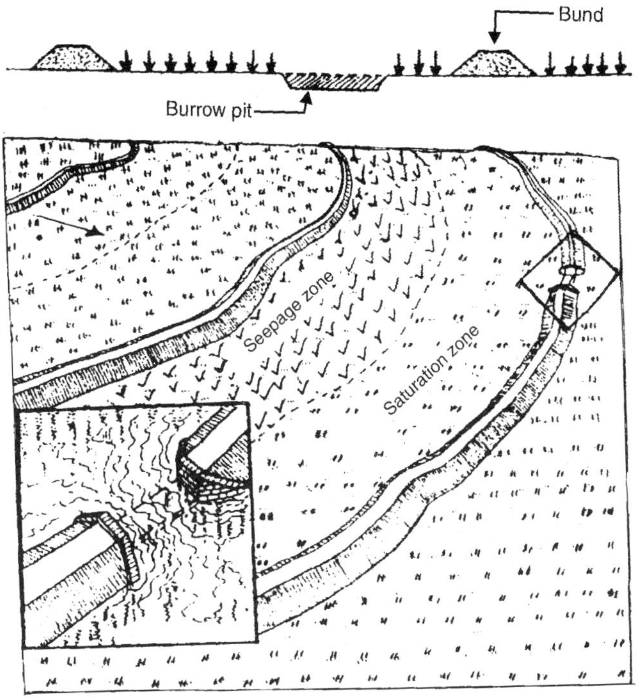

Fig. 5.11. Contour Bund

overland flow. These bunds are suitable for slopes of 1^0 to 7^0 and are frequently used on small holdings in the tropical regions where they form permanent buffers in a strip cropping system. The bunds are generally placed at 10 to 12 m. intervals. There are no precise specifications for their design and deviation in their alignment up to 10% is permissible. The studies at Central Soil Water Conservation Research and Training Institute (CSWCRTI), Deradun and its research stations, have shown that contour-bunding is suitable for low (<600mm.) rainfall areas and permeable soil to serve both as soil and water conservation measure and not recommendable in low permeable soils as such. The size and spacing of bunds depends primarily on the slope (Hurni, 1948).

Table 5.5. Size and Spacing of Bunds in Relation to Slope

	Bund size (in mt.)		Spacing	Storage	Predicted soil
	Height	Width	(mt.)	capacity (m^2)	loss ($kg/m^2/year$)
6°	0.20	1.90	20	0.19	3.0
14°	0.20	0.70	15	0.07	11.5
27°	0.20	0.40	10	0.04	11.5

Construction. After due consideration is given to the alignment of bunds at planning stage, the bund lines are drawn on the ground, keeping in view the following factors:

1. The construction should start from upslope and proceed downwards.

2. No intermediate bund is kept unfinished or half finished.
3. If any area, in upslope, is not bunded for any reason then diversion drain is to be provided to divert the runoff from causing overflow damages to the bunded area.
4. Soils for construction of bunds may be taken from burrow pits located on up-stream side or on either of the sides in case of shallow medium soils or medium deep soils respectively.
5. The planning and design of these structures should meticulously correlate with cultivation practices starting from seeding/transplanting to harvesting.
6. Excessive disturbance of topsoil should be avoided.
7. If the land situation and crop allow, it may be implemented along with contour ditches.

Area Lost. The area occupied by a bund, which is the base area of that bund, is not available for cultivation purposes. On an average about 6 percent of the area is lost through bunding and it will be increased with increase of slope. However, bunds can be used for growing grasses or suitable crops and thus compensate for area lost. Fodder, green-manuring plants, leguminous trees etc. may be planted on the bunds and thus understand the potentiality of farm forestry on contour bund.

Maintenance. Due to lack of maintenance, much of the benefits are lost and in certain cases damage may occur, which leads to destroying of the bunds. It is, therefore, of paramount importance that the following works are attended before the onset of rains.

- Plugging of all breaches firmly;
- Filling up of all sags in bunds and restoration to proper height and section;
- Plugging of all holes (created by rat, fox etc.);
- Repairing of all outlets;
- Plantation of grasses or plants on bunds;
- If necessary, reconstruction of a portion or entire bund.

(b) Contour Stone Walls. It is a type of contour bund made up of stones and gravels. In slopes where stones and rocks are available in plenty this structure is used

Fig. 5.12. Contour stone wall.

in place of contour bund. It is also used as initial treatment for construction of terrace. It effectively slows in the runoff and collects the eroded soil on its upslope side (Fig. 5.12).

(c) Graded Bunds. This method is used where excess water is being removed safely out of the field placed in a bunded area. It is constructed in relatively high rainfall (< 600 mm) areas but can be recommended even in areas where rainfall is less than 500 mm and clay content of the soil is high like the deep, black soils. In contrast to contour bund it drains out the runoff water, though safely, out of the reach of small farmers; thus is not at all a better option in dryland farming. Contour ditches/ trenches are much better alternative to graded bunds.

As it essentially diverts excess water from cropped land, suitable channels and outlets are required for safe removal of water. Grassed or vegetated watercourses and outlets are compatible with graded bunds. Where grass and vegetation would not establish easily or large quantities of water is required to handle, structural measures like drop structures may be adopted.

(d) Contour Ridges. It is the making of ridges along the contour, usually at a slight gradient, at regular intervals determined mainly by slope steepness. These ridges are positioned across the overland flow of runoff water and dispose it slowly along the contour. Thus it increases the time of holding and also the flow path of runoff providing more time for in-situ sedimentation of eroded soil as well as percolation of runoff water. However, contour ridging is generally ineffective on slopes steeper than 4.5°. There are many variations in this method depending upon the situation.

A system of alternating 1mt. wide broad beds/ridges and 0.5mt wide sunken furrow is found to be effective in the soils of Hyderabad in India. In the well-drained soil, greater storage of water and more effective control of erosion can be achieved by contouring the ridges with crosstie over the intervening furrows. This system forms a series of rectangular depressions, which gets filled up with rainwater. This practice is called *tide ridging*. A similar technique called *range pitting* is sometimes used in contour ridge system on grazing land or horticultural fields where a series of pits are dug in between two ridges. These pits are about 50cm.× 50cm., 75cm. deep and 40cm. apart. The size of pit may vary according to the situation of the site.

(e) Tied Ridges. Greater storage of water and more effective erosion control can be achieved by connecting the ridges with cross-ties over the intervening furrows,

Fig. 5.13. Tide Ridge.

thereby forming a series of rectangular depressions which fills with water during rain. This practice is known as tied ridging and should only be used on well-drained soils. It will decrease the run-off and increase the storage while graded ridges alone generally leads to an increase of surface run-off. Compared to ridging, tied ridges enhance water logging, create anaerobic condition in rooting zone, excessive leaching of nutrients and rise in water table in lower slope areas.

Trenching/Pitting

(a) Contour Trenches. These trenches may be continuous or staggered along the contour with a spacing of 5mt. to 10mt. horizontal intervals. The trenches may be 0.6 to 1.0m in width at the top, 0.4m at the bottom and 0.4m in depth. These dimensions may vary depending upon the slope, type of soil, type of rainfall and predominantly upon the type of crop or vegetation to be planted.

Fig. 5.14. Contour Trench.

The contour trenches are generally used for the slope more than 33% even when they are in badly eroded condition. Always ridges should back the trenches from which soil has been excavated to construct them. Trenches should always be connected with the artificial or natural waterways for disposing off the sudden outburst of clouds particularly in high rainfall (< 1000mm.) dry hilly areas.

(b) Staggered Trenches. The staggered trenching is generally used for individual plants. Different crops based on individual plants, particularly the horticultural plantations, are very much suitable for adapting this technique in dry land farming.

With more than 8% gradient, to check and hoard runoff water alongwith eroded soils, the trenches measuring 2mt. by 0.6mt. by 0.5mt. is very much useful. The trees and grasses are planted on the bream (the pile of excavated soil on the down-slope).

Fig. 5.15. Staggered Trench.

(c) Graded Trenches. Here the trenches are laid with some gradient along the contour and act as a waterway to convey runoff at the time of peak runoff time to a point where it will be discharged without any erosion.

They may be used in soil slopes ranging from 10 to 30 % where soil depth is less than 75cm and the average rainfall above 1000mm. It is very much useful when the lands in between two trenches are used for annual crops, which cannot resist the soil erosion.

(d) Range-pitting/ Scooping. Sometimes in grazing/grass lands or in fallow lands a series of 50cm x 50cm x 7.5cm deep pits are dug 40cm apart throughout the area. It collects soil carried by rainwater and stores good amount of water for necessary percolation. The capacity of the pits on an acre thus treated is roughly 1000 cubic fit, and they will hold an additional 0.3 inch of rainfall. This is a tremendous advantage in areas where the moisture from sporadic rainfall is usually lost through run-off.

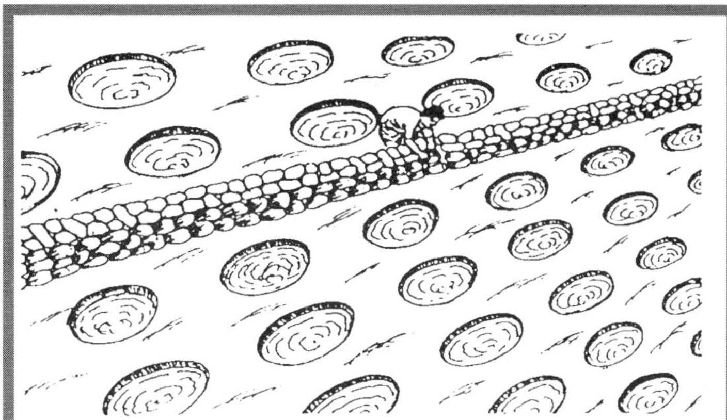

Fig. 5.16. Range Pitting or Scooping.

Terracing

(a) Level Terrace. Level terraces are used to retain the rainwater by storing it in the terrace itself. They are therefore un-graded and generally designed with the capacity to store runoff volume with a ten-year return period without overtopping. However staggered outlets are designed in each terrace for safe disposal of high intensity storm. The level terrace is generally implemented in soil slopes between 1 to 10% and in areas of low rainfall. Actually these types of structures are very much useful for field crop cultivation in dryland farming.

(b) Graded Terrace. It is very much similar to the level terrace in every aspect except the gradation. Here the terraces are designed with a slight gradient, usually 1 : 250, to the contour for channeling the intercepted water across the slope to a suitable outlet. This is particularly adapted in the dry areas where annual rainfall is high (<1000mm.).

(c) Bench Terrace. It consists of a series of alternative shelves and risers, and is employed where steep slopes up to 30^0 needs to be cultivated. The construction of bench terrace involves converting the ground surface into steps like fields. They are constructed by cutting and filling for reducing the degree of slope substantially. The riser is vulnerable to erosion and is generally protected by a vegetation cover and sometimes faced with stones or concretes. There are commonly three types of bench terrace: ***level top, inward sloping*** and ***outward sloping***. The basic design of different bench terrace system can be modified according to the nature and value of crop grown. More valuable crops like costly vegetables and fruits can be grown in terraces in dry undulating areas with proper water conservation measures. The width as well as the vertical intervals of terraces mainly depends on the steepness or gradient of slope and on the crops. The characteristics and purpose of the three types of terraces are: (1) Level top type- the name 'level or table top' describes the basic character of this type of terrace. Actually the upper surface of these types of terraces is provided with 0% slope. It is generally practiced where rainfall is good, slope less then 4.5^0 and soils are permeable. These terraces are useful for implementation of different water conservation measures. (2) Inward sloping type- It is suitable in high rainfall areas with deep permeable soils and is not fit for dryland farming. This type of terrace is always provided with a safe waterway to carry the excess water towards the natural outlets. (3) Outward sloping type – It is recommended for relatively low rainfall and shallow soiled areas. In fact this method is very much suitable for dryland farming, particularly in small land holdings in the area of undulating topography.

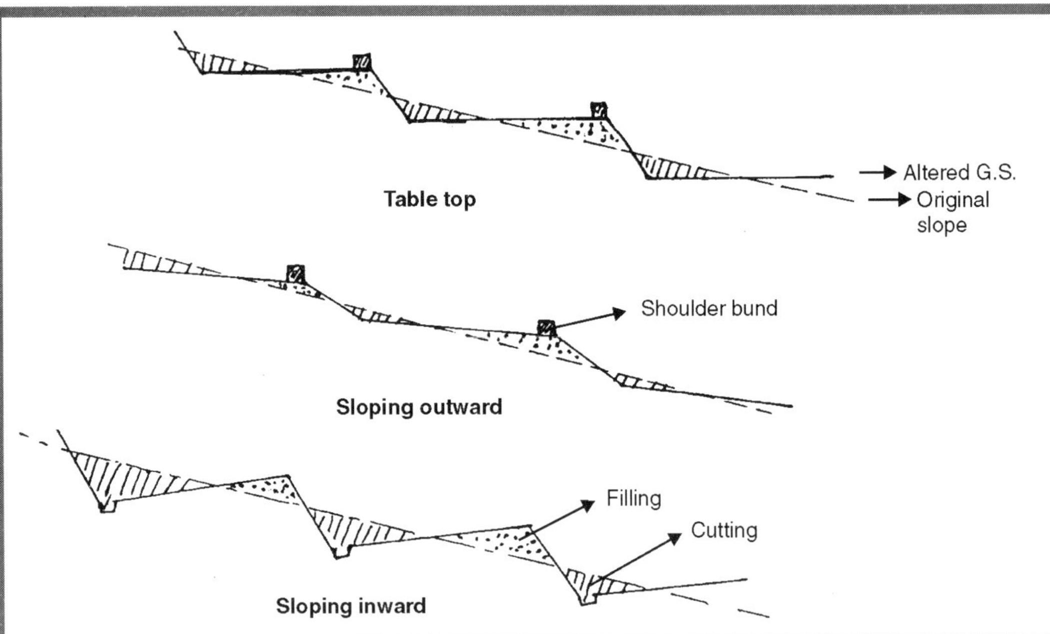

Fig. 5.17. Different Types of Terrace.

5.2.1.3. Tillage and Other Intercultural Practices

Conventional Tillage. It is the standard practice of ploughing first with disk or *mouldboard plough,* then one or more disk harrowing, then a spike-tooth harrowing and finally surface planting. Ploughing carried out with the mouldboard or country plough in stony land or with soils, which do not fall cleanly of mouldboards, the *disk plough* is often used. Ploughs invert the plough furrow and lift and move all the soil in the plough layer usually to a depth of 100 to 200mm. Secondary cultivation to form the seed bed and remove weeds is carried out by either disc or tine cultivators. Ploughing produces a rough cloddy surface with local variations in height of 120 to 160mm. *Secondary tillage* reduces the roughness to 30 to 40mm whilst drilling and rolling decrease it still further (Evans, 1980).

Frequently the conventional tillage causes problems in dusty, fine textured soils, particularly when dry, in very heavy, sticky soils and in structureless soils. Conventional tillage, on the onset of rainy season in degraded lands, creates serious problem of soil erosion that automatically increases the runoff loss. Actually pulverisation of soil near the surface and creation of compacted layer at plough depth increases the runoff by reducing infiltration. This finally increases the soil loss through erosion. In dry areas it also increases the wind erosion by enhancing the processes of saltation and surface creep of the pulverised soil particles.

Contour Tillage/Cultivation. Ploughing and subsequent cultivation practices (harrowing, seeding, weeding etc.) along the contour reduces soil loss from slopy land up to 50% compared with ploughing across. It is done by creating multiple mini barriers in the form of very small furrows and ridges across the slope i.e. across the flow path of sheet run-off. Naturally it improves the status of in-situ water storage of the soil and in turn increases the scope of more infiltration/percolation of rainwater.

Fig. 5.18. Contour Tillage.

Conservation Tillage. To avoid the above mentioned problems certain tillage systems may be adopted depending upon the situation of the farm, where tillage operations are restricted either by cutting down its number or by carrying out as many operations as possible in one pass and allow at least 30 % of residues to remain on the soil surface.

Numerous studies have been undertaken in recent years to examine the effects of different type of conservation tillage on soil erosion rates, soil conditions and crop yields. The results show that the success is soil specific. It also depends upon how well weeds, pests and diseases has been controlled. Generally, the better-drained, coarse and medium-textured soils with low organic matter content respond best. The systems are not successful on poorly drained soils with high organic contents or on heavy soils, where use of the mouldboard plough is essential.

Strip-Zone/Skip Tillage. In this system the tillage operations are concentrated only on the plant rows where plants grow leaving the inter-row areas untilled allowing the crop residue to be deposited. Depending upon the crop to be grown and the land situation its planning may be designed.

Mulch Tillage. In this system large percentage of plant residue (leaf, stubble, stalk, crown, root) is left on or near the surface as protective cover. Mulch tillage is accomplished with implements (sweeps or blades) that undercut the soil surface, thereby retaining most of the crop or plant residue on the surface. One of the main functions of this system is to break or disturb the continuous capillaries through which the evaporation loss of soil moisture from root zone takes place. Mulch tillage can also be given with a *chiesel plough*, but rotary rod weeder may be needed to improve the weed control. The problem of weed control occurs when mulch tillage is performed in moist soils or when precipitation occurs soon after tillage.

Minimum or no Tillage. This is the method of preparation of land or seed bed with minimum disturbance of soil. The practice of no tillage, where drilling of seeds takes place directly into the stubble of the previous crop and weeds are controlled by herbicide, has been found to increase the percentage of water stable aggregates in the soil compared with tine or disc cultivation and ploughing (Aina, 1979; Doglus and Goss, 1982). In this type of tillage all crop residues are maintained on the soil surface, which reduce erosion and enhances water conservation. For no tillage to be effective, adequate plant residue must be available. No-tillage cannot be effective in degraded soils with low plant residue. Where residue ranges from 1.5 to 2.5 tonnes/ha and weeds are controlled by herbicides, water storage and crop yields are similar to fields with stubble mulch tillage.

Crust and Seal Management. Soil crust and seal formation are serious constraints to seedling emergence and also soil and water conservation in alfisol soils. This problem can be reduced by decreasing the content of silt and clay in surface soil. For this, a *non-turning tillage* system is better than tillage with turning ploughs for alfisol soils. Here the problems of crusting and sealing are encountered more during the early part of the crop-growing season when the crop canopy is not fully developed. During this period considerable runoff was observed even when the soils were very dry. The *shallow tillage* imposed as an additional interculture has been found to be effective in breaking up the crust and improving infiltration (Pathak et.al., 1991) thus it reduces soil and water loss.

Hard Setting. In a previously cultivated or loosened soil, hard setting involves the coalescence of some or all of the artificially / newly aggregated structure, formed during and after wetting and hard setting without restructuring during drying. Poor aeration, impeded infiltration and accelerated runoff and erosion contribute to this situation. Chalka (Alfisols) soils of Telengana region of India are typical examples of hard-setting soils. Oxisols and Ultisols of Orissa and Bihar, particularly those with rain-fed agriculture also experience this problem. In order to minimise hard setting and increase crop yields, treatment with organic materials was found helpful.

5.2.1.4. Organic Matter

The role of organic matter on physical soil management (granulation, plasticity etc) is endless and it can solve almost all such problems directly or indirectly. Giving special emphasis on organic matter for dryland farming, the detail of it is described separately {in Chapter-5.4.}.

5.2.1.5. Influence of Root System

Plant roots are the best binding factors of soil. Grass roots improve soil structure largly due to their good canopy and excellent binding force. Roots grow and die in the soil and doing so it supplies food and energy to the soil fauna and micro-flora, which have a strong role in soil management. The roots also provide a mass of living organic matter and on decomposition it adds organic matter not only in the top few inches of soil but to greater depths as well.

The living roots of higher plants physically modify the soil as they push through cracks and make new openings of their own. These roots not only influence the equilibrium of soil solution by withdrawing soluble nitrate, but they also have something to do in a direct way with nutrient availability. Organic acids are formed at the root surfaces and hence become effective solvents.

5.2.1.6. Influence of Soil Stabiliser and Silt

Improvement of texture and structure of surface soil can be achieved by applying soil conditioners and silt. Silt can be applied in porous soils only. Soil conditioners like organic by-products, polyvalent salts and various synthetic polymers are used to bind the soil particles into aggregates.

Polyvalent salts such as Gypsum has been used successfully to improve the structure of sodic soils. Temporary stability, lasting from two weeks to six months, can be obtained in most soils by using soil stabilizers or soil conditioners involving oil or rubber-based compounds. They are too expensive for agriculture use. Soil conditioners are divided into two groups, soil hydrophilic - increased infiltration and decreased runoff and soil hydrophobic- decreased infiltration and increased runoff (De Vleeschauwer, Lal and De Boodt, 1978).

Animal slurries, factory waste, lime, sewage sludge, ferric and calcium sulphate, mixture of natural rubber latex and mineral oil, rice husk ash and many other materials has been tested for use in soil stabilisation. Some of them are too expensive and some of them are unpleasant to handle, but most promising are the organic by-products of crops, which are often available locally.

5.2.2. Chemical Management of Soil

This management is directly related with availability and status of nutrients of soil that directly or indirectly helps the plant growth. Proper management of

(1) nutrient and (2) salinity and sodicity can increase the status and availability of plant nutrients of a soil in every respect.

5.2.2.1. Nutrient Management

Most of the dryland soils are poor in available nitrogen (N), low to medium in available phosphorus (P) and generally adequate in available potassium (K) at their present level of productivity. Micronutrient deficiencies are frequent in the sodic and calcareous soils of dryland areas in which certain elements such as boron, iron and manganese are relatively low. With continious cultivation, zinc deficiency has been prominent. Sulphur deficiency has also been reported in many soils.

Chemical Fertilizer In Nutriant Management.- Due to deficiency of N in soil, response of a wide range of crops to applied N, in the form of nitrogenous fertilizer, has been reported. In high rainfall areas (Deradun) 200kg/ha was observed for both rice and maize, whereas under moderate rainfall (Varanasi, Rewa) response was noticed up to 120kg/ha. In case of areas with rainfall between 800 to 1000mm, response was up to 80kgN/ha for the crops like sorghum and ragi. Studies have indicated that no or less fertilizer is required if wheat succeeds a green-manure legume or grain legume. Only a dose of 30kg P_2O_5/ha is recommended for maximisation of yield of wheat. Splitting the N-fertilizer application reduces the risk of leaching losses and also helps to overcome the aberrant weather.

Response to applied phosphorus (P) is not widespread. Lack of response of crops to P applied to soil having low P_2O_5 status is one of the important problems of soil fertility management of many dryland. The apparent lack of response on vertisols could be attributed to the high fixation of added P caused by the high clay content. Deep placement of P is very useful. During 'rabi' season, when crops are grown on receding soil moisture, placement of P in advance of actual seeding by 10-20 days was found to be a useful practice. Depending on crops, a dose of 8 to 22 kg P /ha may be applied.

Dryland soils are generally adequate in available Potassium (K). However, crop production in some of the dryland regions, particularly in sandy loam and red soils, was found to be limited due to deficiency of Potassium. In such situation, a dose of 10 to 25 kg K/ ha may be applied.

On **micronutrient** front, continuous cultivation of a particular crop depleted a few specific micronutrients from soil thus affecting the yield in due course. In drylands of India, appreciable increase in yields has been obtained by application of zinc, molybdenum and iron in sorghum and pearl millet. In light textured soil boron application at 5kg borax/ha was found useful for groundnut. Responses of groundnut to lime and sulphur and of maize to zinc application are reported from red soils of dryland areas.

Organic Matter in Nutrient Management.-Use of fertilizer in drylands will continue to face uncertain returns due to abnormal rainy season that hasten up the process of non-availability of nutrients. A combined approach, in respect of organic and inorganic source of nutrient, should be practiced in dryland areas. On an average, application of nutrient in a ratio 1 : 1 in respect of fertilizer and organic manure is recommended for dryland farming. Of the several sources of organic materials, suitable for soil application, farmyard manure and compost of different kinds are very much common. Green manures and green leaf manures are also useful source of nutrients in dry areas. Application of crop residues by incorporating them in soil can bring a marked improvement in both chemical and physical properties of soils besides increase in yield. {For detail discussion on organic matter see "management of organic matter" in Chapter- 5.4.}

For Perennial Crops. Nutrient management for perennial or longer duration crops or plants is entirely different than that of annual field crops. The dose or requirement of nitrogen, phosphorus and potassium differs from crop to crop in different soils. However in respect of management of the same in dryland areas the following points may be followed:

1. Except organic matter no fertilizer is required at the time of planting.
2. In vegetative growth stage, fertilizers in combination with organic manure may be applied in two split doses i.e. just ahead of rainy season or after harvesting and before emergence of new leaves/shoots.
3. In reproductive stage, fertilizers in same combination may be applied, to augment fruit/seed development, after emergence of flower.

However, it may vary from crop to crop. Moisture status should be maintained after application of fertilizers and manures. Mulching may be combined with fertilizer application for better nutrient management in perennial crops.

5.2.2.2. Soil Testing

Soil testing is required to know the actual status of available plant nutrients in a field. On the basis of that result and the type of crop to be grown in that field, the application of plant nutrients is decided. Through this testing the acidity / alkalinity and presence of soluble salt can also be determined.

Testing of soil samples for both macro and micro nutrients may be done at laboratory. Laboratory testing provides almost accurate result. The essential macronutrients like nitrogen (N), phosphorous (P) and potassium (K) and pH (for acidity or alkalinity of soil) can be determined in the field itself with the help of portable soil testing kit. It will be an approximate result but adequate enough to take field level decision.

Soil testing of a field or a plot requires a soil sample duly collected from that field. About half a kilogram of soil is collected from a particular field in such a way that it will represent the entire field or plot in respect of soil characteristics. The area of land is determined on the basis of land situation, soil condition and crops to be grown. Depending upon root zone depth of crop the depth of soil collection is determined. The depth for wheat, seasonal vegetables etc are 15cm and for sugarcane, potato etc is 20cm.

How to Collect Soil Samples

1. Select 10 to 15 scattered spots of the plot or field depending on the size. While selecting spots the areas affected by any special situation like presence of residue of organic matter heap, under tree or bush etc may be avoided.

2. 'V' shaped small ditch is dug out with the help of a spade in the selected spots up to the desired depth.
3. A slice of soil is collected from each ditch with the help of spade or 'Khurpi' and put in a clean bucket collectively.
4. Soils collected from all the spots are cleaned and mixed after bringing it in operational stage. Then the mixed soil is spread in a circular fashion on a clean surface under a shed.
5. Divide the spread soils into four equal parts making a cross '+' on it. Discard two opposite parts and reduce the soil amount into half.
6. Rest two opposite parts are then again mixed and spread in the same fashion. Again half of the soil is discarded in the same manner and the process is repeated until the soil is reduced to 500gms.
7. This soil sample is dried properly under shed and packed for testing. This sample should contain the details of land, date of collection, name and address of farmer, last crop and next crop.

5.2.2.3. Management of Salinity and Sodicity

In dryland areas, the accumulation of salt in the soil surface frequently occurs due to very high evaporation demand of dry atmosphere. Soil salinity is invariably associated with drylands as the two most important modulating factors responsible for salt accumulation are least expressed under dry environment. Incomplete weathering of constituent soil minerals ensures a steady source of salts. Also the inadequate rainfall favours the deposition of salts in the soil surface.

The plant growth on such soils is affected because of higher osmotic tensions of soil solutions, poor physical conditions of soil created by sodium, toxicity of specific ions such as boron, and nutrient imbalance in soil solutions. Such soils are generally placed in two broad groups of saline soils and alkali soils.

Saline Soil. Saline soils are recognised by the presence of neutral salts like chlorides and sulphates of sodium, calcium and magnesium. The pH of the saturated soil paste does not exceed 8.2 and the electrical conductivity of the soil extract from that paste is more than $4dsm^{-1}$ at 25°C.

Successful crop production demands a reduction in salt concentration and prevention of salt accumulation in such soils. Excessive soluble salts should be removed from the root zone of the crop through leaching when the water table is deep. If the water table is high, the leaching should be coupled with drainage. The drainage may be performed by perforated PVC pipes placed at 1.5m below the ground surface. The drainage water should be disposed off properly and carefully. Salt tolerant crops viz. sugar beet, cotton, barley, oats, tomato, rapeseed, and grapes etc. may be grown.

Sodic Soil. Sodic soils have very high sodium, which creates problem in the cultivation of crop. The exchangeable sodium percentage (ESP) is very high, generally more than 15% of total exchange capacity of soil. Soil pH is more than 8.5 often rising as high as 10. The electrical conductivity is less than $4dsm^{-1}$. These soils contain large presence of carbonates and bi-carbonates of sodium.

The detrimental effect of sodic soil on plants is largely due to toxicity of sodium and hydroxyl ions. So excessive exchangeable sodium percentage has to be reduced to tolerable limit. High pH value creates insoluble calcium carbonate, so reducing pH level below 9.5 will increases solubility of calcium carbonate. This calcium will replace the sodium from the exchange complex and sodium sulphate will be formed. When this sodium sulphate is leached to deeper layers, the sodicity of the soil will be reduced considerably. The application of readily available source of calcium i.e. gypsum to the soil will reduce the exchangeable sodium by forming soluble sodium sulphate.

Investigations carried out under all India co-ordinate projects of Indian Council of Agriculture Research (ICAR) on water management of salt affected soils and use of saline water indicated that rice-rice-green manure and rice-rice cropping systems recorded higher yield and total income in problematic Vertisols in Karnataka. Fruit trees like Tamarind, Jamun, Guava and Ber showed complete survival where Pomegranate and Baelpather died in highly alkaline soils due to continuous water logging during rainy season (Central soil and salinity research Institute, 1997). Growing trees and agricultural crops together is a better land use pattern for productivity, sustainability and economics in moderately alkaline soils (Singh et.al. 1997). Grass like Rhodes grass, karnal grass, coastal Bermuda and paragrass are reported to be highly tolerant species to sodic soil conditions (Kumar and Adrol, 1982).

5.2.3. Management to Reduce Soil Loss

Soil is lost from field due to erosion caused by water or wind. Arid areas have severe wind erosion and semi-arid regions are subjected to sheet and gully erosion. Dryland agriculture will not be remunerative when the rainwater is lost by runoff that causes heavy soil loss. Therefore, it is necessary to minimise the soil loss by adopting different erosion control measures.

Measures to control soil erosion caused by water and wind can be broadly grouped into four: (1) agronomical, (2) mulching, (3) mechanical and (4) forestry measurés. A brief outline of these measures is narrated below.

5.2.3.1. Agronomic Measures

Agronomic measures have to be considered in combination with other measures for soil and water conservation programme. These measurers are generally effective on gentle slope up to 9%. These measures are as follows:

Land Preparation. land preparation including post-harvest tillage influences infiltration of rain water in soil and obstruct the surface flow, thus reduces erosion. Deep ploughing or chiselling have been found effective to reduce erosion. Rough cloddy surface helps in controlling wind erosion.

Minimum Tillage and Stubble Mulching. Minimum tillage conserves crop residues and reduces soil pulverisation to minimise soil loss by wind erosion. It also helps in stubble mulching which is effective in wind erosion control and moisture conservation. Stubbles on the surface also trap the soil particles and also help in decreasing run-off. Furthermore, it conserves soil organic matter, which helps in soil aggregation.

Contour Cultivation. Cultivation operations (tilling, harrowing, seeding, weeding etc.) along the contour i.e. across the slope reduce soil and water loss. Planting and sowing along the contour obstacles run-off, increases infiltration and reduce soil and water loss.

Choice of Crop. Close growing crops are more effective for erosion control than inter tilled crops. 'Erosion permitting crops' in row such as Sorghum, maize, pearlmillet etc. are not as effective as 'erosion resistive crops' like cowpea, groundnut, greengram, blackgram etc.

Generally legumes provide better cover and protection to soil by minimising the impact of raindrop and acting as obstruction to rill and sheet erosion. This can be termed as *Crop mulching*. For control of wind erosion the cover crops should be drought resistant with enough canopy to cover the soil. Mono-cropping of erosion permitting crops accelerate both soil and water loss. Intercropping with erosion permitting and resistant crops or their rotation has been found effective against soil loss. The suitable crops should be sown in time to develop adequately before commencement of peak rate of runoff.

Strip Cropping. To prevent runoff and soil loss, a few rows of erosion resisting crops can be grown alternated with erosion permitting crops. These rows in the form of strips should be planted across the slope as well as along the contour. It can effectively reduce both soil and water loss from the field. For controlling wind erosion crop strips should run straight and at right angle to the prevailing wind.

Organic Manures and Fertilizers. It is an indirect method to control soil loss. Manures improve physical conditions like soil structure, aggregation etc that increases rate of infiltration and reduce runoff in heavy soils thus reducing soil loss. Application of fertilizer improves vegetation growth (canopy) that aids in controlling soil loss.

Vegetative Barriers Including Windbreaks and Shelter Belts. Closely planted grass or shrub rows along the contour in agricultural lands are used for control of soil loss/erosion. They check the velocity of runoff and retain the sediment by acting as barrier to runoff. 'Khus' (*Vittveria zelanica*) is the most recommended plant for this purpose.

Windbreak is a barrier for protection from wind usually associated with soil erosion. Shelterbelt is a longer barrier and is a combination of shrubs and trees. {For details of windbreak and shelterbelt please see Chapter-5.3.8.}

5.2.3.2. Mulching

Mulching, the best method of soil covering, can protect soil from direct impact of rain, thus reduces soil loss considerably. A minimum ground cover of 30 percent with crop residue or other plant material is necessary to keep soil loss within acceptable limit. Mulching of soil surface also reduce moisture loss and soil loss due to wind. Residues on surface absorb wind force to reduce erosion considerably. Mulching is practiced in different ways with different materials depending on the area, crops, agronomical practices and availability of mulching materials {For Details of mulching please see Chapter-5.5}.

5.2.3.3. Mechanical Measures

Mechanical measures for reduction of soil loss are 'bunding', 'terracing', 'trenching', 'check damming' etc. Except check dam all other structures are already discussed in this chapter. {For detail of check dam see chapter – 4.3.11.}

5.2.3.4. Forestry Measures

Trees, grasses, and *legumes* can be successfully grown in degraded land of dry areas to reduce further soil loss.

Tree species found economical are *pinus patula, Pinus kesia, Eucalyptus camaldulensis, Moliodora spp., Acacia spp.*

Grasses and legumes for soil management should be perennial in nature, drought resistant, rhizomiferous, and should have good canopy, deep root systems, prostrate habit and usefulness in cottage industry. Some of the useful grasses and legumes are:

Grasses : *Cenchrus ciliaris, Chloris Guyana, Cynodon dactylon, Dicanthium annulatum, Heteropogau contortus, Iseilema laxum, Panicum virgatum, Eragrostis curvulla.*

Legumes : *Stylosanthus hamata, Centrosema pubescenee, Atylosia Scarbaceoi.des*

5.2.4. Management of Soil Organisms

Soil organisms are so important to maintain the soil health that without them the soil will be sterile. The soil organisms, both plant and animal, are so minute that they can be seen only with the aid of microscope. The number of larger organisms from smaller earthworms to the larger rodents is comparatively very small. However, all are significant items and are to be handled for better soil management. Though most of the soil organisms directly or indirectly benefitted plants, a few act as pest of the plants. They are herbivores and the examples are fungi and bacteria of few soil borne plant disease, parasitic nematodes, larvae of few insects which attack plant roots, snails and slugs.

The first and foremost job of soil organisms is to decompose the soil organic matters that finally supply nutrients to the plants. Actually the food chain of soil organism's ecosystem starts with organic matter. As soon as any organic substance touches the soil, it is subjected to a coordinated attack by microflora and by micro and macro animals that live on dead and decaying plant tissue. If moisture is present, the bacteria and fungi initiate the attack. They are joined by Larvae mites, snails, beetles, millipeds, woodlice, springtails, earthworms, and enchytraeid worms. These animals chew or tear holes in the tissue, opening it up to more rapid attack by microflora.

5.2.4.1. Role of Soil Organisms

Cultivation of soil brings a drastic change in the soil environment. Agricultural practices generally tend to reduce the species drastically and, especially with soil fauna, the total organism population. Soil organisms have several beneficial roles on soil that influence growth of higher plants. The roles of soil macro-animals are:

1. It pulverizes, mix and granulate the soil and incorporate organic materials into lower horizons.

2. It directly adds organic matter to soil by leaving partly digested organic residues, excrement and their own dead bodies.
3. It increases aeration and drainage of soil.

The role of specific flora and fauna of soils are dependent upon many factors. Apart from climate and vegetation of the area, temperature, acidity and moisture status of soil are the factors which govern the activity of both type of soil micro-organisms. The activities of specific groups of soil micro-organisms are commonly identified by (a) their numbers in the soil, (b) their weight per unit of soil biomass, and (c) their metabolic activity. The numbers and biomass of groups of organisms, which commonly occur in soils, are shown in the following table.

Table 5.6. Numbers and Biomass of Soil Microbes Commonly Found in Soils

Values Common in Surface Soils

	Number per square meter*		Biomass** (lb/AFS)
Microflora			
Bacteria	10^{13}–10^{14}	10^{8}–10^{9}	400-4,000
Actinomycetes	10^{12}–10^{13}	10^{7}–10^{8}	400-4,000
Fungi	10^{10}–10^{11}	10^{5}–10^{6}	500-5,000
Algae	10^{9}–10^{10}	10^{4}–10^{5}	50-500
Microfauna			
Protozoa	10^{9}–10^{10}	10^{4}–10^{5}	15–150
Nematida	10^{6}–10^{7}	10–10^{2}	10–100
Other fauna	10^{3}–10^{5}		15–150
Earthworms	30–300		100–1,00

* Generally considered 15 cm (6 in.) deep. but in some cases (e.g., earthworms) a greater depth is used.

** The biomass values are on a live weight basis. Dry weights are about 20-25% of these values.

Maximum of these microorganisms utilize the soil as their habitat and do the most important job of addition or making available of organic matters to soil thus indirectly benefiting the higher plants. In any case, all these organisms are part of perhaps the most intricate biological cycle in the world.

5.2.4.2. Management of Soil Organisms

Moisture and organic matter are the two vital items required for better proliferation of different soil organisms. Unfortunately dryland areas are inherently dearth of these items. However, successful dryland farming is highly dependent on these two items, which automatically takes care of the soil organisms.

From the farming point soil organisms (Soil algae, soil fungi, soil bacteria, soil actinomycetes, protozoa, nematodes and other soil fauna) affect the crop in two completely different ways. Beneficial organisms change the soil eco-system causing favourable growth of crops and the harmful ones cause damage of the crops.

A few organisms provide direct benefit to the crops by fixing nitrogen, making phosphate available etc. {for details see chapter 5.4.4.}. Other acts for decomposing organic matter, manipulating soil structure etc. indirectly helping the crop growth in dryland areas.

The following aspects of dryland farming can efficiently manage the soil organisms.

1. Increased vegetation cover,
2. In-situ soil-moisture conservation practices,
3. Application of sufficient organic matter,
4. Mulching, particularly with organic residues,
5. Legume based cropping system,
6. Use of grass and forestry in farming practices,
7. Vermiculture

5.3 CROP/PLANT MANAGEMENT

Crop management is the most versatile part of management of dryland farming. Starting from crop selection to post-harvest stage the management of crop changes its course of action as per the necessity. Commonly one perennial or two to three seasonal crops, in sequence, are generally grown in a particular piece of land in a year. But in a farm particularly in a dryland farm different types of crops are needed to be managed and grown. In dryland farms crop management is closely associated with its water, soil and animal management. It also takes care of energy and bio-mass production and reclamation of degraded lands of a farm. However, the ultimate goal of crop/plant management is to bring dryland, including the marginal, sub-marginal, degraded and degradable land, under production.

The management of crop generally deals with the subjects like (1) crop selection, (2) cropping pattern, (3) planting and seeding technology, (4) agronomic/cultural practices, (5) plant protection, (6) management of wind-break, grass waterways etc.

5.3.1. Tradition of Crop Grown in Drylands

Traditionally in dryland, crops are grown mainly for subsistence and are selected out of the locally available crops on the basis of their own need. Naturally basic needs like food, clothing and shelter dominates over any other economic needs. They are not necessarily the most efficient in regard to productivity, moisture use, economic returns and labour utilisation potential. The lack of technological support in the crop front in the dryland areas is the main reason of this lacking in respect of modern farming. Actually, history says that, the traditional farming as well as cropping systems of dryland areas were very much environment friendly and eco-friendly.

In dry areas, traditional single-crop system involving a long fallow period is commonly practiced on annual basis. Traditionally sorghum, chickpea and safflower are grown as sole crop or inter-cropping combinations in central India, wheat as sole crop or inter-cropped with chickpea in North Central plains and cotton as sole crop or inter-cropped with pigeonpea are grown in some vertisol areas of India. Paddy as sole crop is commonly found in the low lying flooded zones of dry areas of southern and eastern India. In dry areas having comparatively good rainfall, double cropping is frequently tried with a short duration pulse crop utilising the pre-monsoon rain followed by a cereal crop in rainy season. However, the productivity of all these crops are very low and all of them are grown as subsistence crop.

To meet the increasing demand of food as well as money out of decreasing cultivable land on increasing population front a drastic change of this traditional approach is obligatory. Thus the guiding principle in choice of crops or plants for dryland should be the economic growth rhythm that would fit in with the water availability of that area.

5.3.2. Crop Selection

In dryland farms the selection of crops should be done on the basis of the information as under.

(1) Climate, which identified the 'effective cropping season' for grain crops and the 'effective growing season' for perennial crops. Rainfall, temperature, humidity, sunshine are the main components of climate. However, rainfall is the main item considered for determining the effective cropping/growing season of a dry area. The effective cropping season, often called as length of growing period (LGP), of dryland area generally depends upon the natural moisture availability in the soil. However, introducing proper water conservation methods including water harvesting can increase the availability of water.

In India, ICAR on its 23 dryland research centres worked out the effective cropping season for each centre based only on the analysis of long term rainfall data of these centres. The effective cropping season ranges from 8 weeks at Bellary to 51 weeks at Dehra Dun at an average of 26.34 weeks per year.

On an average length of cropping season governs the choice of cropping system in the following way:

Length of Growing Period	Cropping System that can be Adapted
Less than 75 days	Perennial vegetation/monocropping with short duration pulses,
75 to 140 days	Monocropping,
140 to 180 days	Inter cropping,
More than 180 days	Double cropping,

The average length of effective growing season along with the information regarding type of soil, particularly the water holding capacity is required for determining the sequence of cropping in a particular area. These will also serve as a good guide in selection of efficient annual crops, primarily grain crops, their varieties and sowing for given location.

(2) Genetic Capability of Crops to Produce Economically in the Soils of low Moisture Status. The proper selection of crop in consideration of effective growing season should be done by substituting the traditional crops by the economically and genetically superior crops or varieties suitable for the given situation or which can be acclimatised in that situation. In Deccan areas of India, wheat is widely cultivated with a low production potential whereas safflower is a superior substitute in respect of selection from field crops. Similarly Mustard and chickpea are identified to be more productive in Agra and Varanasi region respectively than the traditional wheat crop (Venkateswarlu and Singh, 1982). Since recent past different fruit crops has been proved better option in drylands of India than that of field crops in respect of economic return, risk associated with moisture stress etc.

The crops having the comparative genetic advantages like (1) deep root system with long tap-root over shallow or fibrous root system, (2) thin or small leaf structure over broad leaf structure, (3) deciduous type over evergreen type, (4) climbers / creepers over standing crops, (5) leguminous crops over other non-legume grain crops, etc should be taken into consideration while selecting the crops for dryland farming.

(3) Characters to Choice Crops and its Varieties. Crops and varieties which have the following characteristics and which will enable them to withstand the adverse effects of the environment, are suitable for drylands.

1. They should have short life or yearly reproductive cycle so that they can escape the late season drought.
2. They should be photo insensitive.
3. They should have faster rate of growth in the initial period to withstand aberrant weather conditions.
4. They should have a strong and penetrating root system, which can tap water from large part of soil.
5. They should have low water requirement.
6. Number of flower followed by more number of grains or fruits and also high harvest index of the crop or variety are also desirable attributes.

Among the field crops safflower, sesamum, sorghum, coarse cereals, chick and pigeon pea, black and green gram, groundnut, fingermillet, horsegram, castor, maize, sunflower, gram, cotton, tobacco, and their hybrids (if available and suitable) are recommendable in dry areas.

In India, for example, early maturing groundnut variety JL24 (Phule Pragati) has made significant impact in groundnut production in dry areas of Gujrat and Maharastra. 'Bhima' a variety of safflower specially bred for drylands is performing well in deccan region (Umrani et. al., 1984). The performance of high yielding varieties of horsegram such as 'Sina' and 'Man' is also very satisfactory on shallow and un-retentive soils (Nimbalkar et.al., 1986).

Among the tree or fruit crops bel, aonla, ber, custard apple, cashew, citrus, pomegranate, guava, mango, sweet orange, etc may be recommended depending upon the farm and its water situation. Plants like phalsa, fig, tamarind, jamun, gunda and karondac can also be grown in degraded dry tracts.

Table 5.7. Field/Grain crops with tolerance to various stresses.	
Characteristics	**Crop Names**
High temperature tolerance:	Cotton, Groundnut, Chillies, (favour Jute & Yams only in humid tropics)
Drought resistance:	Common Millet, Barley, Chickpeas, Safflower (lower temperatures) Sorghum, Bullrush Millet, Phaseolus crops, Radiatus (gram mung bean), Cassava, Castor Bean, Sesame, Groundnut (Spanish variety), Pigeon peas, Sunflower
Lower temperatures favour:	Wheat, Potato, Sugar, Tomato, Safflower
Very high rainfall tolerance:	Rice, Cassava, Yam
Wide climatic tolerance:	Maize, Soybean, Groundnut (Valencia & Virginia type), Phaseolus lunatis, Kenaf, Hemp, Sweet Potato, Sugar cane, Tobacco

The final selection of crops and its varieties are done on the basis of few additional information, which are narrated below :

(4) Duration of life Cycle /Growing Period. Duration of life cycle of different annual crops suitable in dry areas is an important information required for selection of crop specially for intercropping. Cotton and mungbean / blackgram / soyabean is an intercropping combination where cotton usually takes 160 to 170 days to mature and the legumes mature in 70 to 80 days, well before the cotton crop. The vegetative and reproductive stages of legumes do not affect the cotton crop. As a rule, the advantage of intercropping depends on the competition free period between the crops. In pearl millet + pigeonpea the competition free period is about 60 to 70 days. In case of sorghum + pigeonpea, the period is reduced to nearly 40 to 50 days and hence is less advantageous.

Short duration varieties of castor are found as a better alternative to other uneconomic crops on shallow soils in Sholapur of India. Safflower usually planted as mixed crop in sorghum was recommended as a sole crop.

(5) Availability of Soil Moisture and Supplement Irrigation Water. Dryland farming, particularly cultivation of crop, frequently suffers from shortage of soil moisture in critical periods of crop growth. An arrangement for storing of protective or supplement irrigation water in different water harvesting structures particularly in 'farm ponds' and application of the same at the critical juncture helps to boost the yield of cereal crops up to 40 to 60 percent (Randhawa and Venkateswarlu, 1980). But due to high cost of the supplement irrigation and poor farm-gate price of field crops frequently established it as an uneconomic venture. Vegetables like chilli, brinjal, tomato, bottle gourd, etc. also yield more with supplement irrigation and economically better option than cereals (Vijayalaxmi, 1987). The protective or supplement irrigation to different perennial fruit crops suitable in dry areas, like pomegranate and *ber* proved highly rewarding.

(6) Storability, Marketability and Commercial Value of Produce. The effectiveness of marketing of agricultural produce rests on technical and economic considerations of pre and post harvest operations like assembling, grading, storage, transportation and distribution. In the context of increasing sales and securing higher prices, scientific grading and storing of agricultural produces has assumed greater importance because of its perishable nature. Demand of agricultural produces has a direct bearing with the economy of farming. At present on economic point of view the medicinal and aromatic crops are getting preference over the traditional crops, particularly in dryland areas. So selection of crop in dryland farming should be done in accordance with marketability and economic importance.

5.3.3. Cropping System

Cropping system is an important component of farming activity and represents cropping patterns practiced in a farm. Depending upon the identified growing season and choice of suitable crops and its varieties the cropping system should be identified. The moisture present in the root zone area and available water for supplement irrigation are the two other aspects that should be considered while designing the cropping system. The cropping systems are classified based on the number, combination and special arrangements of crops grown at a given time or in succession during a year. There are many advantages necessarily been extracted through proper designing of cropping system. However, increase of ***cropping intensity*** is the main goal to be achieved through cropping system. It is very tough in dryland where uncertain and extreme climate coupled with water scarcity makes the thing more difficult. Both annual and perennial crops can be considered for the cropping system in dryland farms, however perennial base cropping system have clear advantage over annual (field crops) base cropping system in respect of water use efficiency and economy.

Mono-cropping, with suitable annual crops (mostly the grain crops) is generally practiced in areas where cropping season generally lasted not more than 20 weeks. This single cropping system is possible either as 'kharif' crop (red and shallow black soils) or as 'rabi' crop (deep black soils) depending on the prevailing land and soil condition of an area. In scientific dryland farming, mono-crop or single crop is generally not recommended, but sometimes high density mono-crop may be practised with any very short duration crop to restrict possible soil erosion in rainy season. Still mixed or inter cropping with deep-rooted crops in the said high density short duration single crops may be more beneficial in rainy season.

Mono-cropping system is very much suitable in dryland areas with perennial crops like orchards, plantations, vine-yards, and any other tree-crops. In the initial growth stage it may be inter-cropped with suitable short duration crops to obtain more cropping intensity.

Multiple cropping is nothing but to cultivate more than one crop in the same piece of land. It is the ultimate goal of designing a cropping system in dryland but is a very difficult job. The method involves either growing two or more crops a year in sequence or grow two or more crops on the same piece of land at the same time, which is termed as intercropping. Multiple cropping have the advantages like (1) it avoids risk of total crop failure due to erratic rainfall or pests and diseases and ensure economic security, (2) it increases the annual net yield per unit area and (3) it can ensure the optimum utilisation of natural resources such as land, water, sunlight and nutrients.

In dry areas with more than 750 mm rainfall and with more than 30 weeks effective growing season, double cropping is assured (Reddy, 1999). One major crop and one minor crop are generally used in dryland suitable for ***double-cropping*** system. In double-cropping system, short duration (70 days) pulse crops (greengram, blackgram, cowpea) are followed by 'rabi' sorghum, safflower, barley, mustard etc. in low rainfall regions. Whereas cereal (rice)/oilseed crops followed by chickpea, mustard or wheat is used in moderately high rainfall areas. In areas with 191 to 210 days of moisture availability period, sorghum-safflower double-cropping system is more common as per the findings.

As per recommendation, regions having 20 to 30 weeks of effective cropping or growing season *intercropping* or *mixedcropping* may be practised. This cropping system is the ultimate goal of dryland farming like the double or triple cropping in humid or wet region farming.

Intercropping is the cropping system where two or more crops are to be grown at a given time in special arrangements e.g. row or line arrangement. It may be annuals with annuals or annuals with perennials that is generally found in orchards. It may be deep-rooted crop in contour lines and shallow rooted crop in strips in between. In principle, there should not be any competition between the two crops used in inter-cropping in respect of different growth factors like sunlight, water and nutrients instead, an interdependent relation may exist there. For these matters combination of cereals and legumes is very much recommendable in dryland farming. The length of root and nutrient balance is very much effective in this combination. They may be inter-cropped in alternative strips, which is termed as strip cropping and the cereals and legumes will act as *erosion permitting* and *erosion-resistance crop* respectively.

Mixed-cropping is the same system as intercropping except it maintains no distinct arrangements in the field. Actually two or more crop seeds are broadcasted at a time, where the crops should have a distinct difference in physiological characters like height, canopy type, colour, root type, life span etc.

Management of Crop in Respect of Intercropping—

- Prepare seedbed as per the base crop;
- Use less competitive and more compatible crops in respect of soil and water conservation;
- Proper adjustment of time of sowing or planting considering residual moisture, canopy cover, water requirement and moisture supply;
- Plant population of all the crops should be fixed considering the spacing required by the individual crops on high-density status;
- Application of adequate nutrients is essential as per requirement of all the crops;
- Weed control should be done manually and properly, particularly in initial stages;
- Integrated pest management should be followed to control pest and diseases;

In light textured soil (Aridisol) pearlmillet and clusterbean can be grown inter-cropped in 2:1 row ratio with cowpea or greengram or blackgram. On heavier soils, groundnut is inter-cropped with pigeonpea. Even in regions where natural moisture availability period ranges from 110 to 175 days, pearlmillet or sorghum or clusterbean may be grown with green gram or pigeonpea.

In India, through dry farming research, several crops have been screened for inter-cropping system. Sorghum, pearlmillet, pigeonpea, groundnut and cotton have been used as base crops. Among the several combinations tested, pearlmillet and pigeonpea proved to be the best combination across several regions. In this case the competition free period is almost 60 to 70 days and during this period pigeonpea makes up enough growth in terms of dry matter which results in higher production.

Suitable row proportions of component crops should be selected to provide enough space for inter-crops without compromising on the population of the base crop and for easy management of farm operations. *Paired and skip row planting* patterns may be adopted in such situation. Available research evidence showed that a 100% density of the base crop plus 75% density of most of the inter-crops generally gives high value of land equivalent ratio (LER).

Crop-Rotation

One of the first principles of dry farming in regard to cropping systems is crop rotation. The practice of crop rotation in humid regions is not necessarily recommended in semiarid lands due to the following differences:

- Only a limited number of crops are adapted to the climatic conditions and the farmer must sow the crop best suited to the moisture conditions encountered at that time.
- Moisture is so limiting, that "soil improving" crops are much less effective than in more humid areas.
- Success with rigid or complex sequences is difficult in the face of widely varying rainfall.

Reasons for Crop Rotation: There are five basic reasons why crop rotation should be practiced:

1. **Moisture Conservation.** Any system of crop rotation should be planned with moisture requirements as the main consideration. For a given set of climatic conditions, a crop may be described as either moisture dissipating or conserving. After harvest of a moisture conserving crop, the soil contains more moisture than at planting. This reserve of moisture can help the succeeding crop. Moisture may be insufficient for both growing a crop and also conserve enough water to ensure the succeeding crop. In such case, if annual rainfall is 10 to 15 inches (250 to 375 mm), utilization of dirt and stubble mulch will be needed at least every other year; if rainfall is 15 to 20 inches (375-500 mm) at least one in every three year. Actually mulching in every year, in any form, is the best way to avert the effects of severe drought.

2. **Pest Control.** Where related crops are successively planted in the same place, viruses, molds, blights, and selective insect pests tend to build colonies up in the soil. Crop rotation, that leaves at least two years in between subject plants in the same location will eliminate the abnormal build up of most such pests for most crops.
3. **Erosion Control.** Plants which are thickly planted or which produce a thick ground cover tend to resist erosion much better than those which are intertilled or tend to be moisture conserving. Loss of soil due to erosion is a significant dry farming problem and erosion controlling crops should be included in rotation, preferably in a strip cropping mode.
4. **Soil Nutrients and Structure.** When related crops are successively planted, specific soil minerals and nutrients are withdrawn faster than they can be replaced by decay or subsoil movement. This selective depletion causes a soil to be "worn out" quickly. Simple rotation with other crops makes depletion more uniform so that soils "wear out" more slowly. However for proper maintenance of soil nutrient status and structure, crop-rotations with legume-cereal, monocot-dicot, shallow-deep rooted crops, grass-nongrass, etc are the immediate solutions.
5. **Distribution of Labor and Risk.** It is generally advisable for the subsistence farmer to grow all crops in the rotation scheme simultaneously, apportioning to each crop the fraction of fields that it requires. This helps the scheduling and distribution of labor at the bottlenecks (planting, harvesting, etc.) so that the entire crop need not be done simultaneously. There is also reduced risk of total crop failure and increased variety/nutrition in the diet.

On the basis of experiments carried out at different locations by 'All India co-ordinated research project for dryland agriculture' (AICRPDA) and 'Central research institute for dryland agriculture' (CRIDA) a location-wise efficient cropping system is identified (Singh, 1987) which is given as follows :

Table 5.8. Important Rainfed Crop Sequence with Different Cereal Crops in India.

Soil Type	Crop Sequence	
Vertisols and related Black soils	Mung bean - safflower Soybean - safflower/chickpea Rice - chickpea/lentil	Sorghum - safflower Maize- chickpea/safflower
Alfisols and related soils	Cowpea - finger millet Rice - horse gram Groundnut - maize	Soybean - finger millet Cowpea - maize Rice - chickpea/linseed
Inceptisols	Maize - chickpea Soybean - wheat Maize - wheat+ mustard/chickpea	
Entisols	Mung bean - mustard Rice - chickpea/lentil Black gram -mustard	Pearl millet-chickpea/barley Rice - wheat

Source : Singh (1987)

However, it is unfortunate to note that none of these research works are associated with any crops other than coarse grain field crops. There are many species and varieties of vegetables, spices, medicinal/aromatic plants, fodder etc, which have immense potentiality in dryland farming. They may be used in inter-cropping system with cereal or grain crops apart from intercropped with tree crops. In better rainfall areas it may be used as one main economic crop in double-cropping system.

A list of recommended fruit and vegetable crops and their varieties are listed below:

Table 5.9. Popular Cultivars of Fruit and Vegetable Crops in Dryland of India.

Crop	Varieties
Fruit	
Ber	Gola, Mundia, Kaithali, Banarasi Karaka, Early Umran
Pomegranate	P-23, P-26, G-137, GK.VK. 1, II HR Selection, Mridula
Aonla	Kanchan, Krishna, Balwant, NA 6, NA 7
Sweet orange	Blood Red Malta, Mosambi, Yuvraj, Pineapple, Valencia
Sour lime	Kadayam, Promalini, Vikram
Custard apple	Balanagar, Mammoth, Island Gem, Red Sitaphal, Arka Sahan
Guava	Sardar, Allahabad Safeda, Kohir Safed, Safed Jam
Papaya	Coorg Honeydew, Pusa Delicious, Pusa Majesty, Pusa Dwarf, Pusa Giant, Co I, Co2
Bael	NB 5, NB 9
Sapota	Kalipatti, Cricket Ball
Fig	Poona, Black Ischia
Mango	Banglora, Neelum, Keshar, Bombay Green
Tamarind	PK.M 1, No.263, Pratisthan, Yogeshwari
Vegetable	
Drumstick	PKM 1
Onion	Patna Red, Nasik Red, N-53, Pusa Red, Pusa Ratnar,Pusa White Round, Pusa White Flat, Punjab Selection, N-40, CIN- 491, NS 7-1, Hisar-1, Hisar-22, Agrifound Dark Red, Arka Pragati, N-24-1
Chilli	Pusa Jwala, NP-46A, Mathania, Sindhur, Pant C-1, Arka Mohani, Titan (USA), Arka Gaurav, Arka Basant, Bharat, Indira
Cowpea	Pusa Dofasali, Pusa Phalguni, Pusa Barsati, Pusa Rituraj, S-203, S-488
Cluster bean	Pusa Sadabahar, Pusa Mausami, Pusa Navbahar, Durga Bahar, HBG-4, ARG-80
Tomato	Pusa Ruby, Pusa Early Dwarf, Pusa-120, HS-102, Sweet-72, S-12, Mangla, Punjab Chhuhara
Brinjal	Pusa Purple Long, Pusa Purple Round, Pusa Kranti, Pusa Anmol, Arka Sheel, Arka Shirish, Arka Kusumakar, Arka Navneet
Amaranth	Col, Co 2, Co 3, Chhoti Chaulai, Badi Chaulai
Okra	Pusa Makhmali, Punjab No. 13, Punjab Padmini, P-7, Parbhani Kranti

Crop	Varieties
Pumpkin	ArkaChandan, Co-1, Co-2
Musk melon	Pusa Sharbati, Pusa Madhuras, Hara Madhu, Punjab Sunehri, Durgapura Madhu
Watermelon	Sugar Baby, Arka Manik, Arka Jyoti, Durgapura Meetha, Kesar, Mateera
Bottle gourd	Pusa Summer Prolific Round, Pusa Summer Prolific Long, Pusa Meghdoot, Pusa Manjari, Pusa Naveen
Bitter gourd	Pusa Do Mausmi, Arka Harit, Pride of Gujarat
Ridge gourd	Pusa Nasdar
Sponge gourd	Pusa Chikani
Round melon	Arka Tinda

Relay Cropping. It is a very useful method where crops are to be grown sequentially in the same piece of land without any gap between the harvesting of one and sowing of another. In this system of cropping generally the harvesting time and seeding time are overlapped. It is a useful method of utilising the residual moisture of soil, which helps the 2nd crop to germinate and grow beyond its tender age under the standing 1st crop. Stubble mulching may be practiced in the relay cropping system. Grams and other pulses are very much suitable as the succeeding crop.

Cover Cropping. This system of cropping is not meant for any direct commercial activity but to protect and nourish the soil or supply of green matter for other purposes like fodder production, hay making, composting/vermicomposting etc. Cover crops are grown as a conservation measure either during the off-season or as ground protection under trees. In India cover crops like cowpea, lucerne, 'dhaincha' etc are generally grown and used as green manuring crops. Few of them are ploughed in their vegetative stage and in case of others the crop residue are ploughed at the time of land preparation after harvesting of that crop. This system of cropping protects the soil from erosion hazards. It also increases the moisture retention capacity and after decomposition it increases the fertility status of the soil. Cover crops are grown in rainy season, under tree crops, provide better water retention capacity and protect the soil from erosion and leaching of nutrients. However, in dry season the cover crops compete for the available moisture and may adversely affect the growth of main crop. Studies in rubber plantation in eastern Java shows that cover crops may reduce the soil moisture by up to 50 percent during the dry season compared with clean-weeding (Williams and Joseph, 1970).

Strip Cropping

It is a system of cropping that involves growing of few rows of a crop of specific character like a strip alternatively with strip of another crop of different character.

Commonly strip cropping is used to control soil-erosion and/or to manage rain and residual moisture. For controlling soil erosion, erosion-resisting (grass, legume etc) and erosion-permitting (corn etc) crops are grown in alternative strips. In case of rainwater and residual soil moisture management shallow-rooted and deep-rooted crops are grown in alternative strips.

This system may be divided into 3 groups: contour strip, field strip, and buffer strip, depending upon the purpose and situation (Troeh, Hobbs and Donahue, 1980).

In most of the cases strip cropping is implemented along the contour line. This system of cropping works better in slopes of 3 to 8.5 degree. The width of the strips commonly varies in between 15 to 45 meters. However, it may further vary depending upon the type of crop, soil, and implements to be used for cultivation (FAO, 1965: Cooke and Doornkamp 1974)

The main disadvantage of strip cropping is that, it is tough to implement in small farms. On the other hand it is also imposing problem to highly mechanized/sophisticated agricultural systems.

5.3.4. Seed Management

Modern cultivation methods in water available or water abundant areas eroded the indigenous plant genetic resources. Farmers of dryland areas all over the world have produced, stored and used their own seeds. Seeds that have been engineered to produce higher yields (high-yielding, hybrid and genetically modified seeds) are not only expensive but also more prone to pest and disease, which often show a decline in productivity, particularly in stress condition. However, if the farmer's knowledge and the agro-climatic condition favour then the hybrid or other such seeds are better purchase from the market every year. When an indigenous variety is chosen, a few simple practices can improve seed harvest, processing and viability. {For details of these practice pleas see Chapter 4.5.1.}.

5.3.5. Sowing and Planting Management

Sowing of seeds and planting of seedlings or plant materials is a very important part of dryland farming. The basic principle is that, the seed or plant materials must be subjected to the appropriate environmental conditions: available water, proper temperature regimes, a supply of oxygen, and sometimes light.

Proper germination of seed is the ultimate objective of sowing. Seed germination begins with *imbibition* of water by the seed increasing its fresh weight. This stage of rapid increase in water uptake followed by a period where there is little water uptake is called the *lag* phase. The fresh weight of the seed begins to increase again as water uptake drives the emergence of the *radical* i.e. the growing point of the root and it emerges from the base of the embryo axis. Once the growth (both root and shoot) begins from the embryo axis, weight of the new seedling plant increases, but total weight of storage tissue decreases. Water absorption increases steadily as new root explore the seedbed and fresh weight of seedling plant increases. At this stage water stress can reduce the germination percentage. Most kinds of seed germinate equally well over a range of available soil moisture, from field capacity (FC) to permanent wilting point (PWP). However, germination of some seeds, particularly those with dormancy problems (e.g. beet, lettuce, etc.), is inhibited as moisture levels are decreased. Such seed apparently contain inhibitors and require leaching by way of dipping into water for a reasonable time depending upon the temperature and humidity. Few other seeds, if exposed to excess water for quick imbibition, gives rise immediate germination (e.g. big sized seeds like grams, groundnut, gourd's seed etc.). Moisture stress strongly reduces the rate of seedling emergence from a seedbed. This decline in emergence rate occurs as the available moisture decreases to a level approximately half way through the range from FC to PWP.

Now to obtain better result of sowing of dry or soaked or leached seeds the following steps should be taken:

1. Sunken line or row sowing should be preferred over broadcasting.
2. Seeds should be placed in proper depth of soil and coverd properly.
3. Sowing should be done either by seed drill or through dibbling method.
4. Mulching is must for all kinds of sowing
5. Sowing of seeds (3 to 4 numbers) of climber / creeper crops in a pit (0.7 × 0.7× 0.5 Mt.) filled with manure and soil (1:3 proportion) and having nominal irrigation facility is one of the best option for dryland farming.

Planting of seedlings or plantlets or any other planting (budded, grafted) materials is always a specific job to be done very carefully, specially in case of long duration crops (Figure 5.19). For planting of planting materials the following steps should be taken in a dryland farm:

1. Entire land should not be cultivated for planting of individual crops.
2. Dibbling (in pits) should be the best option for planting.
3. Quality seedling is always preferable over ordinary budded or grafted plant to encash the advantage of original taproot system in case of severe drought. However, *in-situ budding or grafting* is the best option. Here the seed or stone

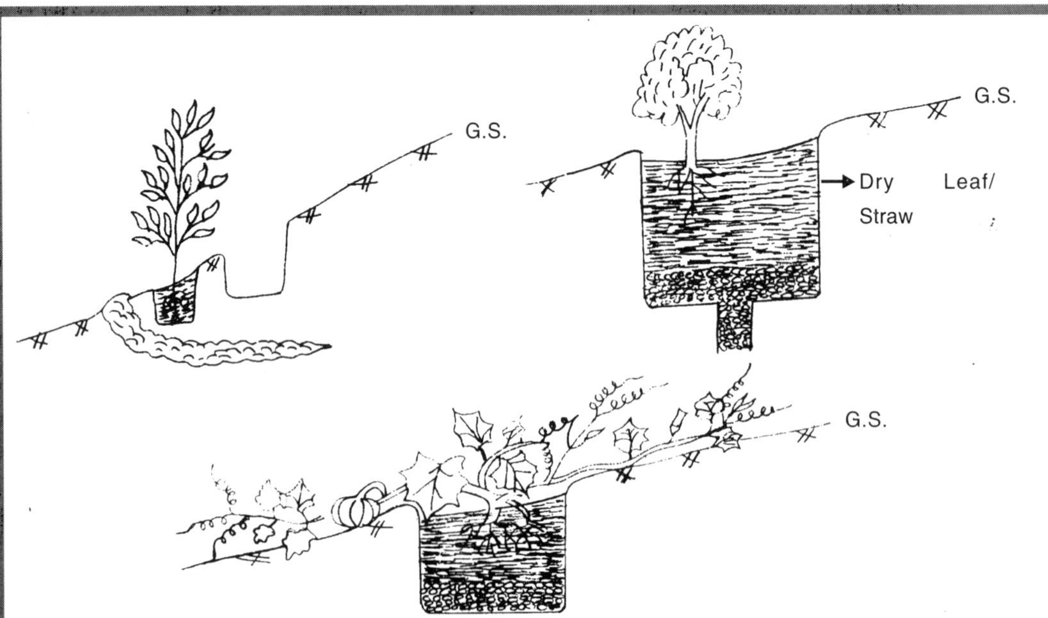

Fig. 5.19. Different System of Planting.

planted in the specific pit and after one or two year the budding/grafting with quality 'sion' will give rise the best suitable plant for a dryland farm.

4. The forest plant/seedlings may be planted in a 1 cubic ft size pit, however direct seeding is preferable. It is better to place the pits on the down side of contour trenches and guarded by 'V' or half moon shaped bunds.
5. The latest idea of transplanting vegetables may be practiced in dryland farming by way of planting the seedlings in 1ft.× 1ft. trenches or in the pits of 1 cubic ft. size.
6. For low-care fruit crops or any hardy shrubs the pit size will be 2ft.× 2ft × 2ft. in the same location as stated in point no. four. Layers of crop or plant residues like straw, dry leaves etc. will be placed at the bottom before filling. The pits will be filled with good quality soil and organic matter in 3:1 ratio.
7. For tree crops and for high value fruit crops the pit size will be 1mt.× 1mt.× 1mt. and will be placed along the contour. If possible, it may be supplemented with a 15cm. wide and 2mt. deep hole at the centre of the bottom of the pit. All the other recommendations made in the earlier points will also be applicable here.
8. All the planting should be done at the onset of the monsoon. Arrangement of irrigation for dry season for first 2 to 3 years is preferable as life saving measure as well as for better initial growth.

5.3.6. Pest Management

The term '**pest**' is used subjectively to include almost any organism that causes economic damage to a farming activity (here we consider only the damage of crop). Different kinds of organisms can be considered as pest in crop production. Study of several disciplines like plant pathology, entomology, nematology, and weed science is necessary to control diseases (caused by fungi, bacteria, viruses etc.), pests (insects, mites and other invertebrates), nematodes and weeds respectively.

Diseases

A plant disease is a disturbance from a pathogen that interferes with pant physiology, producing changes in plant appearance (called symptoms) and or subsequent yield loss. Diseases can be caused by non-infectious agents (like micro and macro nutrients, heat, frost etc.) or by **pathogens**, infectious agents that can reproduce or replicate. We commonly consider the pathogens as the main causal organisms for disease infestation. The infectious agents causing plant disease include Fungi, Bacteria, Viruses, Mycoplasmas, Protozoa and Nematodes. Out of these agents protozoa causes relatively less important plant diseases. With the exception of some fungi, most of the plant pathogens are microscopic in size. Therefore, although diagnosis of disease can be made from symptoms, often a diagnosis cannot be confirmed unless the pathogen is isolated and examined in a laboratory.

Out of these pathogens **fungi** causes maximum number of crop diseases. It causes leaf spot, rusts, root rots, dumping off, mildews, blight, sigatoka of banana etc. The life cycles of many fungal parasites are complex and highly specialised. Many fungi have alternate hosts on which they can persist in the absence of a susceptible crop plant. In general, most **bacteria** are beneficial to many different kinds of organisms. However, a few of them causes serious diseases to those organisms including plants. They are extremely small in size so the identification is very much difficult. Their rapid reproduction rate and short life cycle make them capable of creating rapid colonisation on a susceptible host and did maximum damage. Different species of pathogenic bacteria causes blights, leaf spots, cankers, crown galls, soft rots in different crops or plants. **Mycoplasmas** are closely related to bacteria and they require a vector (it is the organisms that transmit any plant pathogen from an infected host to fresh host) or other mechanical means of transmission to infect a host plant. They cause a few important plant diseases like lethal yellowing of coconut. **Virus** occur within host plant cells and are carried by sap and vascular system within a plant, but cannot move to a new plant host unless transmitted in sap, plant material, or other debris. The transmission may be done through wounds caused by human, tools used for crops and mostly by vectors e.g. aphids, leaf hoppers etc. Some viruses have many weed hosts from which vectors can acquire the virus and move it into crop plants. Some of the important virus diseases are tomato spotted wilt, different crop mosaics, sugarbeet yellows, and citrus tristeza.

Environment Factors Affecting Disease Development in Dryland. Increase of *temperature* upto a good extent generally increase the growth rate of many fungal and bacterial pathogens but very high (above 40°C) and very low (below 10°C) temperature reduces the same. However a particular pathogen always require a specific temperature for their proper development. One of the most important environmental factors affecting progress of disease is *moisture.* Increase of moisture status stimulates the ending of resting stages, germination of spores, growth and penetration into host. Leaves or roots that remain moist for long periods of time are particularly susceptible to colonisation by fungi and bacteria. These growth phenomenons of the pathogens are very fragile and dry out easily when exposed to water stress environment. In dryland areas due to high temperature and water stress situation the infestation of different pathogens are less in comparison to the wet areas.

Nematodes

Nematode pests are plant parasites, which live in soil around plant roots and feed on root tissue. The mouthpart of nematod is a needle like stylet used to puncture plant cells and withdraw fluid and nutrients. At summer temperature its life cycles are rapid and the reproductive rates may be high. Winter is important in interrupting the life cycle and population growth of nematodes. In most cases the damage, caused by nematodes, is indirect. The most common symptoms are reducing yield and quality, stunting, yellowing, erratic stands and nutrient or water deficiency. Soil or root samples are required to undergo laboratory diagnosis to determine the nematode problems in these cases.

Insects

Insects are the largest and most diverse group of organisms, and certainly many different kinds are represented in most agricultural fields. However, non-insect arthropods like spiders, millipedes and terrestrial isopods are often included in the concept of insect and required to manage. Some of the methods used for managing insect pests may also be applicable against snails, slugs etc which may cause damage similar to that of other insects.

After a young insect hatches from an egg, it passes through several stages before it becomes an adult and in every stage, depending upon the species, they may cause damage to the crops. In few insects the immature insects, called *nymphs*, are directly hatched from eggs, resemble the adults except that the wings are not fully developed or expanded. This is called *simple metamorphosis*. In other insects the appearances of young and adult stages are different. Here the young insects are generally converted into the adult insects which is called *complete metamorphosis*. The immature stage is the *larva*, which may be caterpillar, maggot, grub or worm- like individual, which may cause severe damage to the crop. The next stage is *pupa*, a resting stage that doesn't feed and may occur on the host plant or in soil or debris. Then the last stage is adult, which may cause severe damage to the crop.

From agricultural perspective, it is important to recognise that the food sources of larvae and adults of a same species that may be different, which prevents competition between parents and offspring. For instantce, some of the caterpillars are damaging the crops whereas its adult butterflies or moths are beneficial to the plant. Also, it is important to recognise the extent of feeding and damage to the crop. A large 5th-instar larva may be five times as severe as that of a 1st-inster larva.

Other Pests

On worldwide basis, rats and mice cause extensive losses to rice and other grain crops (at pre- and post harvest stages). Birds (including bats) often act as a severe pest to the grain crops like maize, sorghum, millets, peas, rice, wheat etc. and also to the fruit crops. Apart from these two, many other vertebrates like rabbits, squirrels and other rodents etc., even the deer, elephants sometimes seriously damage the crops in particular areas.

5.3.6.1. Methods of Pest Management

Methods for management of plant diseases are directed toward three areas: (1) Improving the quality and ability of the plant host to withstand the damage caused by the pests; (2) Reducing or eliminating populations of the pathogen or parasitic nematodes or insects; and (3) Establishing environmental conditions unfavourable for development of different kinds of pests.

On the basis of the said three criteria, the methods of pest management may be classified in five different ways. The outline of the methods to control different kinds of pests is placed as follows:

Methods	Means	Remarks
1. **Biological**	Disease: parasites	• Versus bacteria or fungi;
	Cross protection	• Versus viruses;
	Nematodes: Parasites	• Bacteria, fungi;
	Predators	• Mites, nematodes;
	Insects: Pathogens	• Viruses, bacteria, fungi, protozoa, nematodes;
	Predators	• Least or highly specific biological agents;
	Sterile male-	• Applicable in location specific situations;
2. **Chemical**	Disease: Fungicides-	• For reducing or eliminating fungi;
	Bactericides	• For reducing or eliminating bacteria population;
	Insecticides-	• Against insect vectors of plant virus;
	Nematodes: Nematicides	• Act as fumigants or non-fumigants;
	Insects: Insecticides	• Natural & synthetic chemicals targetting insects;
	Acaricides	• Natural & synthetic chemicals targetting mites;
	Repellants	• Natural & synthetic, for specific situations;
	Attractants	• Use of feromones;
3. **Physical**	Disease & Nematodes:	
	Soil sterilisation-	• Raises temperature to kill pathogen & nematodes;
	Soil Solarization-	• Raises temperature using solar energy;
	Hot water-	• To kill pathogens & nematodes of plant materials;
	Burning	• To kill pathogens & nematodes of crop residues;
	Fallow-	• Deprives of a host of pathogen & insects etc.;
	Pruning-	• To downsize the population of pests;
	Insects & others:	
	Soil sterilisation, burning, barrier etc.-	• Useful in location specific situations for specific insects or vertebrate;
	Hand-picking/banishing-	• Limited, but effective in small scale & specific cases;
	Traps-	• Efficient on systemic and community approach;
4. **Preventive**	All types of pests:	
	Quarantines	• Stop entry of new organism by government;
	Certification program	• Disease and pest free seeds & plant materials;
	Sanitation	• Reduce the chance of contamination & spreading;
5. **Cultural**	All type of pests:	
	Resistant cultivars	
	Crop nutrition	
	Cover crop & trap crop	Very much useful in proper combination.
	Planting time	
	Plant population/density	

5.3.6.2. Integrated Pest Management (IPM)

To avoid problems of chemical management of pest and considering its future threat, a right combination of cultural, physical, biological, bio-chemical and judicious chemical methods may be suggested to control the pests effectively, economically and obviously ecologically. This environment friendly composite system of pest management is generally termed as integrated pest management (IPM).

For prope implementation of IPM a few basic aspects of crop husbandry is essential to be looked into before adapting or selecting this control method. These basic aspects are as follows:

1. **Soil and crop health.** Good soil, in respect of crop production, is a prerequisite for raising healthy crop. Incidence of pests is always less in such cases.
2. **Crop diversity.** The greater the crop diversity in an agricultural system, lesser is the incidence of pest. It also increases the diversity of predators and as most pests are host specific they are in trouble where the crops are diversified.
3. **Observation.** It is very important to spend some time in the field to observe or inspect carefully the type of infection or damage caused by a specific pest (insect/fungus/bacteria/ virus etc.). Though it is not always easy to identify the causal organisms, continuous observation and subsequent study will increase the efficiency of the farmer in this aspect.

   Monitoring of insect population or number of infected plants or plant parts is an integral part of any pest management. A pest becomes a problem only when its population or infection exceeds a critical level. If by the system itself or with any external help the infection is maintained below the threshold level the damage caused to crop remains minimal. This actually helps to maintain the balance necessary to maintain the sustainability of the agro-ecosystem. Careful observation to distinguish pests from predators will help to evolve a strategy to control pests.
4. **Knowledge on pests and their life cycle.** Insects go through various stages of development (egg, larva/nymph and adult) during their lifecycle. Any one or more of these stages become destructive. The damage causing fungus, bacteria and other microorganisms maintain a specific life cycle depending upon the environmental situation. High humidity along with a moderate to high temperature always encourages proliferation of pests.

Considering the above basic ideas of pest management the proper combination of pest control methodology should be evolved for subsequent adoption for a particular dryland farm. A brief outline of these methods are narrated below :

Cultural Method

1. **Tilling of soil.** This can effectively control many soil borne insects and diseases by exposing the organisms present in the soil to the sun and predators.
2. **Timing of sowing/planting.** To sidetrack the attack of some pests the timing of sowing/planting may be changed in some crops. Even leaving the land

'fallow' reduces the severity of infestation of certain pathogens or insects in the following season by way of depriving them from the proper host.

3. **Crop planning.** Since many pests are host specific, changing the crop sequence and cropping pattern disturb their lifecycle and food supply resulting in death. It would be ideal if root crops, leaf crops and grain/fruit crops are grown simultaneously or sequentially in the same plot.
4. **Management of crop residues and infested plants.** Most of the crop residue and infested plant parts harbour the egg/larva/pupa of pests. Destroying these items will break the lifecycle of such pests.
5. **Using resistant variety.** In case of chronic infestation locally adopted resistant crops or varieties may be used to reduce the incidence of infestation.

Physical Method

1. **Hand picking.** In small plots, pest control may be effectively done by collecting and destroying the pests in different stages of their development regularly. Hand picking in the initial stage of infestation reduces the severity of damage effectively.
2. **Use of traps.** Devices like light, sticky surface, certain odour, sticky or poisonous food etc.allure many pests and thus entrap and kill them.
3. **Mass destruction.** Burning of infected plants or plant parts, hot water treatment/ sterilization/ solarization of soil are some methods of mass destruction of pests.

Biological Method

Methods of pest control by introduction of parasites and predators naturally or artificially are termed as biological control. Basically, biological control involves the manipulation of quantity of predators, parasites and pests to maintain their populations in the field in such way that the loss of the crop remains below the economic injury level. Some important predators and parasites useful for biological control are -

1. 'Ladybird beetles' belong to the order Coleoptera. Both adult and young, feed on *mites* and many other insects including *aphids, whiteflies, scale insects, thrips* and *mealy bugs*.
2. 'Green lacewings' (aphid lions) belonging to order Neutroptera feed on *aphids, mites, whiteflies,* and other *soft-bodied insects*. But, it is itself an easy prey to other predators.
3. 'Syrphid Fly' preys on *aphids, mealy bugs* and *leafhopper.*
4. 'Soldier Beetle. This is relatively a large insect and feeds on *cutworms, cankerworms, slugs, snails* and *gypsy moth* larvae.
5. 'Praying Mantis is another big insect which eats up almost all small insects.
6. 'Braconid Wasp' of Hymenoptera order is a small insect the young larvae of which are parasites on *aphids* and on larvae of many *butterflies* and *moths*. The adult wasps deposit eggs directly on the host. The larvae that emerge, feed on the host to complete their life cycle finally emerging as an adult to act as

predator. In the process aphids are killed thus reducing its number. The cycle goes on and prevent subsequent damage of the crops.

Bio-chemical Methods

If these IPM methods fail to contain the economic loss of the crop, application of pesticides, preferably of vegetative origin, would be necessary. Often these are the plant extracts chiefly having repellent and antifeedant properties. Chemical pesticides usually possess only one or a few toxic elements. It is easier for the pests to develop resistance against such pesticides. Whereas, vegetative pesticides possess a number of toxic chemical in them and pests find it difficult to grow resistance against so many chemicals. Vegetative pesticides have also broad-spectrum effect. Some examples are given below:

1. 'Neem' (Azadirachta indica) - The seed kernel has high percentage of azadirachtin content that has a storage life varying between 3 to 8 month. Extract of leaf, oil and oil cake are used as bio-pesticide and are effective against rice plant hopper, aphid, cutworm, Colorado beetle, locust, leaf miner, mite, potato jassid, cabbage worm, white fly, fruit fly, moth etc.

   Preparation of leaf extract- One kilogram of leaves is soaked overnight to prepare 5 litres of extract. Grind the soaked leaves and filter the extract properly. Then add 1 ml of emulsifier (khadi soap solution) to 1 litre of extract and spray the solution as bio-pesticide.

2. 'Persian Lilac' (*Melia azedarach*)- It is closely related to neem, but twice as potent. The method of preparation of extract is similar to neem. It acts as an antifeedant, stomach poison, contact poison, repellent and growth inhibitor.
3. 'Sitaphal' (*Anona squamosa*)- Its unripe fruits (ripe fruits are edible), seeds, leaves and roots act as repellent, contact and stomach poison, antifeedant, larvicide. It is mainly effective against aphid, rice plant hopper, red pumpkin beetle, grasshopper and diamond-back moth.
4. 'Karanj' (*Pongamia pinnata*)- Extracts can be made from the leaves, seed kernel, cake and oil, which act effectively against citrus aphid, citrus leaf miner, rice weevil, maize stem borer, root knot nematode, brown plant hopper etc. as contact and stomach poison or repellent.
5. 'Garlic' (*Allium sativum*)- This spice crop has both medicinal and pesticidal properties to human beings and crop pests respectively. The extract of its bulbs is mainly repellent, anti-feedant, insecticide, bactericide, fungicide, nematocide and active against ticks.

   Preparation of extract- 100 grams of finely ground garlic are soaked in 2 tablespoonful of liquid paraffin for 48 hours. Then 30 grams of khadi soap, duly diluted in 1/2 litre of water, are mixed thoroughly with the entire solution. 15 ml of this solution can be mixed with 1 litre of water for field application.

6. Tobacco' (*Nicotiana tabacum*)- The extracts of commercial tobacco is a strong poison. It acts as a stomach and respiratory poison, insecticide, repellent and fungicide. It effectively controls aphid, mite, thrips, leaf miner, stem borer, caterpillar and many other insect pests and pathogens.

Preparation of extract - 250 grams tobacco is boiled in 4 litres of water for 30 minutes. 30 grams of khadi soap is then mixed with it thoroughly. It is used as spray after diluting with 4 parts of water.

7. Many other plants like Tulsi (*Ocimum sanctum*), Chilli, Marigold, Turmeric, Aloe, Onion, Lantana etc. have pesticidal properties. But details of preparation of extract and the dilution have to be standardised by scientific trials.

Chemical Methods

If the pest infestation is sudden and devastating, some chemical pesticides, preferably with less residual toxicity, may be used initially. But these substances that pollute the environment, should subsequently be replaced by the above mentioned methods.

5.3.7. Plant Nutrients Management

In general, 90% of dry weight of a plant consists of Carbon, hydrogen and oxygen and the rest 10% includes essential elements like nitrogen (N), Phosphorus (P) and Potash (K) (Table – 5.10).

Table 5.10. Presence of essential nutrients in plant on dry weight basis.

Nutrient Element	Chemical Symbol	Classification	% of dry weight
Carbon	C	Non-fertilizer	–
Hydrogen	H	Non-fertilizer	89.0
Oxygen	O	Non-fertilizer	–
Nitrogen	N	Primary macro-nutrient	4.0
Phosphorus	P	Primary macro-nutrient	0.5
Potassium	K	Primary macro-nutrient	4.0
Calcium	Ca	Secondary macro-nutrient	1.0
Magnesium	Mg	Secondary macro-nutrient	0.5
Sulfur	S	Secondary macro-nutrient	0.5
Iron	Fe	Micro-nutrient	0.02
Manganese	Mn	Micro-nutrient	0.02
Zink	Zn	Micro-nutrient	0.003
Copper	Cu	Micro-nutrient	0.001
Boron	B	Micro-nutrient	0.006
Molybdenum	Mo	Micro-nutrient	0.0002
Sodium	Na	Micro-nutrient	0.03
Chloride	Cl	Micro-nutrient	0.1

Unlike the plants in more humid and fertile land, here, plants grow in a restrictive environment in dryland condition, and the soils of such area is not capable enough to buffer the short supply of plant nutrients. If the soil of such dryland is sandy in nature, the importance of fertilization, including both macro and micronutrients, increases manifold. Thus, in comparison to humid area, the dryland crop requires more fertilization particularly in case of horticultural crops. The best method of application is through drip irrigation system, wherever possible.

In general soil application of organic matter, macro and micronutrients are adapted in dryland farming. However the nitrogen and potassium may be applied through irrigation water through drip irrigation system, wherever possible.

Manures and Fertilizer application Schedule:

Organic matter @ 5-10 mt/ha in the form of compost or FYM may be applied in soil at the time of preparation of land. Along with organic matter, to correct pH level of the acidic soil, finely ground **limestone** is applied to the medium to raise the pH level. Dolomite limestone is best for this purpose because of presence of magnesium as essential nutrient along with calcium. The rate of application varies with type of root medium and magnitude of the pH correction needed. However lime should be applied one month before the application of nitrogen.

Where pH of root medium is neutral, instead of limestone, **Gypsum** (Calcium sulfate) should be incorporated into the soil every year. This is sufficient to supply entire calcium and almost entire sulfur to the plant throughout the year. If the magnesium status of the crop (perhaps through foliar analysis) is low, occasional application of this nutrient through Epsom salt (magnesium sulfate) dissolved in sufficient water is done to treat soil.

Next is the status of **phosphorus** in the soil, which is determined through soil test. Unless it is high in the root media, application of single super phosphate is needed. Single application at the time of preparation of land or just after planting/ seeding will provide necessary phosphorous to the soil for the next year.

Micronutrient mix, containing iron, manganese, zinc, copper, boron and molybdenum can be applied directly to the soil medium once in a year. Otherwise liquid application of commercial mix of micronutrients can be made through drip irrigation system 3 to 4 times during a year.

The rest two important macronutrients viz. **nitrogen and potash** may be applied in multiple doses maintaining a fixed schedule (weekly/fortnightly/monthly) or in accordance to the crop requirement. When ever applicable it may be applied through drip irrigation system with the same schedule. Commonly to determine the dose of nitrogen and potash in each application, the number of application is to be divided by the total requirement of nutrients, after that the necessary adjustment or addition or reduction may be done according to the necessity. However, more frequent application of these two nutrients is always beneficial to the crop, particularly for the long duration continuously growing crops. This is illustrated in the following figure.

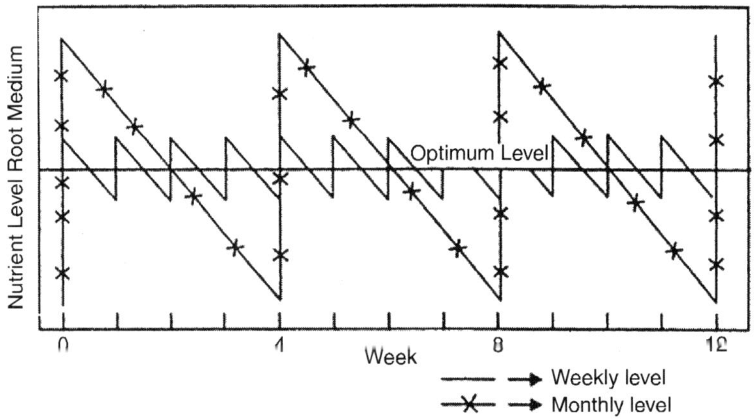

Fig. 5.20. Nutrient level of root-zone during weekly and monthly programme of fertilizer application

5.3.8. Tree, Wind-Break, Grass Waterways and Live-Hedge Management

5.3.8.1. Management of Tree

Trees should be incorporated in dryland farming as water and soil conservation measures, as wind-breaks, as supplier of fuel, fodder and bio-mass and as shade/ ornamental purposes. This type of tree planting in a farm is called ***farm-forestry*** and ***agro-forestry***. The tree farming systems have immense potentiality for dryland farming particularly for its sustainability. Apart from wind-break the requirement of fuel and fodder is so important in dryland farming that a separate *fodder-fuel plantation* can be established, as in parts of Gujarat (Tejwani,1981).The other important contribution of these trees are to provide sufficient bio-mass at the time of lopping and pruning. This bio-mass along with dry leaves can be converted to organic matter through different composting methods. ***Alley farming*** is another system of tree farming where crops are grown in alley formed by hedgerows of trees or shrubs. It involves three critical considerations, namely choice of crops, alley width and pruning heights of trees or shrubs.

The ecological interaction between trees and crops in dryland farms is an important aspect to be considered. Trees control soil loss and reduce run-off. It improves soil properties through its root action and efficient nutrient recycling by adding organic matter. A combination of annual crops and trees/bushes can increase bio-mass production without competing for water due to difference in root depth. There are types of trees/perennial shrubs and herbs that are suitable for extreme soil and climate and ensure return. However, all varieties of trees are not suitable for all these purposes stated above.

For on-farm production of bio-mass and subsequent compost, a mixture of 2 to 4 species of green manuring trees like *cassia siamea, C. spectabilis, Gliricidia sepium* and *Derris indica* are suitable. For tree farming or farm-forestry in dryland areas *subabul* and *maharuk* were found promising for fodder and paper pulp while *paradise tree* is good for oil.

For *agro-forestry* system few important tree species are generally used depending upon the agro-climatic situation. *Desleucaena leucocephala* is a quick growing fodder tree but also provide wood for fuel and pulpwood. *Prosopis julifera* and *prosopis chinensis* are drought resistant and provide wood for fuel and poles. *Acacia albida* is well adapted to sandy soils and produce good fodder, *acacia nilotica* and sisbania *grandiflora* are the other species found suitable for this purpose.

5.3.8.2. Management of Windbreaks and Shelter Belts

Wind-break is a barrier placed at right angles to wind direction to reduce wind velocity and by spacing them in regular intervals, the length of open wind blow can be broken up. The longer version of live (made up of plants) wind-break is frequently called *shelter belt*. It is designed in a way that it raises sharply on the windward side and having two tree rows and upto three shrub rows, one of which is placed in windward side. Sometimes wind-breaks or shelter belts are made up only with shrubs/ hedge and are called *hedge belts*. The density of the belt should not be very thick like an impermeable barrier nor sparse as transparent, it should be translucent. The suitable density of the barrier is equivalent to a porosity of 40 to 50 percent. The *area of protection* increases with the square of the length until the latter is about 24 times the height of the windbreak. The recommended belt length is about 24 times the barriers height. *Widths* of the belt vary from about 9 m for a two-row tree belt with associated shrubs to about 3 m for only hedge belts. These width mean that belts can occupy about 3 percent of the land area they are protecting.

Shelterbelt gives the greatest overall reductions in wind speed over a distance 30 times the barrier height (Skidmore and Hagen, 1977). However, because of variable growth and poor maintenance of the trees, a distance of about 10 or 12 times the height of the belt is more realistic (Khimina,1973). The denser wind–breaks are effective for shorter distance. Using wind tunnel studies, Woodruff and Zingg (1952) develop the following equation for determining shelter belt or windbreak spacing:

$$L = 17H\ (V_t / V) \cos \mu$$

Where L is the distance apart of the belt,

H is the height of the belt,

V is actual wind velocity measured at a height of 15m above ground surface,

V_t is the threshold velocity for particle movement, taken as 34 km h^{-1}, and

μ is the angle of deviation of prevailing wind from perpendicular line to belt.

The greatest effect of shelterbelt is found where a collective belt planting is executed with farmer's collaboration. Within this the individual farmers are free to plant hedges along the boundaries. The plant species selected for wind-breaks should be rapid growing, tolerant of wind and light. Preference should be given to local rather than imported species. Apart from breaking wind-speed it has an air conditioning effect on the surroundings and can decrease evapotranspiration from crop field effectively.

5.3.8.3. Management of Grass Waterways

Grass waterways {For 'Waterways' please see Chapter-4.3.6.} serve as outlets for channels of field (terrace) and diversions. They are usually constructed on natural depressions and one or two seasons ahead of the construction of channels or diversions so that by this time the vegetation in it is well established to take care of the onslaught of the incoming flowing water.

For designing a grass waterway the soil characteristics, land use/vegetative patterns, topography classes in that area and peak runoff of the proposed waterway is to be determined. Then the approximate cross section is to be worked out. However, grass waterways are generally constructed or transformed in parabolic shape whose side slopes are kept very flat e.g. 8:1 or 10:1 that also permit easy crossing of farm implements. On fields with alternating gentle and steep slopes a grass waterway with drop structures on the steeper sections should be used.

The grasses/vegetation to be established should include mixtures of quick growing annuals and hardy perennials. They should be sown/planted in the onset of monsoon and with application of fertilizers/manures and lime for quick estublishment. The vegetation in waterways should be periodically trimmed, bushes and shrubs to be removed and sedimentation should be periodically cleared. The selection of grasses for planting should take account of soil, climate and need of density of cover. Commonly used grasses are Bermuda grass (*Cynodon dactylon*), Kentucky bluegrass (*Poa pratensis*), Smooth bromegrass (*Bromus inermis*) and Napier grass (*Bromus inermis*). It is, however, recommended to seek local agronomic advice before making final selection.

5.3.8.4. Management of Live-Hedge

For making boundary of a dryland farm or a terraced area and sometimes for making contour lines, few specific plant species are being utilised. These non- browsing, hardy and drought resistant plants can simultaneously protect soil and crops from erosion and grazing respectively. Apart from these two vital works they help in many other aspects of dryland farming. The importance of live-hedge are listed below:

- Protection against stray animals;
- Act as a barrier to sheet or rill erosion;
- Act as windbreaks;
- Can create micro-climate;
- Can act as green manure plant;
- Can provide food, fuel-wood, and fodder as economic activity;
- Can have medicinal and aromatic value in commercial outlook.

Selection of Live-Hedge. For dryland areas *Drought resistant capacity* is the prime factor of selection. *Non-browsable* character is the second factor to be considered. Then the nature of soil binding, soil & water conserving capacity, usability for mankind and economic return are the other factors generally considered for selection of live-hedge.

On the basis of the said characteristics *Agave Spp.* are found to be the best live-hedge species for dryland areas and details of them are narrated afterwards. *Agave sisslana* (Sisal), *Agave cantala* (Agave), *Acacia rugata* (Shikakai), *Lantana camara* (Lantana), *Caesalpinia crista* (Molucca bean), *Erythrina variegata* (Coral tree), *Euphorbia sp.* (Thor), *Jatropha carcus* (Jatropha), *Pandanus odorastissimus* (Kewada), *Perkinsonia aculeate* (Horse bean), etc are the important live-hedge plants that can be effectively used in dryland farms.

They are generally planted behind bunds or trenches placed right angle to the slope. Where the boundary lies along with the slope, the planting of live-hedge may be done in jig-jagged manner. Planting or seeding should be done just ahead of rainy season and try to provide life-saving irrigation in following dry season, if necessary, for the first year only. Maintenance, by way of gap filling, proper trimming/cutting, maintaining of bunds or trenches, is the most vital aspects for extracting maximum benefit out of live-hedge.

Sisal/Agave is a widely distributed plant and very much suitable for live-hedge in dry areas. The economic return of this plant appears to be very much encouraging. This easily manageable plant is non-browsable, drought-resistant and can effectively check soil erosion. It is frequently used for gully reclamation purposes for its strong soil binding property. About 10000 can be planted on contour bunds in a hectare. Agave fiber is strong, coarse, superior and more flexible than *Manila hemp* and it ranks next to cotton in USA. Its fiber is used in ship cordage, webbing and sacking, paper making and hardware production. Each plant yields about 250 to 300 leaves during its life span (7-8.5 years). These leaves yield 27 to 35 kg of dry fiber. By way of research and improved culture this yield can be increased to a great extent. Only about 2000 tons of sisal fiber is being produced in India annually against a demand for about 23000tons.

5.3.9. Post Harvest Management

Commonly the food grains are the primary item produced in farming system of any specific area. Apart from grains, fruits, vegetables, and other food or non-food crops are also produced side by side in that system. In dryland farming the scope of economic production of non-grain crops is more. Now on the basis of perishable nature of produce the subject 'Post Harvest Technology' can be broadly divided into: (1) post harvest technology of grains or non-perishable produce and (2) post harvest technology of perishable produce. However, post harvest management basically starts just before the crops are attaining the stage of maturity or ripening or the stage of harvesting. The above group-wise points of post harvest management are narrated below:

For grains and Non-Perishable Produce

- The physiological stage of harvesting should be selected properly to avoid dislodging or over-ripening or any kind of loss of crop.
- After harvesting, the produce are properly cleaned or thrashed for separating main economic parts from the plant or pods. However, the dried plant parts, which generally cover the harvested produce, preserves them better.
- The produce should dry properly, preferably in sun. Then clean the produce and separate them into different marketable grade.

- The dry, clean and graded produce are then stored in bulk or in gunny bags in a room/chamber made up of bamboo / straw / well-burnt soil / tin / bricked or masonry structures, in most cases above the ground. Some underground storage structures made up of clay or brick are also used.
- The temperature, moisture, oxygen, carbon-di-oxide concentration of the storage space should be considered.
- Protection against stored-grain- pest should be adopted meticulously. Proper storing and regular inspection of the storage site can avoid any application of pesticide. If required instead of chemical pesticide 'neem leaf' or such other things may be used against the stored-grain-pest.
- Periodical drying is essential for long storage.

For Perishable Produce

In principal, perishable produce like vegetables, fruits, etc. require completely separate technology for post-harvest management. Here the main goal is to maintain the quality as fresh as possible and to reduce the decaying process. However, the procedure of post-harvest management of perishable agro-produces are as follows :

- Harvest the crop at night i.e. after sunset or well before sunrise and remove the field heat.
- Clean and separate them from unwanted materials, particularly from rotten/ damage materials.
- Grade the crop as per market demand and pack them in well ventilated container with a height not more than 1 to 1.5 feet depending upon the type of produce.
- The produce should reach in the market or processing plant at early morning.

5.3.10. Medicinal, Aromatic and Dye Yielding Crops

Medicinal, aromatic and dye yielding plants provide a more potential alternative to many of the coarse grain or other crops already recommended for dryland farming. High cost, toxic effects, environment and eco-hazardous synthetic chemicals used in different drug, essence and colour have created renewed interest in natural or plant products. Many of these plants are suitable for dry areas, where the quality of the produce is superior in comparison.

Certain complex chemical substances like alkaloids, steroids, essential oils etc. accumulated in different plant parts are known as secondary metabolites. The concentration of these metabolites was reported to be higher under water stress conditions probably providing drought resistance to certain plants (Yaniv and Palevitch, 1982). Some of them are highly tolerant to both biotic as well as abiotic stresses. Thus, dry regions offer superior niches for cultivation of medicinal, dye and aromatic crops.

As per study, it was suggested (Pareek and Gupta, 1993) that Psyllium, senna, periwinkle, asgand, lemongrass, palmarosa, vetiver, indigo, mesta and henna can be grown over low fertility soil in dryland tract of India. Solanum, basil, annato and safflower are profitable to grow in medium fertile soils of dryland using moderate inputs like irrigation, manure and fertilisers etc. However the promising medicinal, aromatic and dye yielding plants are given in the following table:

Table 5.11. Promising Medicinal, Aromatic and Dye Yielding Crops with Some Details

Botanical Name	Common Name	Economical Part	Alkaloid & its Use
Medicinal Plants			
Catharanthus roseus	Periwinkle	Leaves Roots	Vinblastine, Vincristine- Anticancer Ajmaciline-
Cassia angustifolia	Senna	Leaves. Pods	Sennosides-Laxative
Aloe barbadensis	Aloe	Leaves	Aloin-Cosmetics
Withama somnifera	Ashwagandha	Roots,Leaves	Withanoides
Solatium Khasianum	Solanum	Berries	Solasidine- steroids
Muccina purieans	Muccina	Seeds	L-dopa-Parkinson
Aromatic Plants			
Cymbopogon martini	Palmarosa	Flowering herb	Geraniol, Geranyl acetate. perfumery in cosmetics, soaps Foods
Cymbopogon flexuous	Lemon grass	Leaves	Citral- Perfumery. Flavour Vitamin A can be isolated. The spent grass can be used as cattle feed, fuel in distillation unit, manure
Vetiveriazizanoids	Vetiver	Roots	Vetiverol, Vetiveryl acetate
Cooling			
Ocimum basilicum			
O gratisimum			
O. cannum			
O americamim	Basil	Inflorescence and leaves	Methyl chavicol Eugenol Linalool Methyl chavicol
Dye Yielding Plants			
Indigofera lincioria	Indigo	Leaves	Blue dye–textile dye
Bixa orellana	Bixa	Seed	Orange–textile & food colourant
Larwsonia inarmis	Henna	Leaves	Orange-cosmetic & textile
Hibiscus sahdarifa	Mesta	Calyx	Red-food colour
Tagents errecia	Marigold	Flower	Food color
Carihamus lincioria	Safflower	Flower	Food color

5.4 MANAGEMENT OF SOIL ORGANIC MATTER AND BIO-FERTILIZER

Low moisture and high temperature along with erosion hazards reduce the organic matter content of a dryland soil. This is an inherent problem of all dryland areas in relation to crop production. To overcome this problem it is necessary to understand the detail about *soil organic matter,* its origin, action, loss, replenishment, production and management.

5.4.1. Soil Organic Matter

Soil organic matter is that component of the soil, which has come from both plants and animals. Plants are the primary source of soil organic matter while animals and numerous microorganisms are considered as secondary source. Their pertially decomposed body as well as the waste materials add organic matter into the soil on decomposition. Certain organisms like earthworms, centipedes and ants play an important role in translocation of plant residues in the soil.

There are three types of soil organic substance; (1) *litter,* the un-decomposed to partially decomposed or fermented organic remains of plants or animals that still maintain identity of their original form. (2) *organic matter,* the partially to fully decomposed or fermented organic remains that have lost identity of their original form and have acquired a fibrous or laminated structure. (3) *Humus,* the dark colloidal remains of well-decomposed organic matter, which resists further decomposition.

The amount of organic matter present in mineral soils varies widely. Mineral surface soils contain a trace to 15 or 20 per cent of organic matter. There is no satisfactory direct method of determining exact quantity of total organic matter present in soil. The usual procedure is to find the amount of organic carbon first, which can be done quite accurately. Then this figure is multiplied by the factor 1.7. This will give the approximate amount of organic matter present in that soil.

5.4.1.1. Function of Soil Organic Matter

Effect on Soil Colour

- It imparts brown to black colour that absorbs more heat.

Influence on Physical Properties of Soil

- It encourages granulation structure to the soil.
- It reduces plasticity and cohesiveness of the soil.
- It increases water holding capacity of the soil.

Increase Cation Adsorption Capacity of Soil

- In organic matter this property is two to thirty times more than that of mineral colloids.
- Soil organic matter has 30 to 90 percent of adsorption power than that of mineral soils.

Supply and Availability of Nutrients

- Presence of easily replaceable cations increases availability of different plant nutrients in soil.
- It supplies nitrogen, phosphorus and sulfur to plants.
- Extraction of plant nutrient elements from minerals by humus.

Enhanced Soil Microbial Activity

- It increases microbial population in soil.
- It increases atmospheric nitrogen fixation through increased population of nitrogen fixing bacteria and their improved activity.

5.4.1.2. Resource Potential of Organic Matter

Apart from naturally present organic matter in virgin soils, there is possibility of adding more by collecting from human settlement. The major sources are listed below:

1. **Livestock and Human Waste**
    - Animal shed wastes including dung/drops/excreta and urine.
    - By-products of slaughterhouses and dead body of animals.
    - Human excreta.
2. **Crop Residue, Tree-Tops and Aquatic Weeds**
    - Crop wastes like stalks, leaves, shells and other parts.
    - Plant materials available out of pruning /clipping /lopping etc.
    - Aquatic floating or rooted-floating weeds or plants like water hyacinth, etc.
    - Blue-green algae, azolla, etc.
3. **Green Manure Crops**
    - Sunhemp (*Crotolaria juncea*), 'Dhaincha' (*Sesbania aegyptiaca*), Cluster beans, Cowpea, Horse gram etc.
4. **Urban and Rural Wastes**
    - Rural and urban solid and liquid wastes- dustbin collection, sewage, sludge.
5. **Agro-Industries by-Products:**
    - Oil-cakes
    - Husk, bran, shell, bagasse, press-mud, sawdust etc.
    - Fruit, vegetable wastes, cotton, wool, silk, jute wastes and tea, tobacco wastes.
    - Fishmeal.
6. **Tank Silts or Pit Soil.**

5.4.1.3. Composition of Plant Residue

Compounds present in plant tissue are many and varied, but the elemental composition is Carbon (44%), Oxygen (40%), Hydrogen (8%) and Ash (8%). All inorganic elements, including nitrogen and in lesser amount Sulfur, Iron, Phosphorus etc. are represented in the Ash. The *Carbohydrates* ranging in complexity from simple *sugars* to *celluloses*, the *Fats and oils* (waxes, tannins), the *Proteins* in both water soluble and crude forms and the *Lignins*, generally present in older plants, are the basic compounds present in plant tissue.

5.4.1.4. Decomposition of Organic Materials

When fresh organic substance is added to a soil, different microorganisms utilize it in their lifecycle through feeding etc. This biodegradation process is carried out by different groups of heterotrophic microorganisms, bacteria, fungi, actinomycetes and protozoa. The role of cellulolytic and lignolytic microorganisms in decomposition of crop wastes and residues is of prime importance. Microorganisms, involved in the process, derive their energy and carbon requirements from the decomposition of carbonaceous materials and for every 10 parts of carbon, 1 part of nitrogen is required for building up of their cell protoplasm. Fungi are more efficient in carbon assimilation thus more active in biodegradation than bacteria and actinomycetes.

Aerobic and *anaerobic* are two types of decomposition processes involved in biodegradation. When organic materials are broken down in presence of oxygen, the process is called aerobic decomposition. It is most common in nature and generally occurs in arable soils where organic residues are stabilised into humus, with different groups of microorganisms. On the other hand, anaerobic microorganisms, particularly few groups of bacteria, break down organic materials by a process of reduction in absence of oxygen. This type of fermentation takes place in bio-gas plants, waste ponds, marshy soils, buried organic materials etc. In both the processes the organisms use nitrogen, phosphorus and other nutrients to develop their cell protoplasm.

Rate of Decomposition

Organic matters vary to a great extent in their rate of decomposition. It is due to the compounds present in that organic matter, which decompose differently. Out of these compounds sugars, starches, simple proteins and crude proteins are *decomposed rapidly* and hemicelluloses, cellulose, lignins, fats, waxes etc are *decomposed slowly*. All these compounds are usually decomposed simultaneously when fresh plant tissue is added to a soil. The rate of decomposition, however, decreases when it moves from sugar to wax. Thus sugars and water-soluble proteins are examples of readily available energy source for soil organisms. Lignins are a very resistant source of food, although they eventually supply more total energy in due course.

Fresh organic matter generally passes through the following process or reaction when added to soil.

1. Bulk of the matter in presence of carbon dioxide and water undergoes enzymatic oxidation and evolve heat (sometimes up to 65 to 70°C).
2. Nitrogen, Phosphorus and Sulphur as essential plant nutrients are released and/or immobilized by a series of specific reactions relatively unique for each element.
3. Either through microbial synthesis or through other reactions, compounds (modified compounds of original plant tissue) resistant to microbial action are formed and incorporated into the soil.

The rate of decomposition of an organic substance and release of nitrogen from that substance increases with its higher to Carbon: Nitrogen (C : N) ratio. The organic manures with C:N ratio of 20:1 or less, as in case of green manures, animal exereata, etc. are good source of organic nitrogen immediately required for crop plants. Though nitrogen release from organic materials with C:N ratios of 30:1 or higher may be too slow to benefit crop plants but its slow releasing nature helps soil to improve its organic status sustainably. So the choice of organic amendments for soil may be done on the basis of C:N ratio, which is important in determining the quantity & period of nitrogen availability. A few relative C:N ratios of some common organic materials are given below :

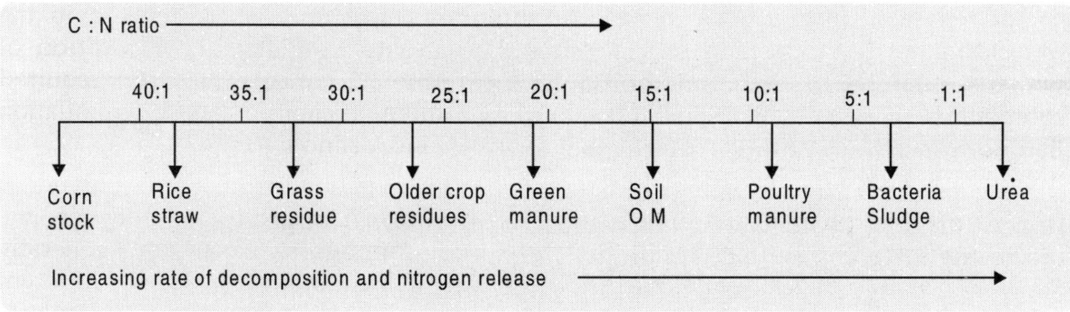

Supply of Nitrogen

Demand of crop and rate of decomposition- Supply of nitrogen to a crop can be provided through application of organic manure, but the percentage of nitrogen in organic sources varies widely. As the C:N ratio increases the nitrogen percentage decreases.

Now the rate of decomposition and release of nitrogen in the form of Nitrate (NO_3^-) should match with the growing period or demand of crop for maximum utilisation of organic manure as well as for maximum crop growth (Figure 5.21).

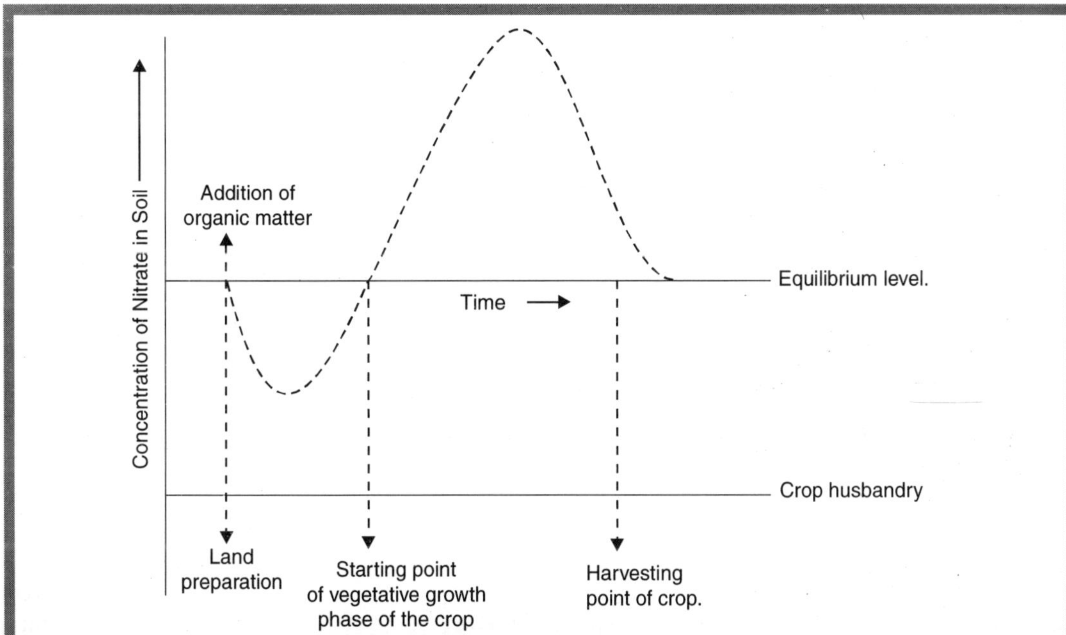

Fig. 5.21. Position of nitrate available in soil after addition of organic matter and its relation to crop husbandry.

5.4.1.5. Humus

It is a complex and resistant mixture of brown or dark brown amorphous and colloidal substance modified from the original tissues or synthesized by the various soil organisms.

As the decomposition of fresh organic matter occurs, two major kinds of organic compounds tend to remain in the soil: (a) modified compounds of plant origin, which is resistant to decomposition but increases their reactivity; and (b) new compounds, which are synthesized by microorganisms and held as part of their tissue. The compounds of microbial origin are not insignificant in quantity. Studies have shown that up to one third of the organic carbon may be present in this form. However, these two groups of compounds provide the basic framework of *humus*.

At the time of formation, the processes permit nitrogen to become an integral part of the humus complex. Newly formed humus is quite resistant to further microbial attack and thereby its nitrogen and other essential nutrients are protected from ready solubility and dissipation.

Nature and Characteristics of Humus. Humus is a highly colloidal substance. It is amorphous and not at all crystalline like other minerals. Its surface area as well as absorption capacity is in excess of those exhibited by any of the clays. Naturally the cation exchange capacity (CEC) of well developed humus ranges from 150 to 300 mille equivalent per 100 grams, whereas in silicate clays it is about 8 to 150 meq/100 gm. Humus adsorbs water equivalent to 80 to 90 percent of its weight from saturated atmosphere, in case of clay it is only15 to 20 percent. The physical properties like plasticity and cohesion is low in humus, which reduce stickiness and increase granulation in fine textured soil.

Influence of Soil Organic Matter/Humus on Soil Aggregates. Organic matter plays a very important role in binding individual soil particles into soil aggregates. Humus created out of organic matter is adsorbed on the surface of clay particles through the process of orientation of organic molecules (generally long chain molecules). This process is effective in moist condition.

The effect of soil organic matter on aggregation is more pronounced in the soils containing less than 25% clay. The organic matter has much higher correlation with aggregates greater than 0.1 mm diameter than the aggregates greater than 0.05 mm diameter. It denotes that organic matter is conducive to the formation of relatively much larger water stable aggregates. A rapid increase in granulation occurs after the addition of organic matter, interestingly application of fresh organic matter is more effective for this initial effect. But that initial effect on granulation decline if that organic status decline in due coarse. It emphasizes the importance of frequent/regular addition of organic matter into the soil to maintain proper soil aggregates, which automatically maintain good soil structure. However, action of more resistant and stable humus components provided most long-term aggregate stability to a soil.

5.4.1.6. Factors Affecting Soil Organic Matter

Climate. Climatic conditions, especially *temperature* and *rainfall,* exert dominant influence on the amount of organic matter found in soils. In general high temperature accelerate decomposition of organic matter but enhance the loss of the same. So if one moves from a warmer to a cooler climate, he will find that the organic matter of comparable soils tend to increase. If the moisture conditions and vegetation remain same the average of total organic matter and nitrogen increases two to three times for each 10^0C fall in mean temperature. Generally, under comparable conditions, the nitrogen and organic matter increases as the effective moisture become greater, however, the C/N ratio become wider, especially in grassland areas.

Vegetation. Naturally, vegetation increases the organic matter status of a soil, but the nature of vegetation is a factor for the same. Studies have shown that in those climatic zones where the natural vegetation includes both grassland and forest, the organic matter is higher in soils developed under grassland than under forest. The organic residues of grassland and their mode of decomposition may encourage a higher organic level than it is found under forest.

Texture and Other Factors of Soil. Due to the lower moisture content and more oxidation, the lighter soils have usually less organic matter and nitrogen than the finer textured soils. On the other hand the heavier textured soils, because of their high moisture contents and relatively poor aeration, are richer in organic matter than lighter equivalents.

The *lime content* and *erosion status* of soil are other factors that influence the accumulation and activity of soil organic matter and its nitrogen.

Cropping. Usually the status of organic matter is much lower in cropped land than in comparable virgin areas. It is due to the continuous removal of plant materials i.e., crop and its residue from the field and relatively less return to the land. Mono-cropping after a fallow and multicropping, especially with leguminous crop, help to maintain soil organic matter status of the soil.

5.4.1.7. Management of Soil Organic Matter

The inherent capacity of soils to produce crop is closely and directly related to their organic matter and nitrogen contents. In majority of farm soils, particularly in dryland areas, it is difficult to maintain or increase the level of these two constituents up to a satisfactory level.

The Approach. The amount of crop residues to be returned to the soil is invariably dependent upon the level of organic matter in a soil. This approach, as far as practicable in cropped soil, in combination with application of foreign organic materials can maintain the organic status of the cropped land similar to that of comparable virgin land.

A good crop of oats, maize and sugarcane will perhaps produce 5000, 8000, and 15000 pounds of dry matter per acre respectively from its aboveground parts. Now, if the roots left in the soil when such crops are harvested then the added organic matter in soil will be even half of such weights. Many do not appreciate the fact that the maintenance of a satisfactory supply of organic matter in arable soils is possible only because of root residues added in this way. In arable soils, rootlets of crops spread rapidly and as the life of these roots are very short they continually contribute to the supply of readily decomposable organic matter. So production of bumper crop, through proper liming and fertilizer application, can maintain the organic matter supply by increasing the crop residues particularly the amount of *roots*.

Apart from this, there are several methods of production or collection of partly / fully decomposed or un-decomposed organic materials for application of the same into cropped soil. These materials are called as *organic manures* and are apparently used for supply of plant nutrients though it performs many other important functions in soil.

Under the arid tropical conditions, the soil organic matter is quickly lost. Fresh application is necessary to replenish this loss in order to maintain soil fertility. According to present recommendation 12.5 tones of FYM in areas of medium to heavy rainfall (about 1250 mm/year) and 5 to 7 tones in areas where rainfall is low (up to 500 mm) should be applied per hectar for intensive agriculture. In extremely dry areas, where annual rainfall is below 500 mm, the application of 2.5 tones of FYM/ha is suggested to increase the yield. Now considering the scientific and economic approach of dryland farming the recommendation for application of FYM or any other organic manure will be for increased quantity in favour of low rainfall areas where every drop of available water will be utilised for crop growth. The details of organic manure and its production are narrated hereafter.

5.4.2. Organic Manure

Organic manures are mostly by-products of farming and allied activities, derived from plant and animal sources. Few other human activities presently produce organic manures such as town or municipal wastes, factory wastes etc. Most of the partly or fully decomposed organic materials are bulky in nature and supply small quantities of plant nutrients. These organic materials are termed as *bulky organic manures*, e.g. farmyard manure, rural and town compost, vermicompost, night soil, green manure etc. Organic materials which contains higher percentage of major plant nutrients like Nitrogen, Phosphorus and potash are called *concentrated organic manures* e.g. oil-cakes, blood, bone and meat-meals, fish-meal poultry manure etc. These concentrated organic manures are not available in well-decomposed form but many of them quickly degrade when added to soil. Higher residue of oil in oilcakes delay the rate of decomposition and thus solvent extracted oilcakes are quicker acting than the 'ghani'/expeller-pressed oilcakes. The nitrogen of oilcakes is available to plants in about 7 to 10 days after application. Only 'mahua' (*Bassia latifolia*) oilcake takes about two months to nitrify.

Role of Organic Manure. Organic manures play an important role in crop production by influencing fertility of soil and productivity of crop directly as well as indirectly. In addition to nitrogen, phosphorus and potassium, presence of a fair amount of micronutrients such as manganese, zinc, copper, and iron are simultaneously added to the soil through organic manure. But the most important effects of organic manures are the indirect benefits by which the plant growth as well as yield can be sustainably increased. It improves granulation of soil thus improving the structure, seed germination status, water holding capacity, drainage situation, base / cation exchange capacity etc. of that soil.

Most importantly it helps augmentation of beneficial microbial population and their activities such as decomposition, biological nitrogen fixation, solubilisation of insoluble phosphorus and availability of other plant nutrients. It has been reported that organic matters check proliferation of plant parasitic soil nematodes. For energy, nitrogen-fixing *symbiotic bacteria* (Rhizobium, Azospirillum) use the carbohydrates of the hosts that grow better with organic matter. On the other hand, nitrogen-fixing *free-living bacteria* (Azotobacter)use the soil organic matter directly. They fix nitrogen and either pass it on to the infected host or make it a part of their own tissue. When both are decayed in soil, part of this nitrogen is available to higher plants. This is one of the most vital aspects of management of dryland farming.

Production of Organic Manure

Successful dry land farming largely depends upon the supply of well-decomposed organic matter i.e. manures. This supply can be sustained through several methods depending upon the prevailing situation i.e. availability of raw organic matter. Though many kind of raw organic matters are available in and around every farm, they are neither in applicable form nor in sufficient quantity to augment the organic status of the soil.

Approach. The First job will be the production or collection of raw organic matter and next the selection of suitable methods for production of organic manure. Dung and refuse of farm animals, fresh and dry crop residue and green biomass (leaves, twigs, young branches) collected from green manure or other tree/plant growing as part of farm or agro forestry may be used as raw organic matter for production of organic manure. Depending upon the situation and requirement the ratio of availability of these materials may vary in dryland farms. Accordingly, the production methods of organic manures be selected and practiced. However, at the time of selection of different raw materials for mixture of manure pit to produce organic manure the C:N ratio should be considered and effort should be given to bring down the average C:N ratio nearer to 20 : 1.

Brief descriptions of few production methods are given below.

5.4.2.1. Farmyard Manure (FYM)

The collection of different organic wastes of farm, generally dominated by animal refuses, and apply the same in the crop field is a very simple but not efficient way to increase the organic matter status of that soil. The decomposition process of FYM is not uniform and generally being exposed to sun and rain, reduces its efficacy in relation to compost.

This is the traditional organic manure available through decomposition of the liquid and solid excreta of the livestock stored in a pit or tank. This raw organic material is mixed with varying amounts of straw or other litters and deposited layer by layer. For uniform decomposition the pit or tank should provide a minimum air circulation to the decomposing organic materials. At the same time a shed may be provided over the pit or tank to avoid rain and sun. This will increase the quality of manure.

On an average well-rotten farmyard manure (FYM) contains 0.5% Nitrogen (N), 0.2% Phosphate (P_2O_5) and 0.5% Potash (K_2O). Nitrogen present in FYM is slow acting and less than 30% of it is generally available to the first crop. About 60 to

70 percent of the Phosphate and about 75% of the Potash become available to the immediate crop. The rest of the plant nutrients are carried over to the subsequent crops.

5.4.2.2. Compost

Compost manures are the decayed form of plant residue like leaves, twigs, stubble, grain residue, hedge clippings, all kinds of town and village wastes, water hyacinth, saw dust, bagasse and any kind of organic materials. A large number of microorganisms feed on these wastes and convert them into well-decomposed manure called compost. The process of decomposition or *composting* is hastened by adding few other materials like wastes from animal shed, night soil, urine or fertilisers.

This is an improved or modified method of traditional production of FYM. Here farm waste dominates in raw materials and a uniform decomposition is assured. At the time of decomposition loss of nutrient is reduced in composting. To get good manure through composting one should follow the following steps:

1. Dig a pit at a suitable corner of the farm, preferably in a shady place.
2. The base of the pit should be overlaid with a thick polythene sheet or by any material to reduce the leaching of nutrients drastically.
3. Spray the bulky farm/crop/ plant waste (including whole or parts of green manure plants) and mixture of concentrate wastes in alternate layers of 16 cm. and 8 cm. thickness respectively. By manipulating these two types of raw materials much higher quality (in respect of C:N ratio) of compost can be prepared.
4. Decomposed dung or fresh compost may be added with concentrate wastes, for inoculation, which will supply the microbes necessary for composting.
5. Rock phosphate or SSP may be mixed with the bulky organic material.
6. Plaster or cover the upper most surface with mud or any other materials regularly after every turning and keep it for 2 to 3 months for adequate decomposition.
7. Spray water over the pit at 10 days intervals and turn the entire mass of the pit at monthly intervals during the composting process.
8. Several pits are required for utilising the entire farm waste throughout the year, thereby getting the supply of compost round the year.
9. Both compost and FYM must be incorporated into the soil directly to reduce nutritional loss.

5.4.2.3. Vermicomposting

Earthworms are considered as the friends of farmers. They are living composters because they help to speed up the conversion of fresh bio-mass into manure. They actually feed almost all farm wastes and fresh biomass like leaves, pods, shells etc, and excrete a mud like material whose manure quality is quite high. Apart from nitrogen this manure provides potassium, magnesium, calcium, phosphate, molybdenum in available form and few hormones and vitamins. Its humus content is also high. The dead earthworms, on decomposition, release about 25% nitrogen as nitrate and ammonia and other soluble approximately 3% of organic compounds.

Unfortunately large-scale use of pesticide chemicals in the field has drastically reduced their population. Besides, in dry areas the presence of this unique beneficial creature is less due to lack of moisture and organic matter. Its presence in arid regions is almost negligible. Now with the development of science and technology one can cultivate suitable species of earthworms (*vermiculture*) in controlled atmosphere and utilise them for speedy composting of raw organic materials. So vermiculture creates an opportunity for dryland farmers to produce compost. The cultivable earthworm species, popularly used all over the world, are *Eisenia foetida, Pereonyx excavata* and *Eudrilus enginae.* Few other species like *Lumbricus rubellus, Dendrobiana rubaida, Lampito mauritii* and *spp* of *Pereonyx* and *Dychogaster* are sparingly used for culture.

Earthworms can be reared in masonry tanks or pits to prepare organic manure (Fig. 5.22). They convert organic wastes into rich manure, called vermicompost, quite quickly. Apart from the nutrient contents of vermicompost (0.66%: 0.995: 0.4% of N: P: K: respectively), it improves the soil conditions and its characters. The tank method of vermicomposting is well suited to dry land farming in general.

Fig. 5.22. Home Scale Vermi-compost Unit.

Scientific vermiculture for production of vermicompost requires following inputs for a 1.5 cubic meter sized tank (with a tapped outlet) for a period of 2 months:

1. Dry organic wastes - 100 to 120 Kg.

2. Dung slurry - 30 to 40 Kg.
3. Water - 8 to 10 litre in every 3 days.
4. Numbers of earthworms kept in old dung - 1000 to 2000 (they multiply 20 to 25 times in every 70 to 75 days).

Steps of Vermi-Composting

- Fill the bottom of the tank with a layer of broken roof tiles or coconut husk.
- Fill half of the tank with chopped leaves and other farm wastes mixed with cow dung slurry, if not available with mud slurry.
- Maintain the moisture content to 40-50 % by sprinkling water and allow them to decompose for 15 days.
- Release the earthworms on that half-full tank.
- Fill the next layer by organic waste or by any other biomass.
- Cover the tank with wiremesh to prevent birds from picking the worms.
- Do not pack the pit tightly for facilitating proper aeration.
- Arrangement to protect from ant and other insects should be done properly.
- Sprinkle water every three days to maintain adequate moisture which also maintains the body temperature of worms.
- Collect the liquid fertilizer from the outlet, which is at the floor of the tank, at regular intervals and apply to the crop.
- Compost is ready in 6 to 8 weeks depending upon the quality of the waste.
- About 20 days before removing the compost, mix organic waste, dung slurry in a separate site, and allow the mixture to decompose.
- When the vermicompost tank is emptied, transfer the 20 days old semi-decomposed mixture and start the process again.

The approximate market value of the vermicompost is about Rs.1 to 3/Kg. Thus this activity, i.e. the production of vermicompost, is an economic enterprise.

5.4.2.4. Biogas Slurry

Biogas slurry is a potentially high quality manure for the dryland farming. It is produced in a plant by using excreta of livestock or human as raw material. A smokeless gas is obtained as by-product, which is an efficient fuel for cooking and lighting. This gas can indirectly save the cowdung, vegetation and crop wastes, which are usually used as cooking fuel for home. The spared organic materials can now be used for making additional organic matter.

Besides, the slurry has higher manurial value with a narrow C:N ratio as compared to ordinary F.Y.M produced from the same dung. During anaerobic digestion of cattle dung 15 to 30 percent nitrogen is mineralised to ammoniacal form. The average nitrogen, phosphorus and potash present in digested catteldung slurry is about 1.5%, 0.9% and

0.8% respectively. Slurry contains considerable amount of other plant nutrients in dissolved form. Hence when used directly in diluted form, these nutrients are absorbed readily by the crop. This results in quick growth and greater yield. This is, therefore, usually used for growing of high value crops or vegetables in the dryland areas. Besides, water contained in the slurry augments moisture requirements of the crops in dry areas.

There are several models of bio gas plant depending upon the constructional design like KVIC, Janata, and Dinobondhu etc. The quantity of dung, available daily, determines the size of the biogas plant. An average 1kg of fresh dung can produce $0.04 mt^3$ of biogas and 0.25 to 0.3 mt^3 gases is required per head per day for cooking. Normally 4 heads of cattle is needed to supply dung to the smallest unit.

5.4.2.5. Green Manuring

Sometimes the immature, succulent crops especially legumes (grown for fixing atmospheric nitrogen through *Rhizobium* symbiosis) are ploughed down to incorporate the same into the soil. However, application of green plant directly into the soil is called green manuring. The common green manure crops are Sunhemp (*Crotolaria juncea*), Dhaincha (*Sesbania aculeate*), Cluster beans (*Cyamopsis tetragonoloba*), Cowpea (*Vigna catjang*), Horse-gram (*Dilichos biflorus*), Senji (*Melilotus parviflora*), Berseem (*Trifolium alexandrinum*) etc.

Green manure has several effects on soil depending upon the conditions of the soil. It increases the organic matter content of the soil. It may collect nutrients likely to be deficient in the surface soil and leave them there in a relatively available form. However, green manuring can't provide both organic matter and nitrogen at a time. The mature or nearly mature plants are able to improve the organic matter status but not the nitrogen level to that extent. On the contrary the younger plants can provide nitrogen to the soil but can not improve the humus status of the soil to the same extent.

The green manuring is not as effective (per unit carbon) as farmyard manure (FYM) in increasing the organic matter content of the soil because of greater decomposability. Thus the green manuring has, in most part of the world, been applied more effectively to increase the available nitrogen supply rather than to increase the humus.

Green manuring is a time-tested strategy to create useful organic matter and additional nutrient supply. But, its acceptance lags far behind the benefits it is acknowledged to produce. A critical review on green manuring suggests that in order to make it practicable a few basic conditions need to be fulfilled. A green manuring crop should readily fit into a cropping cycle from a farmer's point of view. In dryland farming the situation is further confounded by the limits on cropping seasons imposed by the length of rainy season. No doubt about 25 percent rains generally occur beyond the rainy season, 15 percent on post-rainy season and about 10 percent on pre-rainy season. So the pre and post-season rains can be gainfully utilised to raise a quick-growing, drought-tolerant green manure crop and at the same time use of off-season tillage can also be possible. The biomass thus generated can be incorporated in the soil at flowering or any physiological stage depending upon the requirement of main crop. However, the green manuring can be practiced in dryland farming with certain conditions eg,.- as catch crop between two main crops, in mild winter with moist climatic conditions and it should not compete with the main crops in any way, in particular for water.

Another approach, best for dryland farming, for generating of green biomass could be the raising of fast growing trees, mostly leguminous (*Leucaena leucocephala, Gliricidia maculata, Cassia auriculata, Cassia tora, Thespesia populnea, Sesbania spp.,etc.*) or shrubs on wasteland in the part of a holding or in the form of rows in an alley-cropping system. When grown in separate land or in alley cropping, the tree/shrub should be pruned regularly to avoid shading of the main crop and this pruned material may be incorporated in soils directly. Otherwise this pruned or clipped biomass may be used for composting to create quality organic matter.

5.4.3. Bio-Fertilizer

What is Bio-Fertilizer ? The inoculants that contain living microorganisms, which on application can enrich the nutrient status of the soil by various means, are termed as bio-fertilizer.

This supply of nutrients is not as simple as in case of chemical fertilizer but is more complex and more beneficial. Bio-fertilizers enrich soil by supplying nutrients by way of fixing nitrogen from air and by enhancing availability of several nutrients through solubilization or by increased adsorption, stimulating plant growth through hormonal action or antibiosis, and by decomposition of organic residues.

The primary group of organisms acting as bio-fertilizers, called nitrogen-fixing organisms (Rhizobium, Azotobacter, Azolla, Blue green algae, and Azospirillum), can fix nitrogen efficiently directly from air. Another group of heterotrophic microorganisms (bacteria, fungi etc) can convert unavailable phosphate to available form from mineral phosphatic compounds with the help of enzyme formed in their body. *Mycorrhizae* is another microorganism that have a potentiality to improve plant growth.

Rhizobium can fix up to 120 kg of aerial nitrogen per acre through a leguminous crop; *Azotobacter* can collect 100 kg of nitrogen from air. Azolla in combination with *Anabaena azolle* can add 16 to 24 tonne of organic matter along with 40 to 60 kg nitrogen per acre. Some of these microbes not only add nitrogen or phosphorus to soil but also release hormones, antibiotic substance, vitamins and enzymes and vitamins that are beneficial to crops.

Wide acceptance of bio-fertilizer is not reported due to various reasons. In this situation the strategy to extract the benefits of *in-situ* indigenous bio-fertilizer organisms in drylands is to concentrate on crop and soil management practices.

Types of Bio-Fertilizer

5.4.3.1. Nitrogen-Fixing Bacterial Inoculants (BNF)

This bio-fertilizer offers an economically attractive and ecologically sound means of reducing nitrogen input, and improving the quality and quantity of internal resources of nitrogen. The N-fixing organisms relevant to dry areas are *Rhizobium,* a legume-Rhizobium symbiosis; and *Azotobacter* and *Azospirilium,* the non-symbiotic and associative symbiotic organisms. The need to inoculate the legumes to be grown in dryland farms must be assessed by considering three interacting factors like plant, bacterial strain and environment. Crop history has a significant influence on the population of native rhizobia present in the soil and inoculation response of legumes to be cultivated. However, some inoculant strains succeed in forming more nodules despite the presence of competing indigenous rhizobia, e.g., NC 92 on groundnut (Nambiar et al., 1988). Application of *Azotobactor* and *Azospirillum* inoculants improve plant growth and yield not only through N fixation but also through production of phyto-hormons.

5.4.3.2. Phosphate Solubilising Micro Organisms

Bacteria like *Bacillus megaterium, B. circulans, B. Subtilis, Pseudomonas straita, P. rathonis,* fungi like *Aspergillus awamori, Penicillium digitatum, Trichoderma sp.;* Actinomycetes like *Streptonyles sp.* etc. are the organisms falling in this group. They solubilize the unavailable forms of inorganic-phosphates like tri-calcium, iron, aluminium and rock phosphates into soluble forms by release of a variety of organic acid like succinic, citric, malic, formic, glyoxalic and gluconic acids. So far only Phosphate solubilising bacteria (PSB) like *B. meghatherium* and *P. straita* have been commercialised as phosphate solubilising bio-fertilizers. Farmers' acceptability of products like 'Microphos' and 'Biophos' has increased particularly in crops like groundnut.

5.4.3.3. Plant Growth Promoting Rhizobacteria (PGPR)

A group of rhizosphere bacteria (rhizobacteria) that exert a beneficial effect on plant growth is referred to as PGPR (Schroth and Hancock, 1981). *Actinoplanes, Agrobacterium, Alcaligenes, Amorphosporangium, Arthrobacter, Azotobacter, Bacillus, Cellulomonas, Pseudomonas, Rhizobium, Streptomyces, Xanthomonas* are several of the genera that belong to PGPR. Out of them *Bacillus spp.* proves to be an ideal inoculant for dryland areas because of their endospore producing ability.Currently *Pseudomonas spp.* are receiving much attention because of their multiple effects on plant growth promotion.

PGPR are believed to improve plant growth by colonizing the root system and pre-emptying or suppressing the establishment of Deleerious Rhizosphere Micro Organisms (DRMO) on the root (Schroth and Hancock, 1981). Field trials has been conducted with wheat, potato, sugarbeet and zinnia on experimental and commercial scale, where 40 out of 63 trials showed significant results with yield increases varying from 7 to 136% in different crops over the control.

5.4.3.4. Vesicular-Arbuscular Mycorrhizae (VAM)

The symbiotic association between plant roots and fungal mycelia is termed as mycorrhizae. These fungi are associated with majority of agricultural crops. VAM occur over a broad ecological range from aquatic to desert environments. These fungi belonging to the genera *Endogone,Glomus, Entrophosphora, Gigaspora, Acaulospora, Scutelispora* are obligate symbionts and the inoculum is prepared using infected roots and soil.

Crop responses to VAM inoculation are governed by soil type, host variety, VAM strains, temperature, moisture, cropping practices and soil management practices.

5.4.4. Prospect of Organic Farming

5.4.4.1. What is Organic Farming ?

It is basically a new terminology carrying an old concept as well as practice of farming. Different people of different countries tried to define the term "organic farming" but a unanimous single definition or explanation is yet to be finalised. As a result of these efforts many other terminologies like 'natural farming', 'eco-farming', 'bio-farming', 'recyclic farming' etc., almost of similar conception to organic farming have emerged.

Now, we may try to generate a conception about the term "organic farming" for the common people particularly for the farmer. According to the necessity of present world and to avoid the detrimental effects on modern agriculture, 'Organic farming' may be conceptualised as "a process of farming where its produce do not carry any kind of residual toxicity or phyto-toxicity". However, initially a minimum level of this toxicity may be set to provide time frame for changing over from existing practice to organic farming. For dryland areas it will be an easier process than that of other regions.

The toxicity in agricultural produce generally occurs due to application of (1) chemical fertilizers, (2) plant protection chemicals, (3) weed control chemicals, (4) Off-farm waste like town/industry waste. So avoiding the application of chemicals may be the way of organic farming, but it is not so simple. To maintain the present yield status so as to replace the chemically supplied N, P, K, an unusually large amount of organic manures will be necessary. If the chemical fertilizers are withdrawn without replacing it by organic manures the production will be reduced drastically. Therefore, 'Integrated farm input approach (IFIA)' along with 'integrated pest management (IPM)' can ensure the economic security and can initiate the practice of organic farming. Now with this experience the chemicals, till used in the farm, can be replaced gradually with organic substitutes and this will convert the farm to an ideal organic farm in a few years.

5.4.4.2. Whether Dryland Farming can be Converted to Organic Farming?

Generally, the dry land soils are devoid of plant nutrients and poor in organic matter status. Thus, the inherent fertility as well as productivity level of these soils is very low. It is almost impossible to increase the production further only through application of chemical fertilizers. Hence, application of organic matter or manure is must in dryland farming. In absence of adequate quantity of organic manure to feed the crops a proper combination of organic manure, bio-fertilizers and to some extent chemical fertilizers, which may be termed as Integrated Nutrient Management (INM), can sustainably increase the productivity of a dryland farm without any residual toxicity.

In the plant protection front, due to low humidity and high temperature the infestation of insects and diseases in dryland areas are generally less. In this advantageous situation integrated pest management (IPM) coupled with application of suitable bio-pesticides/fungicides may successfully protect the crops from different pests. {A brief description of the same is available in Chapter-5.3.6.2.}.

If dryland farming, as described in this book, can follow the above said practices in plant nutrients and plant protection front then it can be said as organic farming.

5.4.4.3. Steps to be Taken For Organic Farming in Dry Areas

- Produce huge quantity of organic manure from on-farm waste throughout the year;
- Application of bio-fertiliser whenever possible;
- Collection of off-farm organic matter for producing organic manure;
- Use of bio-pesticides/fungicides produced on-farm and off- farm.
- Development of cropping plan on the basis of ecological and recyclic nature of crops;
- Inclusion of animal husbandry, fishery, forestry and other components to introduce composite character into the farm.
- Efficient management of water should also be linked with the principle of organic farming.

5.5 MANAGEMENT OF MULCHING

5.5.1. What is Mulching ?

Mulching is covering of the soil surface with suitable materials primarily to *reduce the loss of soil moisture* through evaporation and also to *minimise the erosion of surface soil* by wind and water.

Mulching has other beneficial effects too, like *maintaining of soil temperature, conservation of soil nutrients, building up population of soil organisms, improving soil organic matter content, improvement of soil structure, controlling weed growth, improving germination of seed and establishment of young plants*. All these effects can be used in a farm to derive positive effect on the yield and quality of the crop. Hence, mulching is an important operation in farming, especially under moisture stress situation.

5.5.2. What is Mulch ?

Mulch is the material used for mulching. From soil to paper, husk to plastic, saw dust to straw, any of these material can be used as mulch. Mulching materials can broadly be divided into three groups: (1) organic mulch, (2) synthetic mulch, and (3) natural (other than organic) mulch. Crop /plant residues generally dominate in the lists of organic mulch. On the other hand, different plastic materials are used as plastic mulch. Dust, pebble / stone and loose soil are generally used as natural mulch Tillage that makes a layer of pulverised soil on the surface acts as a good natural mulch.

Use of mulch is dependent on (1) climate, (2) soil, (3) crop, (4) agronomical practices, (5) availability of mulching materials and finally the (6) purpose of mulching.

How Mulch Works

Mulching slows down the ***rate of evaporation*** from a bare wet soil considerably. It shields the soil from solar energy and thus the lower proportion of solar energy absorbed by the soil reducing the rate of evaporation. As long as the wet soil is exposed to air, the evaporated water is rapidly removed by the general turbulence of the air. In mulched soil the water vapour slowly diffuse through mulch, which keeps the air almost stationary, and the rate of diffusion limits the rate of evaporation. In saline soils this will reduce the rate at which salts accumulate in the soil surface due to evaporation of soil-water.

Mulching affects both the diurnal and seasonal fluctuation in ***soil temperature***. Its principal effect on diurnal temperature is to reduce the midday maximum temperature of a bare soil under hot and dry conditions, without having any large effect on minimum temperature. On seasonal effect, mulch increases the soil temperature during winter, and reduces the rate at which the temperature rises during spring or early summer.

The crops received 160 and 230mm of rainfall in each of two 3-months growing seasons. The mulch reduced the soil temperature at a depth of 10cm (4 inch) resulting in increased root length and weight.

Organic mulch increases the permeability of a ***poorly structured surface soil*** and helps to produce good stable structure. The primary effect of mulch on poorly structured soils may however be to increase the length of time of anaerobic condition in the surface soil. It reduces the rate at which it dries after rain or in the spring and also helps to keep the soil surface permeable.

Mulching can have an important ***effect on weeds***, as weed growth is slow through mulch mainly because of reduced sunlight. Black plastic or properly managed heavy organic mulch can reduce the weed growth effectively. However, transparent plastic mulch increases the growth of weed.

Mulching has certain general effects on ***decomposition of soil organic matter***. The moist condition under the mulch favour micro biological activity, thus, increase organic matter decomposition. It also favour improvement and stabilisation of soil structure through this activity is brought about.

Mulching enhances the **supply of plant nutrient** when nutrients present in organic mulch are returned to the soil as it decomposes. As organic mulching materials contain low nitrogen, due to it consisting of mature plant residues, its decomposition locks up ammonium and nitrate, so lowering the level of available nitrogen. Mulching also increases phosphate uptake by crops, partly because it contains phosphates but principally because it encourages the surface rooting of the crop and keep the surface soil moist for a longer time. The effect of mulch on the potassium supply is usually considerable because the organic residues used for mulch are high in potassium. The potassium uptake of crop is generally increased on mulching condition.

Mulching by organic material/residue, particularly in dryland areas, is an effective way to **replace cover crops**, which competes for moisture and nutrients with main crop in dry areas. The best result can be obtained by covering 70% to 75% of the cropped area (not possible in field crops).

5.5.3. Different Mulching Practices

A minimum ground cover of 30 percent with crop residue or similar material is necessary to keep soil loss within acceptable limit. Singer and Blackard (1978) in a study of the effects of oat-straw, oak-leaves and redwood-litter as mulches found that soil loss did not begin to decrease (from a plot $0.37m^2$ in size) until a 50% cover was provided. Afterwards it declined rapidly with increase of cover.

The details of effectiveness and usability of various mulches are described below:

5.5.3.1. Mulching With Organic Materials and Crop Residues

Mulching with crop/plant residue such as straw, crop stalks, husks, shells, dry leaves and other kinds of residue as well as standing stubbles of the previous crop can trap considerable rain or of previous crop irrigation water in the soil. This effectively reduces the velocity of runoff facilitating greater percolation. Thus, in dryland farming mulching with organic materials plays an important role. The benefits of mulching with organic materials are listed below:

1. Retains rain or irrigation water in the surface layer of soil.
2. Allows the water more time to percolate into the root-zone.
3. Maintains favourable soil temperature for plant growth.
4. Reduces evaporation loss of soil moisture.
5. Increases the microbial population and their activity in the soil.
6. Provides organic matter to the soil after decomposition.
7. Checks weed growth.
8. Improves efficiency of irrigation, particularly the drip method.
9. Reduces soil erosion.

In the long run organic mulch increases the permeability and helps to produce good structure of soil and stabilizes it by encouraging the activity of soil microbes, earthworms and other soil biota. The bio-soil-mulch can considerably increase water supply (Table 5.12) to the crops grown with it.

Table 5.12. Effect of Mulching with Banana Leaves in Coffee Plant on the Soil Moisture Percentage.

	After Prolong Drought			After a Rainy Period		
Depth of soil in cm.	0-15	15-30	30-45	0-15	15-30	30-45
(1) Mulched plot	27.5	35.0	37.5	14.0	39.0	38.0
(2) Clean weeded plot	15.5	22.0	26.0	30.0	33.0	32.5

Gillbertare (1945)

This type of mulching also helps plant growth by controlling the extreme soil temperature to a good extent. The average daily maximum and minimum temperature from June to September in the top 2.5 cm. of soil under maize plant was 31.2°C and 17.9°C when clean weeded, and 23.6°C and 19.7°C under a mulch of 20 tones / ha of straw respectively (Meealla and Dulay, 1946). The result shows that mulching moderated the soil temperature in favour of plant growth. Experiment on the effect of different types of mulch on soil temperature (Table 5.13) shows that transparent polythene and straw (wheat) mulch creates highest and lowest mean temperature respectively against the bare soil.

Table 5.13. Effect of Different Mulch on Soil Temperature (°C)

Depth (cm.)	Bare soil	Transparent Polythene	Black painted 2.5cm.gravel layer	Aluminum painted 2.5cm.gravel layer	Wheat straw 10t/ha
1	29.1	34.4	27.4	25.9	23.8
4	28.0	31.8	27.1	25.4	23.6
16	26.8	29.0	26.0	24.2	23.0

Source : Hanks et. al (1961)

Mulching reduces the first stage of drying of soil which helps to improve soil moisture status, reduce soil temperature besides checking seedling mortality and crop emergence (Table 5.14). In West Bengal of India pointed-gourd and off-season coriander (for leaf) have been found to germinate better under straw mulch. It also protected the emerging seedling from impact of raindrop. This will also protect the young green part from get in touch with wet soil, thus reduce mortality significantly.

Table 5.14. Effects of Mulching on Soil Moisture Status, Temperature and Cowpea Seedling Stand

Treatment	Mean Moisture Content (%)	Mean Max. Temp. (°C)	Post Emergence Seedling Mortality (%)
Year	**1982 – 1983**	**1982 – 1983**	**1982 –1983**
No – Mulch	6.5 – 5.2	41.3 – 42.7	80 – 80
Mulch	7.8 – 6.3	37.2 – 38.0	45 – 22

The mulch maintains a higher vapour pressure near soil surface by restricting air movement. It also shields the surface soil from detrimental effect of solar energy.

To achieve the 70% to 75% cover of the soil surface with straw (Fig. 5.23) an application rate of 0.5kg/m^2 is sufficient. The less coverage reduces its effectiveness, whilst a greater covering suppresses the plant growth particularly in initial stages. Denser mulches are sometime used under young tree crops to check the weed growth. Once these plants have attained maturity they can successfully control the weeds below by cutting the sunlight.

Mulching with organic residue can reduce soil loss effectively. On a silt-loam soil with 7^0 slope, annual soil loss is about 2.46 kg/m^2 from uncultivated bare land, but the loss will be only 0.11 kg/m^2 in land covered with straw mulch applied at 0.5 kg/m^2 (*Borst* and *Woodburn*, 1942).

The experimental evidence shows that, of the different rates of mulch application (3,6,9, and 12 T/ha) tested, the most effective application rate was 9 T/ha which is equal to 0.9 kg/m^2 or 0.2 lb/ft^2.

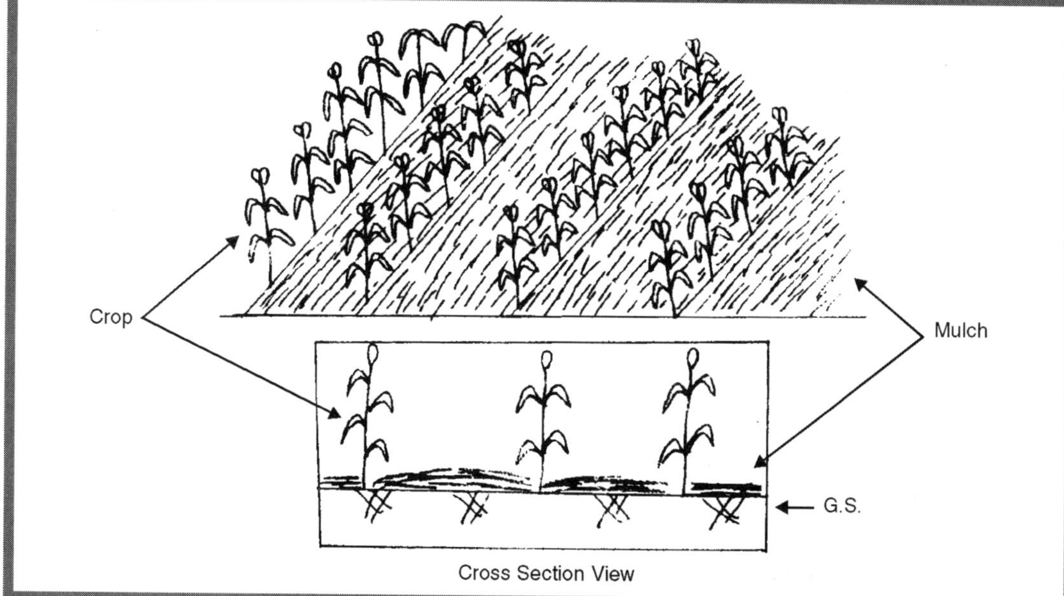

Fig. 5.23. Mulching With Organic Residue.

Limitations. However, organic mulching has some limitations particularly in humid and sub-humid areas, as stated below:

1. It may attract pests and diseases specially termites.
2. It is not suitable in wet areas (induce gleying to crops and anaerobic condition to soil).
3. It is difficult to practice in steep lands.
4. It may create weed problem if the plants used for mulch regenerate through seeds or sets.
5. It poses extra problems in tillage, other cultural operations and sometimes in germination process. But, when combined with conservation tillage; many of these problems can be overcome.

However, organic mulching has high potentiality in dryland areas for successful farming.

5.5.3.2. Synthetic Mulching

There are two types of synthetic mulching materials; Polythene sheet and Bitumen in water emulsions are coming into use. Polythene sheet is spread on the soil surface between rows of vegetable crops, pineapples, melons and other high valued crops. Different kind of polythene sheet in respect of colour (white, black), thickness etc is generally used for mulching purposes. The selection of material is dependent on requirement of climate and crop.

Bitumen emulsion is sprayed over the rows after emarge of seeding. Sometime polythene sheet, punctured at recommended spacing for a particular crop, is spread on the prepared field. The seeds or seedlings are planted through the openings (Fig. 5.24) However, In dry season the openings may be placed in furrow and in wet season it may be placed on the ridge.

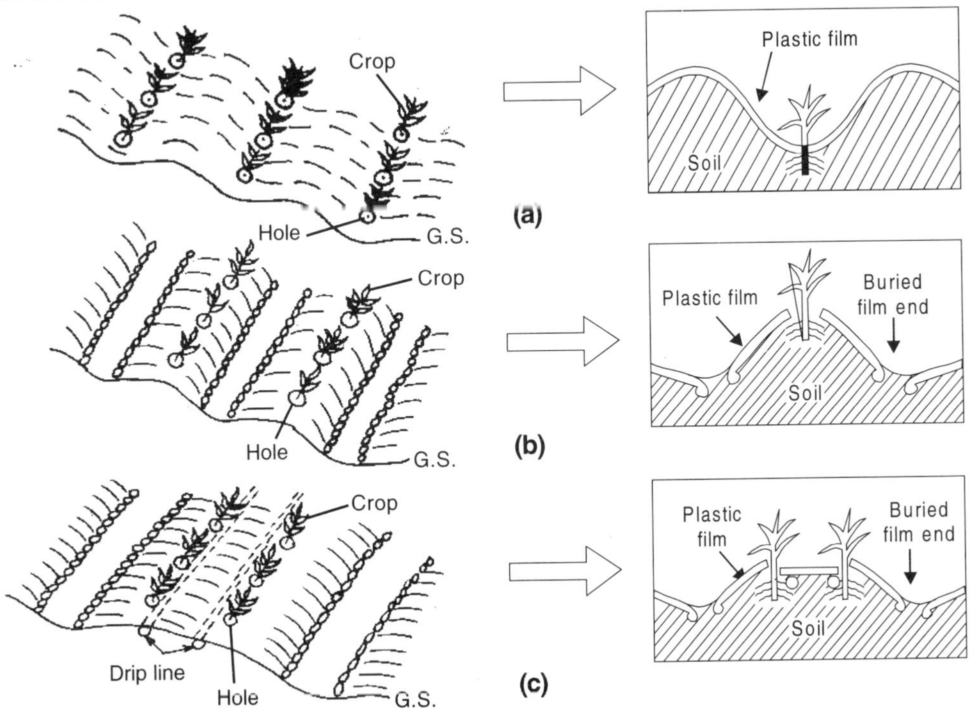

Fig. 5.24. Plastic mulching : (a) for rainfed & dewfed, (b) for furrow irrigation, (c) for drip irrigation.

Being expensive polythene mulching is used in high value crops including fruit trees. Apart from being effective in soil and water conservation, this method helps early harvest of summer vegetables thus catching high market price. In some cases, increase in yield has also been reported.

Besides, polythene and bitumen emulsion mulching raises the soil temperature which has a positive effect in cooler climate. This may also increase the nitrate content of the soil. The heating effect of transparent polythene is greater than bitumen film. High temperature under polythene kills the germinated weeds and soil born pathogen and egg, larvae of insect pests, especially during summer. This is known as soil 'solarification'.

In case of tree crops polythene sheet may either be spread over the soil surface or it may be used round the lower portion of the trunk covering the root zone area. It looks like a conical shaped cover where sufficient opening is provided between trunk and polythene for air passage to avoid the problem of stem girdler. Polythene mulching is very effective when used with drip irrigation system.

Plastic Mulch Films

Plastic mulch is generally 1.1 to 1.5 millimeters thick, 4 to 6 feet wide, available in rolls of 1,000 to 4,000 feet long. It is available in a multitude of colors ranging from clear (transparent) to opaque (black or brown). Recently, colored mulches have been investigated for their influences on insect control and plant yields.

Plastic mulch functions to warm the soil, conserve moisture, and prevent nutrient leaching. Clear plastic has the highest soil warming capability (8° to 14° F over bare soil), but weed growth underneath can be extreme. An herbicide is necessary to keep weeds under control with clear mulch. Black mulch will prevent weed growth by prohibiting light transmittance to the soil and will warm the soil 3° to 5° F higher than the bare ground.

In small farms or in small fields, black, brown, or wavelength selective mulches are often the preferred way to eliminate the use of herbicides. Crops that respond best to mulching are those that respond to higher soil temperatures (muskmelon, watermelon, cucumber, squash, tomato, pepper and sweet corn).

The use of black plastic for mulching vegetables and flowers has become quite popular.

Advantages of Using Black Plastic Mulch

1. Reduces soil moisture loss by evaporation.
2. Prevents weeds form growing due to lack of light.
3. Warms the soil (warm season crops benefit from the warmer soil).
4. Hastens maturity of crops.
5. Keeps the fruit cleaner (especially tomatoes).
6. Prevents compaction of the soil.
7. Prevents the soil surface from drying out and cracking.
8. Reduces the amount of hoeing and cultivating.
9. Reduces the amount of root injury due to cultivation.
10. Facilitates the development of roots in the top few inches of the soil where moisture, oxygen, temperature and nutrients are most favorable for early growth.
11. Reduces the amount of hand spading. In small gardens, hills can be worked up and then covered with plastic mulch without working all of the soil between the hills. This is especially helpful when you are trying to convert a lawn into a garden.

Disadvantages of using Black Plastic Mulch

1. The increased growth of plants grown on black plastic results in greater loss of water by the plant through transpiration. Therefore, on lighter, well-drained soils during dry seasons, it will be more necessary to irrigate mulched crops than crops that aren't mulched.
2. On heavier, wetter soils with high water tables, black plastic mulch does not always produce favorable results because the moisture content of the soil under the plastic is maintained at too high a level for adequate aeration.

Laying of Plastic Mulch

Apply plastic mulch after fields, i.e. beds have been leveled, smoothed, and fertilizer has been applied, and when there is good soil moisture. Initially the soil should be fertilized, using 4 pounds of 5-20-20 fertilizer per 100 square feet (for vegetables) or as recommended by the results of a soil test. At least two-thirds of the fertilizer should be broadcast before ploughing or spading and the rest should be raked in lightly before applying the plastic. When the amount of moisture is at an optimum level, dig a trench with a hoe and bury about 3 inches of the plastic in it and cover it up with soil to anchor it down. Then figure out where the other side of the plastic will be, dig a trench and anchor it down. Do the same for both ends and pack a soil down again after anchoring the plastic. Wind will pull up an entire sheet of plastic if it isn't anchored properly.

In the case of black mulch, good uniform soil contact is essential as the soil is warmed by heat conduction. Commercially, the simplest way to apply mulch film is with a *mechanical mulch layer*. Hand application is an option, but applying more than a half-acre can be difficult and time consuming.

Planting through the Plastic

Tear or cut holes in the plastic about 6 inches across wherever you want the plants or hills to go (probably 3 to 5 feet apart for tomatoes and vine crops). For transplants, dig a pit through the hole, place the transplant in it and cover with soil and then use a starter solution high in phosphorus to water the plants in. For seeds such as squash, tear a hole in the plastic and then plant the seeds in a hill (3 to 6 seeds per hill). Be sure to anchor the plastic down around the hill or the wind may catch it. Clods, stones or loose soil will hold the plastic down. Large plants should be planted in the center of the plastic mulch at recommended spacing. Smaller plants, such as peppers and eggplant, could be planted in two staggered rows on the plastic. Some individuals use tin cans with sharpened edges to punch through the plastic. Leave the mulch down until after the crop has been harvested. Then the mulch can be taken up and burned or saved for another year.

Watering of Plants

Drip irrigation system is the best option to irrigate crops covered by plastic mulch. However, plants can be watered with a hose by placing the hose under the plastic through the holes which were torn or punched in it. Common garden sprinkling devices could also be used. After a rain, you may want to check the plastic to see where the water stands. Then punch a hole through the plastic with your finger or stick wherever you find water standing. This will allow the water to run through the plastic during future rains.

Biodegradable Plastic Mulch

Researchers have been looking for degradable plastic mulch alternatives for decades. Some of the first commercialized degradable mulches were photodegradable, and would break down when exposed to light. Many growers that used these products reported uneven and incomplete breakdown, particularly after tillage buried the plastic fragments at the end of the season. However, biodegradable mulches made from plant starches that are broken down slowly by soil microbes into carbon dioxide and water is now available.

Woven black plastic mulch

It may be a good alternative of Plastic-film mulch in dryland areas. Increase of temperature in summer, under plastic-film mulch, is one of the main problems and to solve this problem application of woven plastic mulch may be recommended. Black/green woven plastic mulch (75% agro shade-net can be used) effectively reduces the soil erosion and reasonably offers all the other advantages as the film-mulch provided, except the increase of soil temperature.

In 1997, a five-year project was initiated to evaluate the effects of woven, black polypropylene mulch on plant mortality, growth and production of select vegetables under limited irrigation.

5.5.3.3. Pebble Mulch

In stony areas a uniform layer of pebble on the soil surface in between two rows will prevent transfer of heat from the surrounding air to the soil, reducing evaporation losses. It also helps to check the run-off water effectively. This method may be used in dryland areas where soil is gravely. It is also favourable for trees in the initial stage of growth.

In the process of cooling during the night, the moisture that evaporates from the soil condenses on the lower surface of the stones instead of being lost to air. As this type of mulch creates problem to cultivation and the fact that it reflects sunlight and holds and radiates heat, we suggest that this method of mulching can only be used for perennial crops, especially trees. Relatively low-lying and tender plants/crops can suffer from the heat absorbed and radiated by the pebbles. In case of annual crops, pebbles and sand can be used as mulch.

In arid north-west China, a 5-16cm thick gravel, pebble and sand mulch is being used on melon field and other annual crops for past 300 years.

5.5.3.4. Tillage Mulch

It is the tillage operations with implements (sweeps, blades or chisel plough) that undercut the soil surface retaining most of the crop residues in the surface. This is basically a type of soil mulch, which combines with organic mulch in the form of stubble mulch. It breaks weed, insect and disease cycles but its main function is to break or disturb the continuous capillaries through which the evaporation loss of soil moisture from root zone takes place. Application of loose soil also does the same thing by way of capping of the capillary channels. When the soil surface gradually dries up the underground water rises to the surface through the capillary pores and gets lost through evaporation. This moisture loss continues till the ground water as well as soil moisture gets completely exhausted. Mulch tillage disturbs this capillary movement and thus prevents the loss of underground water.

5.5.3.5. Live Vegetative (Cover Crop) Mulch

In the light of the basic principal of mulching, creeping or climbing plants that spread along the ground and can cover a good amount of area may be used as mulch instead of crop/plant residue. It is apparently worked as cover crop but the difference is that it does not use the entire land it covers for its own need. Crop used as live vegetative mulch should not use more than 20 percent of mulched land for its own nutrient and moisture supply. This mulch can do almost all sorts of works better than that of the other organic mulch.

The Supplementary Advantages of Live Vegetative Mulch are as Follows :

1. It efficiently protects the soil from the impact of raindrop.
2. Its canopy can retain a good amount of rainwater and provide more time for infiltration.
3. It can control the temperature and humidity of soil and just above the soil in favour of the soil microbes. At the same time it reduces the evaporation loss from soil surface.
4. It also adds fresh organic matter to soil by way of adding dry/dead leaves etc.
5. Interestingly, it may provide moderate to good amount of economic return, which has not happened in case of other mulching materials.
6. It is most suitable for tree crops and orchards.

The Additional Problems Associated With Live Vegetative Mulch Are :

1. Feeding, watering, and other crop management aspects are required to be done properly for better mulching effect, and to avoid compitition with main crop.
2. It increases the transpiration loss of the field, hence increase the surrounding humidity.
3. This mulch acquires and uses a certain portion (less than 20 percent) of the mulched area to spread its root system.
4. Availability of suitable plant is dependent on environment and locality and an early planning is essential.
5. It is not feasible for small area, particularly for elongated small strip like area.

However for tropical and sub-tropical areas, different *cucurbitaceae spp*. having long soft stem and branches like pumpkin, cucumber may be used as live vegetative mulch. These mulches may be planted in cluster or in line as per requirement of the area to be mulched with proper prunning and training.

5.5.3.6. Vertical Mulch

Vertical mulches provide a water pathway to the root zone, and can reduce the area of soil surface that is wet. In this system narrow trenches in farm beds (fig. 5.25) are made and filled with dry stalks of millet, maize, sunflower, amaranth etc. The air space created by these stalks conduct the water quickly down to plant root-zone and reduces its possibility of evaporation considerably. It simultaneously works as sub-surface irrigation system and, on decomposition, supply organic matter to soil.

A laboratory experiment, with a common furrow and a 7cm (3 in) wide and 15cm (6in) deep trench packed with barley straw applied with 5.5cm (2in) of water, shows that after 17 days 42% and 72% of the water remained in the soil in case of furrow and vertical mulch respectively. When the applied water overflowed the vertical mulch and wetted the surroundings, 60% of it was retained after 17 days of evaporation. The subsequent experiment with sorghum crop showed that a similar mulch increased yields 20-40% over control plots with no mulch.

Fig. 5.25. Vertical Mulching

Another experiment in gently sloping Vertisols in semiarid India shows that 15cm (6in) wide and 30cm (1ft) deep vertical mulch with sorghum stubble, placed 2m (6.5ft) apart, gave 25 times more grain yield and 2 times more straw yield of an 'improved variety' of sorghum in a very dry year. The beneficial effects appeared to last 4 years, probably the result of addition of organic matter as the straw partially decomposed after the 2nd year. The only problem with the vertical mulch is the infestation of pest and disease, especially the termites; hence measure against these threats should be taken at the time of implementation of this method.

5.6 ANIMAL & FISHERY MANAGEMENT IN DRYLAND FARMS

5.6.1. Introduction

In dry regions of the tropics where agriculture is always difficult, animals and fish are frequently more capable of supplementing the farming activity utilizing the often abundant plants and crop residues as fodder. Besides some easy growing feed crops, the most important feeds in many dry regions will be those which grow by themselves, naturally and untended. Sometimes improvement of this natural fodder by fertilizing, watering, or selected weeding may be the best solution to increase the yield of animal feed. In addition, appropriate care of animals against hot & dry climate is necessary.

Animals in the dryland farm can be used for a wide variety of purposes. In addition to excellent food in the form of eggs, milk and meat, animals serve as beasts of burden, and can be trained to handle difficult jobs on the farm. The dung is a useful resource for crop production but is also used in plastering walls and floors, and when dry, as fuel. Animal wool, hair, or fur can be used in bedding and clothing.

Basic Problem of Development. Very few scientific information or findings, in ready to use form, are available about the domestic animals suitable for dryland areas. It becomes more obvious when we compare the scientific developments of different domesticated animals of temperate and tropical areas. Neither in breed nor in feed or any other husbandry aspect, in-situ basic technology is available for dryland farming, barring a few established acclimatized imported technologies, which are used sporadically in these areas obviously associated with a number of problems. However, in case of dairy (cow), poultry, piggery, etc some information is available particularly in respect of acclimatization of imported breeds or development of local (in case of cow only) breeds. However very poor and insignificant information, in respect of buffalos, goats, and ducks etc. is available.

5.6.1.1. Effect of Heat on Animals

Dry and hot climate affect the development of livestock in two ways. Firstly direct influence on the animal's growth and its economic utilisation and secondly by indirect influence on the environment of the animals.

Direct Effect. Experimental evidence shows that there are profound differences between species (Findlay, 1954), breeds or types within species (Worstell and Bordy, 1953), and individuals within the breed (Payne and Hancock, 1957) in their ability to withstand direct effect of climate.

All domestic livestocks are homotherms that is they attempt to maintain their body temperatures within a range most suitable for optimum biological activity. The normal range of body temperature of mammals is 37 to 39 degree centigrade(C), while in birds it is 40° to 44°C. However, typical breed body temperatures of some domestic animals are shown below.

Horse – 37.2 to 38.2	Asses – 36 to 38
Cattle – 38 to 39.3	Camels – 36 to 38
Sheep – 38.3 to 39.3	Goats – 38.7 to 40
Pigs – 38.9 to 39.4	Chickens – 41.9
Ducks – 42.1	Turkeys – 41.2

In order to maintain the body temperature while subjected to a wide range of environmental changes, domestic livestock must preserve a heat balance between their heat production or gain and their heat loss to the environment. In general the means by which domestic animals can vary their heat production are much limited in comparison with the methods by which they can dispose of heat.

The thermal balance of an animal can be expressed by the equation:

$$M - E \pm F \pm Cd \pm Cv \pm R = 0$$

where 'M' is the metabolic heat production, 'E' is the heat loss from skin and nose, 'F' is the heat loss/gain for bringing ingested food & water to body temperature, 'Cd' is heat loss/gain by direct contacts between body and surrounding surfaces, 'Cv' is heat loss/gain by convection in between air and skin including linings of respiratory passages, and R is heat loss/gain due to radiation.

The metabolic heat production depends on: (1) Basal heat production for maintaining essential body processes, (2) Digestive heat production that varies with quality and quantity of feedstuff and animal's digestive system, (3) Muscular heat production that varies with movement of the animal, (4) Additional metabolic heat production due to any reproductive/productive process like production of milk, egg etc and physical growth.

In heat production processes, the productive and muscular heat production can be reduced to some extent and digestive heat production to a more limited extent, but they cannot normally reduce basal heat production, as minimal body processes must be maintained. Evaporation is potentially the most important process of heat loss available to domestic livestock. Under normal circumstances, this evaporation loss depends on (1) ambient air temperature, (2) amount of available moisture, (3) area of the evaporating body surface, (4) absolute humidity of air surrounding the animal and (5) the degree of air movement.

Basic Principles to Combat Against Direct Heat. In dryland areas, to maintain the heat balance one have to follow two basic principles.

1. *Reduction of metabolic heat* production when ambient temperature is high. It may be done (a) by grazing or feeding at early night or very early morning and (b) by reduction of physical movement during daytime particularly at summer.
2. *Reduction of heat load* by (a) providing a shady place, (b) arrangement for cool breeze blowing or maximal air movement on and around the animal, (c) sprinkling of water on the body of the animal and obviously by (d) supplying of drinking water particularly in day time when ambient temperature is very high. Lowering of ingested water is known to have more marked effect in heat load than increasing the volume of ingested water.

Indirect Effect. The major indirect effect of hot and dry climate on livestock is on the quantity and quality of the feed available for them. The other indirect effects are on the incidence of disease and parasites and on the storage and handling of animal products.

5.6.1.2. Effect of Hot & Dry Climate on Animal Behaviour

Animal behaviour always changes or is adjusted as per the climate. Naturally in hot and dry climate the animal behaviour particularly in respect of feeding and drinking are different.

1. **Effect on Feeding.** It is suggested that high temperature depresses the food intake of cattle. The effect of high temperature (above 40°C) has very pronounced effect on rumination, particularly in humpless (B.Taurus type) cattle. The high ambient temperature also affected the food intake of other domestic animals according to their adaptability in that situation. High humidity at ambient temperature above 23.9°C also depresses food intake of cattle (Ragsdale et.al.

1953), while increased radiation have some effect on humpless cattle but not on humped (B.Indicus) type cattle (Bordy et.al.). However, unfortunately very few scientific findings about other domestic animals fit for hot & dry climate are available.

2. **Effect on Drinking.** Water is required by livestock at least for two different purposes, firstly, as an essential nutrient and component of body, and secondly, to assist the animal for loss of heat through conductive or evaporative cooling. Ambient temperature has a different effect on the water intake of different types of livestock and also on different breeds, and it appears that acclimatized animals require less water than un-acclimatized ones when managed at high ambient temperature. According to Bordy et.al., increased radiation intensity on cattle increased water consumption.

3. **Effect on growth.** The growth of livestock depends on feeding and drinking. As the drinking and feeding is dependent on ambient temperature and humidity, naturally they influence the growth as well. So proper management of drinking and feeding practices can reduce the adverse effect of hot and dry climate on growth of domesticated livestock. The said management can reduce the adverse effect to a great extent in case of cattle.

   Experimental evidences on effect of climate on milk, butterfat and solid-non-fat indicate that production of all these three items, individually and collectively, is depressed by high ambient temperature. However, independent of management and feeding, it was possible to asses (in Fiji, NewZeland) that the climate had a marked effect on production of milk and butterfat but not to that extent on solids-non-fat.

   Information about growth of goats and sheep in hot & dry areas are scanty. However, the birth weights of kids in dry season is less than the winter season and have a high post-natal mortality due to poor adaptation of high temperature. Higher temperature probably reduce the rate of growth of poultry; lighter breeds can withstand better than heavy breeds.

4. **Effect on Management.** The hot and dry climate affects the animal management in respect of supply of feed, parasite & disease attack and post harvest/production management. In dryland areas due to low rainfall in a specific period followed by a long dry period, the production of high nutritious forage throughout the year is impossible. Naturally, the supply of nutritious forage to the livestock is very much limited and the possibility of low nutrient uptake through hay and water deprivation during the dry season create a major problem to the growth of livestock in dry areas. In dry areas the problems of parasites and diseases are not serious in comparison to the humid region. Still in rainy season, the breeding and infestation of parasites and occurrence of diseases increases.

   In respect of storage and handling of animal products, the dry & hot climate favours rapid deterioration of animal products like milk, egg, meat etc. Naturally it increases the storage, processing, and handling costs and reduces the economic return of the farm, indirectly affecting the animal production or developments in dry areas.

5.6.1.3. Acclimatization

It is a complex process by which an animal adapts itself to the environment in which it has to live. Acclimatization to heat stress may be temporary or permanent and depends upon the animal either by (1) increasing its heat loss or (2) reducing its heat production or (3) tolerance of its tissue to more fluctuating and higher body temperature. These may also act in combinations to obtain best possible results.

Permanent acclimatization to climatic changes in respect of behaviour of the animals or in physiological reactions may or may not be inherited. Natural or artificial selection for morphological characteristics, that assist the animal to acclimatization may also take place. Physiological adaptation may be achieved by changes in hormonal activity. Animals may be selected for certain morphological characteristics that possibly assist acclimatization, such as large skin area in relation to body size, short and light coloured coat, pigmented skin etc.

5.6.1.4. Improvisation in Few Husbandry Aspects

Housing. In dryland areas the growth and production capacity of animals may be increased significantly by way of manipulating the shelter for reduction of the impact of high temperature. The basic designing and site plan of a shelter should be modified in such a way that it should reduce the temperature in and around the animal's house than ambient temperature. Due to poor economic condition of this area any capital-intensive infrastructure oriented technology (Air cooler etc.) is always not recommendable. But it can be done effectively in the following ways.

Firstly, the shelter surroundings should be sufficiently covered by dense green vegetation. The height of the canopy should be more than twice of the roof height. It will create a microclimate in and around the shed, which will reduce the heat and increase the humidity in dry season. These canopy prevented hot wind to enter directly into the shed. In general the dense canopy always reduce the ambient temperature sufficiently under it particularly in hot and dry areas.

Secondly, the eastern and western sides of the shelter should be covered up to one meter height and the roof and walls should be white outside and coloured inside. The application of white colour in outside wall reduces the surface temperature of inside by 1-2° C when compared to unpainted walls in places where temperature remains above 37°C.

Thirdly, sprinkling of water inside the house at the hottest period of the day should be done after covering all the openings by temporary thick semi-porous or porous structures/curtains.

Feeding. The major differences in the feeding of the animals in tropics is that, heat produced within the body generally has to be dissipated, whereas in temperate zone it can often be conserved in order to maintain normal body temperature. Thus it might be expected that energy requirement for maintenance would be lower in tropics (Kehr I954).A significance difference is not apparent in practice, but the workers of Indian Dairy Research Institute have suggested that animals need less energy during the hottest period of the year. It has also been suggested that the protein requirements of animals may be lower in hot climates, but it is likely that mineral and vitamin requirements will be the same at similar level of production.

To maintain thermal balance in favour of proper development of the animal, night-feeding in hot summer periods is recommended. In case of grazing animal, night-grazing is more important during hot dry season when poor quality and decreased water content of fodder force the animals to move further and further. The experiment in Serere in Uganda, by Joblin(1960) shows that facility of night-grazing is critically important for B. indicus cattle during the periods of very dry condition with shortage of forage. In that environment restriction to night-grazing led to a significant decline of live weight gain and it was to the extent of 30 percent. This experimental finding has since been verified by other workers. Field data reflects more accuracy than climate chamber data on assessing the overall effect of climatic environment on feed intake of livestock.

Types of Feed. Out of the four types of feed group i.e. *succulents, roughages, concentrates* and other *locally available feeds,* the roughages, particularly different straws are the major source of feed in dryland areas. As the roughages possess low food value the quality of feed in dryland areas should be enhanced by cultivation of succulents, as fodder. The roughages should be mixed with succulents and concentrates in required proportion to enrich its food value as per requirement. The materials like different oil cakes, broken food grains or grain parts etc. are generally used as concentrate feed. In market ready-made concentrate feeds are available for different animals for different growth stages, which supply necessary important nutrients like protein, vitamin, minerals etc.

Apart from the common feeds, few locally available feeding materials may be used suitably to enrich the feeding staff in dryland areas. In dry tracts many wild and locally adaptable trees and bushes, particularly of leguminaceae family, can add high nutrition to the roughages available there. They are of major importance in dry areas, as many of them remain green in the dry season long after the ground vegetation has dried up. The green parts of these plants generally possesses high nutrient, which can enrich the low value roughages up to a desired level. The number of trees and bushes specially planted for forage production is very limited, however, this kind of created vegetation may increase the quantity and quality of feed for live-stocks in dry season apart from providing cover to the bare soil.

Among the locally available feeds, different usable by-products of industry as well as agriculture can be judiciously used as supplementary feed for livestock. The by-products like molasses, distillery (also country types) wastes, agricultural wastes like skin, shell, any unused green parts etc. may be used as feed.

Water Requirements. Water is the more vital material for maintenance of animal life than any other feed component. The water content of the animal's body varies with age. Newborn animals consist of 70% to 80% water. The average animal, probably contain 55% to 65% water on maturity. Water requirement increases with growth (Table 5.15), with an increase in productive processes such as lactation, laying, fattening. It also increases with physical exercise and when the animal is subjected to heat stress. The water requirements vary between different species, between breeds or varieties within species and between individuals within breeds.

In addition to maintain the body, water plays an all-important role in the animal's thermo regulatory mechanism. Evaporation of water from lung and skin surface helps the animal to dispose of unwanted heat and the specific heat of the body. Water assists the animal to accommodate itself to large changes in heat production with little change in body temperature.

Animals take water in three ways, mainly by drinking free water. They also utilize the water that forms part of the food they consumed (green fodder contains 80% to 90% of water), and occasionally by manufacturing of small quantities of metabolic water obtained by the oxidation of fat within their own body. Some wild animals, particularly those whose habitat is the desert or desert fringes, appear to survive without access to free water for prolonged periods presumably due to the said ways. However different species of domestic animals differ in their respective abilities to withstand dehydration and their demand for free water. Sheep and goats are very tolerant to dehydration and partial dehydration respectively. Buffalos and pigs are relatively tolerant of water deprivation. Cattle in general are not as tolerant as camel or sheep but *Bos. Indicus* types of breeds are more tolerant to dehydration than other cattle breeds.

Table 5.15. Average water requirements of Livestock in the semi-arid Tropics during dry season.

Live Stock	Water Requirement/day (Litre)	Frequency of Drinking
Camel	60 – 80	Every 4 or 5 days or longer.
Cattle	30 – 40	Every 1 to 3 days.
Sheep	4 – 5	Every 1 to 2 days.
Goats	4 – 5	Preferably once a day.

The details of feeds suitable for different animals are described in detail in books, leaflets and in local publications that the dryland farmers are advised to study.

5.6.2. Management of Livestock

5.6.2.1. Cattle

Out of the two basic cattle breeds, humpless or *Bos. taurus* and humped or *Bos. Indicus*, the second one is more suitable for dry areas. The cross breeds between these two, either in stabilized form like Damascus, Siri or in unstabilized form like red sindhi + Jersey etc., are also suitable in dry areas. If the management and feeding is properly done in the second type of crossbreeds, they can be used commercially in dry areas.

The cattle are used in human society for three purposes: for milk production, for meat production and for draught purpose. Out of these three purposes, due to heavy mechanization in recent years, the requirement of cattle for draught or any other physical work is drastically reduced. In recent trends of consuming less red meat, the growth rate of beef production is facing a downtrend. But the requirement of milk is still increasing and will be increased with the increasing population. Apart from these three aspects a fourth aspect is gradually gaining very much importance i.e. the supply of organic manure.

Dryland areas, in particular, is very much dependent on the organic matter supply in respect of crop production. Among the different livestock, cattle are the biggest supplier of organic matter to boost up the commercial and sustainable agricultural production, particularly in dryland areas.

Now, the following table may provide breed-wise (Tropical cattle breeds) information for necessary knowledge.

Table 5.16. Breed-Wise Different Information of Cattle

Breed	Age of first calving (months)	Calving interval (months)	Milk production Per lactation (kg) Normal range	Maxi-mum	Length of lactation (days)	Fat (%)
Asian cattle						
Damascus	—	—	1,500 - 3,000	5,000	190 300	4.0-5.0
Hariana	32-72	19 21	635 - 1,497	4,536	260 320	4.0-4.8
Ongole	26-511	16 18	179 - 16,33	3,266	300 330	5.1
Gir	31-51	14 16	1,225 - 2,268	3,175	240 380	4.5-4.6
Red Sindhi	30-43	13 18	680 - 2,268	5,443	270 490	4.0-5.0
Sahiwal	30-43	13 18	1,134 - 3,175	4,536	290 490	4.0-6.0
Tharparkar	24-47	14 18	680 - 2,268	4,763	280 440	4.2-4.7
Hallikar	39-69	14	2.27 - 1,134		180	—
Thai	—	—	260 - 580			—
African Cattle						
N dama	27-72	14 42	150 - 270	450	150 300	6.5-7.0
Dwarf Shorthorn	30-48	12 24	120 - 360		120 180	—
Sokoto	36	15	454 - 1,361	1,940	230 283	5.8
White Fulani	36-48	12 15	635 - 1,225	2,302	190 360	5.0-7.5
Sudanese	24-54	12 24	454 - 2,723	4,659	168 339	4.7-5.5
Boran	36-52	11 14	454 - 1,814	2,641	139 303	4.1-6.8
Small East African Zebu						
(Bukedi)	25-61	11 14	227 - 998	1,941	223 280	4.7-7.1
Ankole	42-60	16 24	318 - 817	898	212 239	3.0-7.0
Angoni	—	—	630 - 800		245 270	5.7
Mashona	—	—	218 - 499		180 270	—
American cattle						
Jamaican Hope	27-33	12 14	454 - 1.814	9.072	250 305	5.0-5.2
Australian cattle						
Australian Milking Zebu	—	—	1445 - 2647	4848	305	4.8 4.9

Few basic points of cattle management for dry land farming.

1. Proper selection of breed;
2. Heifer rearing is the best way to start with;

3. Breeding should be done scientifically and for crossbreeding, the percentage of pure exotic bloodline is to be maintained within a range of 50% to 75%.
4. Daytime rest in a relatively cooler place is necessary in summer.
5. Feeding should be practiced in between evening and morning.
6. Creation of microclimate with the canopy of any tree species surrounding the resting-place is must.
7. Production/ supply of nutritive forage should be assured.
8. Special attention on husbandry of calves, born in the summer, should be given.
9. Sufficient supply of drinking water in summer should be assured.
10. The maximum number of cattle may not be exceeding 15 in a dryland farm.
11. Proper storage and quick transport facility of milk should be arranged.
12. Scientific facility of composting of cow dung and other waste should be with the farm.

5.6.2.2. Buffalo

As a dairy animal the buffalo is found at its best in India, Pakistan, Egypt and Iraq. In unorganized poor village, the milk yield is 2 to 3kg/day, but in well-managed farm with good quality buffalo the yield is as high as 20kg/day or more. The four main milking breeds of buffalo in India and Pakistan are *Murrah, Nili, Surti* and *Jaffrabadi*.

These animals are long lived, can produce calves with low mortality, and also can work up to 18 years of age with full potentiality, if managed properly. They are very much docile and easy to handle for milking or work.

Heat toleration: Though they are profusely found in the tropical and sub-tropical regions, yet it has poor heat resistance and will suffer distress in very high temperature in comparison to the cattle. The milking buffaloes must have liberal drinking water to avoid dehydration at any time. Actually, the heat regulating mechanism of buffalo is less efficient in comparison to that of cattle. The difference may be partly due to heavy coat of the cow, preventing wetting of the skin, but more probably to the fact that the thyroid-adrenal mechanism is less efficient in buffalo. In hot and dry period they frequently require 'wallowing', i.e. rolling or floundering in mud or water.

5.6.2.3. Poultry

Fowls like other domestic livestock are homeotherms. The popular egg production breed introduced and accepted in tropical areas is *white leghorn*. They appear to be more heat tolerant than other commercial breeds. In humid tropical conditions they are not particularly hardy and susceptible to diseases. As per findings of different scientists, both the production and size of egg are depressed when the ambient temperature increases above 30°C.

For meat production as well as egg production i.e. for dual purposes, apart from white leghorn, Rhode Island Red (RIR) and New-Hampshire Red are popularly used in tropical areas. For broilers the experimental data suggested that, those reared at ambient temperature within the range of 18.3°C to 35°C, the size will be somewhat less in lower temperature but the efficiency of feed conversion is high. It is also suggested that the dry area may be superior environment for broiler production than the humid areas.

Out of different commercial breeds White leghorns are less affected by high temperature with low relative humidity, whereas at high relative humidity the Brown Leghorns are less affected by high temperature.

In case of layer, the first effect of high temperature (up to 42°C) is to initiate a partial moult and radically to reduce egg laying, but with continuous exposure the laying is improved, thus demonstrating a considerable degree of acclimatization. The problems of acclimatization have been fully discussed by Hutchinson (1954) and it would appear that the pure exotic breeds that are used in the tropical climate would in general acclimatize quite well, but they are affected by very high ambient temperature.

However, beside these entire technicalities, one simple and obvious by-product of any kind of poultry, i.e. its litter or manure, has a very important role in dryland farming.

Apart from these exotic breeds, no scientific information about commercial cultivation of country type or local breeds is available.

5.6.2.4. Goats

The goat is an important animal in dryland area on account of its unique ability to adopt and maintain itself in harsh environments. Using height at withers as criterion, the goats are multipurpose animal, producing meat, milk, skins and hair. Out of these purposes meat production is the most important in country like India. The meats are commonly known as 'Redchevon' and are usually low in fat. It contributes about 35% to the total meat production (except poultry) in India through sacrifices of 33.94 million goats annually. The milk of goat, contributes 2.9% of total milk production. Goat's milk does not differs appreciably from that of cow's milk and actually superior in respect of nutritional values (Table 5.17). Besides meat and milk the manure they produce is always a very important product for dryland agriculture.

Table 5.17. Comparison of Different Composition of milk between Goat and Other Milking Animals.

Livestock	Fat	Protein	Lactose	Ash	SNF	Total Solid
Indian Cows	4.8	2.8	4.6	0.73	8.1	13.5
European Cows	3.7	3.4	4.8	0.73	8.9	12.7
Murrah Buffaloes	6.8	3.9	5.7*	–	9.6	16.4
Goats	4.9	4.3	4.1	0.89	9.3	14.2

*The figure included percentage of Ash.

Breeds: Jamunapiri, Beetle, Jhakrana, Surti and Barbari are the main breeds of northwest India where the goats are mainly concentrated. The breed named Black Bengal, famous for its tasty meat and extraordinary skin named 'Glesskid', are predominantly found in entire eastern region. The breeds generally found in southern region are not so good.

Goats are generally adapted quite well than other livestock in dry and hot climates. Its water requirement is 4 to 5 liter once in a day. About feed requirement and other husbandry aspects, no scientific developments are there, which can help this activity to match with the others like poultry, dairy etc. in respect of commercial output. However, still it has a good potentiality in dry land farming.

5.6.2.5. Sheep

The sheep rearing is spreading in dry areas due to its meat and wool production capacity, particularly for fine wool, which is being produced in a very increasing scale. Apart from wool and meat, the manure is also a usable material for dryland farming.

Most breeds of tropical area are selected on the basis of their survivability in prolonged drought period and semi-starvation. However, the investigations of Demiruren et. al. (1971) suggested that they will respond quite well to improved nutritional levels, where sheep are subject to face a long dry season, the productivity can greatly be increased by supplementing their ration. Apart from gain in bodyweight through well nutritative feeding the wool-sheep will usually produce a heavier but coarser fleece. When the sheep are fed indoor the feed may consist entirely of succulent fodders or mixture of forage and concentrates.

Average mature live weights and fleece of a number of Indian breeds (of dry areas) are shown in the following table –

Table 5.18. Average mature live weight and fleece of a number of Indian Sheep breed.

Breed	Av. Mature live Wright (kg)	Av. Fleece Weight (kg)
(1) Baluchi	33—44	1.8
(2) Magra	25—33	2.0—2.5
(3) Nali	25—40	2.5—3.5
(4) Marwari	25—40	1.5—2.5
(5) Sonadi	25—40	0.6—1.2
(6) Kathiwar	27—45	1.5
(7) Deccani	20—32	0.2—0.3
(8) Nellor	38—41	nil
(9) Bikaniri	36—64	1.3—2.4
(10) Chottanagpuri	15—20	0.18

Management of sheep in hot season (May and June): As the green grass is not available for feeding or grazing, stubble of 'rabi' harvested crops may provide some feed materials to them. So the supplementary feeding is very much essential and a feed planning for these two or three months should be designed well ahead.

Sheep do not require to drink water daily, particularly if they are maintained on lush, wet-season pasture grazing. But they should be given access to free water all the time. The approximate daily water intake of sheep, managed in dry areas, is 4 to 5 litres.

5.6.2.6. Piggery

In some dryland areas Piggery is becoming popular on account of the pig's high fecundity and growth rate. The pig production can yield a relatively rapid rate of return on the capital employed. Due to high temperature, it is not possible to grow pigs in dry areas up to 70kg.

The primary purpose of farming is the production of meat and it assures the supply of protein to the less protein diet of dryland areas.

Breeds

The indigenous or country type black small breeds are commonly found in the tropical areas. Few British and American breeds like Berkshire for crossing with indigenous pigs and pork type Duroc breed, transformed by selective breeding respectively are frequently used for tropical areas particularly in dry tropical areas. With high standard feeding and management facilities the breeds like Large White Yorkshire, Tawarth, improved Chinese, Hampshire can be used for upgradation of local breeds.

Effect of Climate

(a) **Effect on Sexual Behaviour.** Male sexual libido appears to be affected by the climate. Steinbach (1972 b) stated that refusal to mount and to ejaculate was positively related to the effective monthly temperature. Climate dose not appear to affect the ovulation rate of female pigs. However, it does appear to affect the oestrous cycle. It was found by Omtvedt et el (1971), that heat stress during the pregnancy rate at 30 days after oestrus by 43%, while heat stress from 8th to 16th day reduced it by 21%. The no. of embryos was significantly lower in the stressed group in both the periods. But heat stress during mid-pregnancy (53 to 61 days) did not appear to affect the gilts significantly in any way. On the contrary heat stress towards the end of pregnancy (102 to 110 days) had a very significant effect on gilts. The heat stress in late pregnancy reduce the ratio of alive new born and still born piglets from 10.4:0.4 (in case of non stress) to 6.0:5.2. The heat stress up to 21 days after birth reduces the survival rate up to 17% in comparison.

(b) **Effect on Growth.** The pig is essentially a non-sweating animal and is very sensitive to changes in climatic environment. In case of piglets, during the first two days of life, the ambient temperature should be 32.2°C'+'. As the pig grows, the optimum ambient temperature for maximum live weight gain and efficiency of feed conversion capacity changes. The experiment by Heitman Hughes in early years (1949) shows that both the live weight gain and food conversion capacity is maximum at approximately 24°C for pigs weighing 32-65kg and at approximately 15.6°C for pigs weighing 75-118kg. The figures of this experiment may vary in case of local or crossbreed pigs, but in dry areas the aim should be to raise a porker weighing approximately 54-64kg and not for the large pigs as bacon.

Housing

The housing of commercial piggery is always a special one among the other live stocks specially in respect of flooring. The floor should be hard, impervious to liquid and easy to clean. However, as the cost and availabity of materials vary so widely in different locations, it is not possible to suggest one set of plan to give a generalized idea particularly in respect of size of house. The rooms or pens of bore, furrows, dry sows and gilts, weaners and fatteners should be built up separately.

Feeding

The importance of proper feeding is very great as the cost of its feed represents a very high proportion of the total cost of the production. If fed overgenerous rations pigs fatten very rapidly. This tendency is highly inheritable. It is also uneconomic.

The pigs are omnivorous i.e. can eat all type of food staff and can't digest too much fibre like other domestic ruminants. Pig farmer should use the cheaper, lower grade feeding-staffs to the fullest extent and to supplement them by more expensive nutritious feeds (mainly protein sources) to the point that true economy dictates.

5.6.3. Fishery in Harvested Water

Fishing is an age-old practice, which dates back to pre-historic period. In course of time man started cultivation or rearing of fish besides crops and animals. Fish culture in tanks had its beginning in India since as early as 350 B.C. whereas it is reported that Egyptians were the first to start fish-culture somewhere in 2500 B.C. followed by China.

With the backdrop of the history of fish culture, only at the beginning of eighties rapid development of fish-culture took place with the help of artificial breeding technique and mechanisation. However in India per capita, particularly in rural areas, consumption of fish is very low against the recommended level of total protein requirement per day. As a food item, nutritional quality of fish is incomparable. It is rich in protein (Table 5.19). It contains all essential amino acids required for the human body. The protein conservation ratio of fish is highest among the other sources of protein supply (Table 5.20).

Table 5.19. Average Chemical Composition of Fish.

Contents	% Of body weight
Moisture	74.00
Fat	2.50
Protein	19.50
Ash	1.50
Phosphorous	0.50
Calcium	0.50
Iron etc.	1.80

Table 5.20. Protein Conservation Ratio of different Animal Proteins.

Product	Protein Conservation Ratio
Beef/Livestock	21 : 1
Pork	8 : 1
Poultry meat	5.5 : 1
Milk & Egg	4 : 1
Fish	3 : 1

Now, naturally the plain land or lowland of high rainfall areas, particularly the river-dominated areas, is mostly focussed for development of inland fisheries. In other areas, particularly the dryland areas, where rainfall is not much but intensity is high, it is possible to create small to big water bodies, which are very much essential for any farming activity in that area and may be used for economic fish-culture. The technique of creation of these water harvesting structures in dryland areas are discussed in the earlier chapter {Chapter no.- 4.3 and 4.4}, which is also very much important to determine the type of fish culture to be performed. Actually, apart from big water bodies, maximum of them are seasonal in nature and the life span practically controlled the fish culture activity of each and every water bodies. However, the present technical advancement can create profitable and technically viable proposition of fish culture for a very short period (within six month).

There are two basic types of fishery activity, (1) hatchery and (2) fish rearing/ culture. This can be performed in the harvested water bodies of dryland areas.

5.6.3.1. Hatchery

The traditional device in 'bundh' breeding areas for hatching or rearing of carp eggs is mud hatchery pits. The water harvesting structure may be designed in the form of the hatchery pits. The pits are dug in series and the inside is plastered by red/clay soil. During twitching stages eggs are collected and put into the hatching pits. Process of hatching in mud hatchery pits can be improved by affording shade and running water flow in these pits. This method may further be improved in respect of efficiency by introducing double cloth hatching 'happa', which provides a success rate of 50 to 60 % (Maximum eggs that can be hatched at a time through this method is about 0.8 to 1 lakh.) with highest cost benefit ratio (1 : 21). As it is purely a seasonal activity and as it requires less fixed cost hatchery in harvested water body, it can provide a good supplemental income to dryland farming.

5.6.3.2. Fish Culture

The semi-intensive fish culture of appropriate carps or species is possible even in very small seasonal water bodies in dryland areas. For selection of fish species and to design the entire cultural procedure the seasonality and size of the pond or tank should be considered first. In dryland areas the seasonality of the water bodies is around 6 months, and apart from big /medium dams most of the created water bodies are generally small in size.

Considering the above said situation the culture of hybrid/African Magur and red/nile Tilapia is very much recommendable, for the small tanks (up to 10 cents) in particular. Some spp. of cat fish' is also another option for this small tanks. It may bring a gross profit of ₹ 4000=00 to ₹ 14000=00 from 2.5 to 8 cent sized ponds respectively (see outline of a scheme given below).

Outline of a semi-intensive fishery scheme (for 2.5 cent i.e. 100 sq.m. pond/tank of dryland areas with a production target of 100 kg).

Item	Quantity
1. Hybrid/African Magur fingerling (1 to 3 gm. Each)	300 Nos.
2. Red/Nile tilapia (5 to 15 cm. long)	200 Nos.
3. Mustard oil cake	25 Kg.
4. Cattle/Pig dung	50 Kg.
5. Lime	5 Kg.
6. Fish feed	60 Kg.

1. Stocking, magur @ 3 nos./sq.m.
2. Stocking, tilapia @ 2 nos./sq.m.
3. Assuming FCR (food conversion ratio) of 2.5 for the feed and mustard oil cake.

Return: - Production of magur — 80 Kg.

Production of tilapia — 20 Kg.

For a given unit of food energy in any small water body, tilapias are known to produce the maximum protein with high quality flesh. The major drawback in tilapia culture is their ability to quickly overpopulate the ponds or tanks through high rate of reproduction. As inbreeding is one of the main causes of deterioration of the quality of the breed; seasonal cultivation of tilapia is always preferable. Now to control the population of tilapia the most widely used technically and economically viable method is mono-sex population with male tilapia only. All-male tilapia populations are preferred, as they are generally known to grow faster than the female. Naturally all-male mono-sex culture of tilapia in small pond or tank is technically and economically more viable proposition for dryland farming.

5.7 SOCIAL MANAGEMENT/INTERVENTION

5.7.1. Backward Socio-Economic Condition : its Reason and Remedy

Lack of availability of proper technology is probably the main reason of this backwardness in comparison to the non-dry or water rich areas. Next, lack of capital, lack of nearby market and marketing facility, lack of credit have made them averse to take any risk towards any technological development. Due to this chronic backwardness most of the dryland farmers are very poor and very little cash reserve is available to them to try anything new. Actually the total farming activity is of subsistence type. So, available information or technology for cash crops suitable to dry area is nil.

Extension services are comparatively poor since the majority of extension personnel prefer to concentrate on the technologies and crops of non-dry areas, as good result can be demonstrated easily in a very short period. Due to lack of knowledge about dry land, its prospect and possibilities, the extension service has never taken a proper shape as most of the extension workers are from non- dry areas. The scientists are also working with the same reasoning, consciously or sub-consciously. Research or demonstration on long duration crops like 'ber' etc or water harvesting in dry areas are always tougher jobs than that on the vegetables like potato etc or shallow tubewell in non-dry/water rich areas. Due to this reason the availability of works on dry land areas is less and as a result the knowledge about dryland farming technology and socio-economic condition of the farmers of dry areas are far from satisfactory.

5.7.1.1. Farmer's Attitude Towards New Technology

The farmer's reaction towards any change in common farm practice is always negative. The chief reason of this attitude is the probability of risk, as assumed by them, involved with that change. The second vital part is the cost associated with that change. In dryland areas the risk in farming is always high and the economic condition of the farmers is relatively poor. These two factors should be considered properly before doing any developmental work, particularly during introduction of new technology in dry areas. Proper dry farming technologies are those, which can reduce the risk of failure and can produce optimum economic return. These technologies can change the conservative attitude of the farmers. The most advantageous aspect of extension work is that the farmers of dry areas are in urgent need of development in farming activity just for their survival in coming days. However, introduction of new technology in dryland areas is very difficult due to the following reasons.

1. Non-availability of enterprising individual who can act as a leader or early adapter in term of extension;
2. Supply points of different modern farm inputs are quite a few and far away from the field as well as from the farmers.
3. Poor knowledge about dryland farming of the external developmental agencies (including government) that are supposed to play a catalytic role in developmental process. This is creating a wide communication gap.
4. The prevailing socio-economic constraints like illiteracy, poverty, lack of credit supply, finally the resultant frustration and inferiority complex of the local people are major blocks against the introduction of new technology in dry areas.
5. Last but the most vital reason is the sub-conscious resistance due to the unforeseen risk towards their subsistence activity.

5.7.1.2. Approach for Introducing New Technology

Immediate solution of all the above problems is almost impossible. However, a long term planning may reduce the magnitude of these problems. The basic principle should be the raising of economic condition of the farmers. Unfortunately the development programmes frequently enrich the mediators instead of the people for whom the programme is meant.

Like non-dry areas, it is not possible to develop the farm activity of dry areas within a year or two by introducing new technology because of the extremity of the nature, particularly in respect of water and temperature. So acceptance of any new technology depends on how far the farmers can take control of the adverse nature in and around his farm to augment the production through that technology. Any change always takes its own time to produce desired result. The positive outcome of these changes results in modified nature of the farm in respect of soil, water, and its relationship with flora and fauna. Considering the social and economical aspect of dry land areas, the following steps may be taken to introduce any new technology.

(a) Increase the knowledge of extension worker or the facilitator about the locality and technologies of dryland farming. It may be achieved by training and by creation of in-situ data of the area.

(b) Developing the community by:
- Transmission of knowledge
- Sharing of achievement and failure among the facilitators and farmers
- Supply of different farm inputs at doorsteps
- Creation of marketing facility of farm products
- Sharing of subsistence activity with each other
- Volunteering towards common goal and common interest

(c) Proper selection of specific technology for individual as well as for the community.

To perform the above, a few tools are necessary which, in combination, can make the jobs successful. They are: -

1. Regular meeting between farmers and extension agencies;
2. Preparation of individual land use plan, which may be interlinked;
3. Training on profitability;
4. Arrangement of inputs supply through co-operatives or other sources;
5. Arrangement of credit through banks or co-operatives or self help groups;
6. Arrangements of proper marketing facility of agriculture produce through public or private agencies.

5.7.1.3. Participatory Technology Development (PTD)

It is a process that combines the knowledge and research capacities of local communities and that of research and development organisations in an interactive learning process. It involves identifying, generating, testing, and practicing new techniques and practices to solve local problems. The ultimate aim is to strengthen the experimental and technology management capacities of local people and communities; thus, farmers play a key role in the entire process.

PTD is a purposeful and creative interaction between local communities and outside facilitators which involves: gaining a joint understanding of the main characteristics and changes of a particular ecosystem; defining priority problem of farmers and farming communities; experimenting locally with a variety of options derived both from farmers (indigenous knowledge and other farmer experiences) and from formal science; and enhancing farmer's experimental capacities and farmer- to-farmer communication.

PTD in agriculture is not a substitute for station-based research or scientist-managed on-farm trials. It is a complementary process, which involves linking the power and capacities of agricultural science with the priorities and capacities of farming communities to develop productive and sustainable farming systems.

The technology development approach is closely linked with a process of general community development. Activities involved in PTD – critical analysis of community-managed changes in the agroecological system, identification and use of indigenous technical know-how, reconstruction of successful local innovations, self-organisation and self-implementation of systematic experiments with selected options – all foster the awareness, self-respect and self-confidence, as well as the diagnostic and experimenting skills of the farmers involved.

PTD approaches also try to foster development of a network of village organisations, to intensify communication on local experimentation and improve linkage with supporting organisations and institutions. Development of village groups and independent relations of exchange encourages the maintenance and further spread of local practices and enhances the prospects for on-going experimentation, improve resource management and self-sustained agricultural exchange.

PTD not only enables the generation of technologies to local environments, but also develops the local capacities, socio-cultural structures and organisational linkages necessary to sustain the process.

5.7.2. Social Development and Dryland Farming

For equitable social development in dry areas, dryland farming should aim at (1) the optimisation of economic returns for all members of the society by using natural resources (2) promotion of different technologies in harmony with nature's regenerating potential and (3) creation of a common platform for the management and maintenance of natural resources. To achieve the said development all round utilisation of both internal (local) and external (imported) resources of the area is essential and for that matter a total social involvement is required.

The basic factor of the social involvement is the strengthening of self-reliance and empowerment of individuals of a society. Unfortunately, at present, all these developmental efforts are externally induced, and mostly unrelated to local reality. Many well-intended schemes to promote this effort frequently failed mainly due to non-technical factors like unorganised individuals, poor or motivated public policies and obviously lack of initiative of the public administration, the only facilitator, due to ignorance or otherwise. So it is not surprising that effects of such programmes, instead of enhancing people's autonomy, often perpetuate dependence and further erode the communities' self development potential. However, now we discuss about some basic conception to develop aforesaid initiatives in a particular society.

Participation : Participation is a mechanism or process where people express themselves and act with mutual responsibility to promote the mutual set of interest, which finally helps the individual. There are a few techniques like participatory appraisal (PA), which are useful ways of initiating the process of participation. The proper participatory process includes (1) information generation (2) analysis of the information generated (3) generation of planning, both community and individual (4) implementation of the planned activity and (5) arrangement of the activity, .The useful participation requires the character of free expression of facts, ideas, opinion and aspirations of every one obviously with a non-selfish attitude.

Dryland farming, which associates many types of constraints, always requires proper participation of all kind of people. *Acceptability* (participation) of all socio - economic groups and external developmental agencies like government, NGOs etc. should be a must for dry land farming. *Woman participation* is a very important area to be looked into in dry land farming. There are many activities in dry land farming in respect of water management, animal management, crop management, organic matter production and obviously the food and nutrition management where women are more important than man. Naturally, the sustainable development of dryland farming is not possible without the active participation of women folk.

5.7.3. Community Activity

In dryland areas many activities are involved, which have common user rights, common responsibility and common liability, on a common property that may be of public or of individual or of common possession. These activities have a very close relation with the successful implementation of dryland farming. The common use of natural properties like forests, grazing lands, streams, rivers, ground water etc, may come under the community activity. This may also include the man made properties like community irrigation tanks, community wells, roads etc. Apart from the creation and management activities of the community few other things are to be performed for sustainable development of dry areas. These things are joint plant protection measures, creation of windbreaks and shelterbelts, cooperative societies for input supply, marketing of produce, providing credit supply to the farmers etc,

Few Purposive Community Activities

1. **Forest.** With a view to develop natural resources, effective development and maintenance of forest for the benefit of the community i.e. for supply of fuel, fodder, biomass and restricted timber, is must. For that purpose a concept of joint forest management (JFM) has come up in recent years. It is nothing but a well-organised community activity on a common property. For proper functioning of this activity through a JFM committee / forum, there must be an unambiguous guideline for everybody on issues relating to use of the forest in every aspect and its protection, management and regeneration. The forest department and/or other government authority may involve themselves in this activity initially, but they should not dictate the proceedings. Rather, they should act as facilitators only. The guidelines should be on the basis of give and take policy and obviously on equity basis.
2. **Water Source.** Community wells/tube wells, 'panchayat' well/bore well with hand pump, community tanks/ponds/dams/reservoirs and other common water sources including river and rivulets etc. need proper management and maintenance to remain efficient and sustainable. For dry land farming the management and maintenance of common water source is vital, particularly where perennial or good individual water resource for every farm is not possible.

Without proper maintenance and management, majority of these common water sources becomes non-functional affecting the people of the area, particularly in regard to farming. Proper maintenance and management of community water sources also improve and sustain the water supply of the water sources possessed by individuals. Users need to be equipped themselves properly to develop appropriate skill to operate, maintain and manage these common properties in a way that would provide equal benefits to all. It may be done by indigenous knowledge and practices, which may have a better chance to be adapted/followed easily by the common people. The outside agencies including government may be involved as facilitator to supplement with modern knowledge for overall benefit of the users. The proper blending of indigenous and modern knowledge / technologies will create confidence among the people and motivate them to respond more readily.

3. **Grazing Land.** Continuous, unplanned and unorganised grazing in the non-cultivable common land suppressed the regeneration of grasses; shrubs and fodder trees thus reducing the productivity and quality of grassland.

   The common lands should not be converted from grazing land to forestland by unscientific afforestation. Village communities should manage this common grazing land to develop common property resource management (CPRM) enterprise. The common right on scientific grazing on grassland rich with nutritive fodder should create an extra interest to participate in CPRM of grazing land. Actually, we put all our thrust towards afforestation ignoring development of grassland, which is very much important for dryland farming.

   In proper grassland management the main job is to fence the area, after dividing it into 6 to 8 segments, then a systematic grazing should be done. To restrict the unauthorised grazing there should be an unambiguous guideline of proportionate grazing facility for every animal. A well layout plan in respect of area of grazing in single phase and number of animal grazed in one phase should be incorporated in the guideline.

4. **Co-operative/Farmer's Club/Self-Help Group/Community Service Centre.** Apart from the discussed community works there are other needs that are related to common interest but not related to any natural resources. The common need involves (1) supply of farm inputs or raw materials, (2) marketing of farm produce and (3) financing.

   These types of organisations should be created through the process of community development activity for the benefit of the entire community. These community organisations should run like business agencies on the basis of benefit sharing and do the jobs for development of the community.

5.7.4. Women's' Participation

In general, particularly in dryland areas, the technical and social decision maker and implementers are all men. The formal and informal training or studying is largely male farmer oriented. Women are utterly neglected in these matters in dryland areas though they, in large numbers, work as worker in different community projects.

Women Participation on Community Lands. Issues related to use and sharing of common resources among various groups in the community hardly receive much attention. In the eyes of society women are landless, do not control finance, have no access to credit and have very little control over resources. They are left out of the planning process of any community work.

Dryland farming frequently require many community activities that require specific knowledge, experience, and training. As a participator if training is offered to women and are given choice to have technical works with a supervisory position, thus can involve them in decision-making actively. Wages difference in case of hired labour is an injustice to women. Actually strength required and apparent quantity of a work is only judged by the society but not the quality and it is unfortunate part for the women.

Women Participation in Family Land. Unlike community activities, here women have, to some extent, decision-making power. As in the drylands men are frequently engaged in outside works and sometimes goes outside as hired labour, the women are forced to take few decisions about farm-planning etc. However, due to lack of information and knowledge they are dependent on their counterparts for decision-making. The men are not always equipped with these information and knowledge but they dominate the proceedings ignoring women as they are invited by women to solve the problem.

So in dryland areas, by imparting training only to women about basics of dryland farming, the process of development of socio-economic situation through farming will increases significantly.

Facilitating women's' participation begins with understanding the community. It is important to create awareness and understanding among the men, so an atmosphere of acceptance can be created in the village whereby women can take an active part in community activities and can take final decision on family land. There are few strategies to facilitate women participation in community activities.

- Impart training only to women about farming, which can be implemented in family lands. This will boost their confidence in decision-making.
- Specify the works for women, which they can perform equally well, if not better, than men.
- Among the implementers and decision-makers women representative will be 50 percent.
- The timings of the meeting etc should be done in accordance with women's convenience.
- Initially as support group, an all-women committee, comprising the women participants can help them to prepare themselves for deliberation in main group meeting. It also helps to build leadership.

■■■

CHAPTER 6

'CHOTANAGPUR' MODEL OF DRYLAND FARMING

6.1. MODEL OF DRYLAND FARMING

Introduction

Agriculture will ever remain the mainstay of Indian economy and also the foundation of agro-based industries like textile, jute, sugar, paper, food processing etc. including many sectors of cottage industry.

The agricultural value of a land has so far been measured by its potential for growing the estimated cereal crops like paddy, wheat etc. The severe food shortage the country faced immediately after independence, produced this mindset in our planners, politicians and agricultural scientists. Recent development in food front, where huge quantity of food grains are lying with the farmer due to lack of space in warehouse and a fall in the market price, have shown that we have carried the 'grow more food' policy too far in respect of grain production. Probably, a subconscious selfishness regarding personal food security among the policy makers, who are mostly purchaser not the grower, is responsible for the present chaos or it may be due to utter negligence regarding the welfare of the farmers.

The cereal growing mind-set of the farmer is responsible for the present plight. They grow their crops even on lands, which fail to give an economic

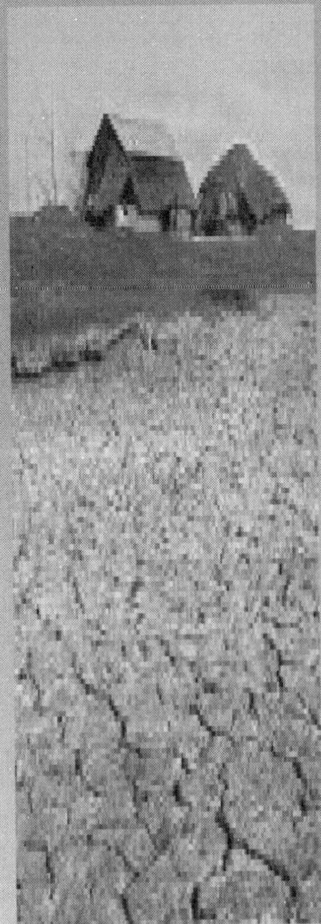

yield. This situation has largely helped the poverty to continue in the rural areas with less fertile lands.

This attitude has to be changed to improve the purchasing power of these people, which in turn will alleviate illiteracy, poor health care and other problems of poverty.

A vast tract of about 8.3 m.h. of such land lies in the eastern plateau of chotanagpur belt.

Ecological Background of This Tract

This tract which consists of red and lataritic zone of Chotanagpur is a plateau lying on the eastern extremities of the Deccan plateau in the newly formed Jharkhand State, North Orissa and South West part of West Bengal. In the near past dense forest predominantly occupied this area, sparsely inhabited chiefly by the tribals who lived mainly as hunter-gatherers. Agriculture, if at all, manifested itself through the primitive shift cultivation.

Later, attracted by the natural resources, people from outside gradually intruded, and destroyed the balance of its ecosystem. Excessive and unscientific milking of these natural resources, mainly timbers and minerals caused serious damage to the land, particularly in respect of water and soil conservation. Quick spread of conventional agriculture, though not profitable, gradually covered the area.

Interplay of these factors made the economy of the local people very weak, but helped a group of people, mostly outsiders, who drained the resources, to become rich. The situation has sown the seeds of discontent which often manifests itself in severe political unrest.

Climate and Land

Climate. The area is subtropical with less humidity. Average minimum annual temperature goes down to 8.5°C during December and January, which are the coldest months. The average maximum temperature during this period is around 23°C. Climate warms up from March onwards and the average maximum temperature during April to June is around 36°C and frequently soars upto 45°. The temperature cools down with the onset of rains, but hovers around 30°C till October.

The average annual rainfall is about 1250 mm of which nearly 80% is received during the period of south-west monsoon which generally sets in from 15th June and continues upto 15th October. The average number of rainy days per year is about 56 days (it includes days receiving 2.5 mm or more of rainfall). There is a frequent 'break' i.e. a halt of rainfall for a week or two during the monsoon.

The maximum humidity veers around 90 per cent during the monsoon months. Other months are more or less dry. The wind velocity is nominal with 4 to 8 km/hr. In summer gusty hot wind (loo) prevails, but cyclones are few and far between.

Land. The landmass.is very much undulating in nature and has a sloping gradient, moderate to stiff, towards easterly direction. The rainwater flows down quickly through natural waterways and finally gets lost in the rivers, carrying away with it the fertility in the form of topsoil, organic matter and nutrients.

The higher altitudes (Dungri & Tanr), mostly consist of alfisols in an advance state of degradation, lacking humus and major plant nutrients, untreated, unbounded and supporting low cropping intensity with poor productivity, if not actually fallow or barren.

The midlands (Baids) are often terraced rice fields with thin soil cover and poor moisture holding capacity. Mid rainy season drought, occurring with unfailing regularity every 2-3 years often damages rainfed crops like rice. A small percentage of lowland (Kanali and Bahal) with high clay content and high moisture holding capacity is good for rice cultivation.

Soil. The soils are of red and lateritic types, the colour is generally reddish or yellowish red and the structure is loamy sandy or often gravelly and sometimes vascular. The upland and mid land soils are mostly dominated by coarse sands and gravels of different size.

In respect of soil profile, on an average, the depth of A horizon is about 15 cm, B horizon is spread from 15 cm to a maximum 150 cm. The C horizon generally consists of weathered parent material of disintegrating coarse grained granite, which may retain the morphology of the rock structure to some extent. The soils are generally of acid in reaction with pH ranging form 5.0 to 6.5. A distinctive feature is the high percentage of acid soluble Fe_2O_3 compared with Al_2O_3. The base exchange capacity varies from 5 to 25 m.e./100 gm of soil depending upon the amounts of clay and organic matter. The available potash in loamy & sandy red soils is sufficient but P_2O_5 (Phosphorus) and N (Nitrogen) status is low. In lateritic soils, the percentages of all the K_2O (potash), P_2O_5 and N are very low.

Land Situation and Land Use Plan

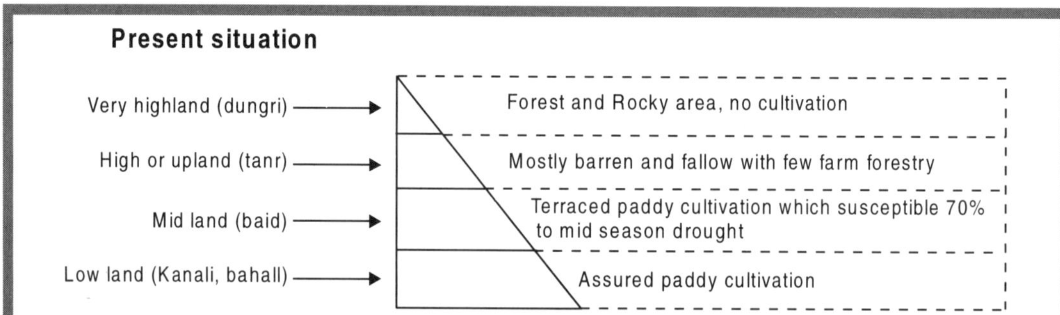

Fig. 6.1. Present situation

Figure 6.1 is self-explanatory for the existing land use planning. It implies that this type of land use pattern is not conducive to the economic development desired for the people. This is actually the subsistence agriculture that the people follow here.

To achieve this recommended land use plan (Fig. 6.2) a few basic and specific but simple technological adaptations are necessary.

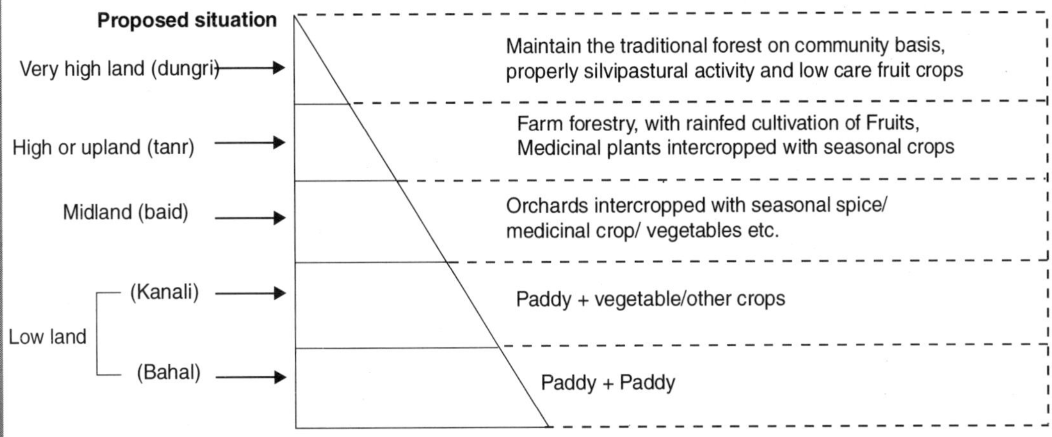

Fig. 6.2. Proposed situation.

Specific Technology to be adopted for sustainable farming in this region :

[A] Water Conservation

Due to land situation most of the precipitation, though high in quantity, is lost through- surface run-off. Technology for storing the rain water either on the surface or into the soil is generally termed as water conservation. There are many methods of water conservation depending upon the land situation, rainfall and cropping pattern.

Storing Rain Water Into the Soil

(a) **Cultural Measures.** Indigenous methods like (i) deep ploughing at the onset of monsoon, (*ii*) furrowing and cross ploughing, (*iii*) tide ridging (*iv*) compartment bunding etc. are used for in-situ conservation of moisture.

(b) **Contouring.** It is the land development work, which is done along the same elevation or height (from mean sea level). Depending upon the land/soil situation and crop need, the following methods can be adopted for contouring:

(i) **Contour Graded Bunds.** It is the technology by which bunds, made up of soil are to be constructed along the contour lines with a small gradient. The size and spacing of the bunds depend on the local situation. The agricultural operations are to be done in the spaces between two bunds. This technology is very effective under 'Baid' and 'Kanali' situation.

(ii) **Contour Trenches.** It is the technology by which the trenches, of continuous or staggered type depending upon the crop, arc to be dug along the contour lines. The dug out soils are arranged downslope in the form of V/crescent to catch more runoff. This technology is useful for 'dungri' and 'tanr' areas.

(iii) **Contour Walling.** It is a porous type of bund generally made up of stones or live hedge along the contour to check the run off velocity and also to collect a sizeable eroded soil. This technology is suitable for stony gravelly 'tanr' and 'baid' areas where planting of orchards, forest trees or any other suitable crop are to be done.

(iv) **Contour Seeding or Planting.** All sorts of sowing or planting on land between two bunds or trenches or walls are to be done along the contour. This technology applies on all lands of the area .

(c) **Mulching.** This technology is used to conserve both water and soil as well as to control weeds. Under this technique soil surface is covered usually with organic residue like straw, leaves, stocks, husks or shells etc. to reduce surface evaporation. It also temporarily traps the runoff water which thus gets more time to percolate into the root zone. The mulching materials on decomposition provide organic matter to the soil which in turn increase the microbial activity, Where mulching with those materials encourages termite infestation, termite control should be adapted. At least 70% of the open soil surface should be mulched to get effective result.

Live Vegetable Mulching. Climbing and creeping crops/plants, that spread on the soil surface, provide good mulching.

Storing of Rain Water on the Surface

A good portion of rain water, which generally gets lost as run off, can be stored in ponds or ditches. A part of this harvested water may be used directly for irrigation, the remaining part percolates down to recharge the sub-surface flow or ground water flow and raise its level to reach the root zone of the deep rooted (tree) crops.

Though a sizeable portion of this surface storage gets lost through seepage, still this technology is recommended in the Chotanagpur plateau due to good rainfall and poor ground water condition. In Kanali and Bahal the seepage loss is less in comparison.

(a) **Farm Pond.** In 'Bahal', 'Kanali' and parts of 'Baid' areas the dug out farm pond is usually constructed at its lower most point in square or rectangular form with a specific catchment area properly bounded. The optimum catchment area of 1 ha can produce 250 m^3 of run-off storable water. For this catchment size a square pond of 12 Mt × 12 Mt × 3 Mt may be constructed with a side slope of 1:1. Every pond should be provided with controlled inlet of 2 Mt width to avoid gullying, in association with a silt-pit. An outlet is also provided with every pond for safe discharge of excess water, if any. The excavated soils may be carried to 'Tanr' areas to improve tho soil condition there.

(b) **Miniature Farm Pond/Tank.** This technology is often termed as '5 percent model' and is suitable for 'Baid' areas. Here 5 percent of the created catchment area or plot is to be excavated at the lowermost point of that plot. The minimum depth of the pond should be 1.2 Mt. Such pond or tank collect run off water as well as sub-surface seepage coming from plots of upper slopes. To increase the water holding efficacy of the ponds, the suitable low cost seepage control measures like (1) Puddling, (2) compaction, (3) application of clay layer, (4) Bentonite treatment and (5) application of soil-cement mixture

(10:1) may be adopted. Size of the catchment of this model obviously depends on the type of soil, slope, and site of the pond. However, 1/4th to 1/3rd acre has been found to be an effective size. The water inlet modified to trap the accompanying silt reduces the maintenance of the pond.

(c) **Percolation Pit in Bunded Plots.** In 'Tanr' areas, with upto 8 percent slope, rectangular plots measuring 100 to 120 sq.Mt. should be bounded by small ridges. A pit should be dug at the lowest corner, where run-off water would be collected. The size of the pit may vary according to the situation. Here the harvested water maintains a seepage line down the slope which may feed the root zone of some crops.

(d) **Making of Staggered Pits and Trenches.** This is an effective and acceptable method of water harvesting for the 'dungri' and upper 'Tanr' areas where the slope is more than 8%. The staggered pits or trenches measuring 2 Mt × 0.5 Mt × 0.5 Mt or 1 Mt × 0.5 Mt × 0.5 Mt are excavated along the contour line with required horizontal intervals. The excavated earth is used to construct small embankment on the down slope creating an additional small catchment area in 'V' or half moon shape. This structure can harvest the run-off water and also prevent the soil loss.

[B] Crop Selection

To implement the recommended land use plan effectively (Figure 6.2) in the areas, where water conservation structures and measures are already implemented, proper selection of crop is vital. Some basic points should be kept in view for this purpose.

1. Crops should have potential to give food, fodder, bio-mass or any other commercial product.
2. The perennial or long duration crops with deep tap root system are always preferable over seasonal crops. They may be planted in single or mixed form with two or three different species as deem fit.
3. Perennials should be inter cropped with non-competitive seasonal crops.
4. Climbers or creepers are always preferable to the bush form, as they provide better ground coverage in relation to their root zone area.
5. The harvested water, stored on the surface, should not be used for paddy cultivation, except Bahal areas, beyond rainy season.
6. After kharif paddy, in 'Kanali' areas, high-yielding or hybrid vegetables or other cash crops may be cultivated, depending upon the quantity of stored water in farm pond.
7. A minimum of 5 to 10 percent of the cultivable area should be covered with fodder crops.

Now the selection of crop and its varieties should be done depending upon the (1) productivity of the land, (2) profitability in the market and (3) capacity of the grower to manage them. From these points of view the following crops are suggested particularly for Tanr & Bad areas.

(a) **Almost no Care**

(1) **Fruits.** Tamarind, wood apple, Jamun, bael. wild ber, ordinary custard apple, Mohua etc.

(2) **Medicinal Plants.** Haritaki, bahera, arjun, kuchila, etc.

(3) **Pulse Seed.** Kalai.

(b) **Some Care**

(1) **Fruits.** Ber, custard apple, in situ grafted mango, seedling guava, Aonla. Pomegranate, phalsa.

(2) **Medicinal Plants.** Aswagandha, ayapana. ghritakumari, tulsi, sarpagandhya, chalmugra, Neem, Turmeric, curry leaf, etc.

(3) **Vegetables.** Drumstick, wild spine gourd/kakora, yam/dioscoria, sword bean, Alocasia/man kachu, Kundri, etc..

(4) **Grasses & Other Trees.** Dinanath, pigeon pea, ritha, bamboo, etc.

(c) **Good Care**

(1) **Fruits.** Mango, Litchi, Guava, Lime, sweet orange, sopota, etc.

(2) **Vegetable.** Tomato, radish, carrot, beet, cowpea, beans, pea, sweet potato, all kind of gourd, cucumber, kakri, amaranth etc.

(3) **Medicinal Plants & Spices.** Isobgul, coriander, fennel, fenugreek, black cumin, ajwan etc.

(4) **Oil Seeds & Cereals.** Mustard, Soyabean.

[C] Few Specific Cultural Practices for 'Dungri' 'Tanr' and 'Baid' Areas

1. Land preparation for sowing or Planting: Except grain crops maximum of the rest should be planted by dibbling i.e. in pits, the size of which will depend on the type of crops.

In Dungri area, for forest spp. 1 ft × 1 ft × 1 ft. pit will be dug just on the down side of the trenches guarded by 'V' or half moon shaped bunds and the saplings will be planted at the onset of monsoon. For no care fruit crops the pit size will be 2 ft × 2 ft × 2 ft in the same locations as stated above. Several layers of crop or plant residues like dry leaves etc. will be placed at the bottom before filling. The pits will be filled with good quality soil and organic matter in 3:1 ratio.

In 'Tanr' and upper baid area pits for big fruits 1 Mt x 1 Mt x 1 Mt and for small fruits and climbing vegetables like pumpkin, cucumber, gourds etc. 0.7 Mt x 0.7 Mt x 0.5 Mt. pits (See fig. 5.14) are to be dug out along the contour. If possible, large pits may be supplemented with a 15-cm wide and 2 m deep hole at the center of the bottom. The filling method should be the same as mentioned earlier.

2. Sowing & Planting Technique

(a) **Sowing in Pits and Trenches.** Pits and trenches, made at suitable sites; should be filled up with good soil and manure liberally. These should be planted more densely with climbing/creeping crops.

(b) **In-situ Grafting/Budding.** In case of mango, ber and sweet orange, the in-situ grafting and budding should be followed. To guard against any chance of failure, two seeds should be planted in each pit in the onset of monsoon. These will be grafted or budded when suitable. If grafting/budding in both the plants is successful, the excess one may be sold out or used in relatively better-soil areas. This method reduces the cost of plant materials to a great extent.

3. Irrigation Methods

(a) **On spot Irrigation.** Here all sorts of water application, mostly from harvested pond or well, covers only the root zone.

(b) **Pitcher Irrigation Method.** The method is very convenient in dry areas. An earthen pitcher of 10 to 15 litre capacity, with a hole towards the bottom, is buried in the planting pit in such a way that the hole remains near the center and the neck of the pot just above the ground level. The hole is plugged with a cotton wick to allow the water of the container to trickle out.

The seeds/saplings are planted above the hole, and the container is filled up. A well-fitted earthen lid is placed tightly on the mouth to reduce the evaporation of water from the pot to the minimum. Fertilizers/systemic pesticides may be added into this water. Soon the roots of the plants reach the hole to enjoy the continuous source of water and grow happily giving a luxuriant health to the plants.

The pitcher irrigation, if necessary, may continue for 3 years or more at the same location, or the pot can be removed and reused at another site. This method is suitable not only for trees, but also for the spreading types of vegetables like gourd, pumpkin, cucumber, melon etc.

4. Organic Matter Production: Organic matter is probably the 2nd most important factor for commercial production of crops, after water, in these areas. It is the only material, which can increase and sustain the productivity of the soil by manipulating the physical and chemical property of the soil. The organic manure may be produced in the following ways:

(a) **Farm Yard Manure Production.** All types of liquid and solid animal excreta should be utilized as manure after proper decomposition.

(b) **Night Soil.** The human excreta, richer than animal excreta in respect of nutrients, is to be used as manure after proper decomposition.

(c) **Composting.** Composting of all available bio-mass, crop residues etc. in the specific compost pits or in field pits/trenches is a very useful method of organic matter production. As bio-mass the pruned leaves and branches, particularly of leguminous plants will create very rich compost. For speedy decomposition mixing of phosphate, with the raw materials, is useful. Another way of quick decomposition is through **vermicomposing,** where selected earthworm spp. is used for breaking down the raw materials.

(d) **Green Manuring.** Seed of sunhemp/dhaincha should be sown, rather thickly, in the onset of monsoon and plowed down at the flowering stage. This is an easy method of enriching the soil with organic matter in 'Baid' and 'Kanali' Areas.

5. Animal Husbandry: The Chhotanagpur region has a very good prospect of animal husbandry. Actually without animal husbandry the commercial cultivation is not possible. Sometimes, animal husbandry is essential in this type of regions not primarily for milk or meat production, but for organic matter. In that case milk and meat will be considerd as by-product which will liberally supplement the farm income.

In this area dairy, goatery, piggery and sheep rearing can be done commercially. Commercial poultry should be added to this list except during the very hot months.

Conclusion

These measures will not only improve the economic condition of the inhabitants through agricultuaral developments but the techniques employed, which reduces the runoff, will tame the rivers down-stream resulting less occurrence of devastating floods. Another benefit would be lesser silting of our rivers and reservoirs.

6.2. BANKABLE SCHEMES ON DRYLAND FARMING FOR UNDULATING 'CHOTANAGPUR' TRACTS OF INDIA

In reference to the conception narrated about 'Dryland farming' the following schemes are prepared to make the farming activity more remunerative and sustainable. To adjust the extreme agro-climatic situation of these vast dry tracts of India the schemes are made. It is basically prepared for the uncultivated or poorly cultivated lands.

To make the schemes of dryland farming more sustainable and more practicable, incorporation of agro-forestry, drip irrigation and animal husbandry in the farming is done. Most interestingly instead of vegetables and othe common inter crops, medicinal and aromatic plants are selected as inter crop, which have a tremendous potentiality in these areas.

Scheme A and Scheme B are the two schemes presently prepared for financing through banking sector. The basic individual costs and the technical assumptions are taken from existing market rates and commercial practice respectively in accordance to the crops used in these schemes.

It is not possible to recommend any model scheme for the dry areas of this kind because of wide range of natural variations particularly in respect of soil, topography and water resource. However, these two schemes may be acted as a guideline both for Bank and beneficiary. But, at the time of implementation the schemes should be altered with a safe bearing in respect of viability after study each and every project carefully. For that purpose the scheme provides ample scope and alternatives

Few Technical Information of the Scheme 'A' and 'B'

1. Choice of Fruit Species and respective Varieties. There are some special problems which have to be carefully considered while selecting the crops in these areas. Except some fruit trees here and there, the area does not grow fruits as plantation or orchards. Paddy is the main crop (in the low lands i.e. in 'Kanali' and 'Bahal') of this area and frequently faces uncertainity. However through trial and demonstration the commercial fruit species like mango, sweet orange, guava, ber, custard apple, limes, lemons have been found suiable. Still the prevailing climate, particularly hot dessicating wind should be considered while selecting the crop species.

2. Planting Procedures and Layout. Due to extreme climatic condition one has to carefully consider the procedure of planting methods of these areas. The steps are to be taken in the following manner.

(a) Planting of all plants should be done on contour lines.

(b) At selected spots the pits should be dug bigger than normal size

(c) The pits are then filled with a mixture of good quality of soil and organic manure.

(d) Then the root stock or stone or seed (for in-situ grafting of mango, sweet orange, ber) or grafted or budded plant materials are to be planted in the pit.

(e) For slopy land, contour or staggered trench should be dug in the upslope of the row or individual pits respectively. The dugout soils should be used for making bunds placed in the downslope.

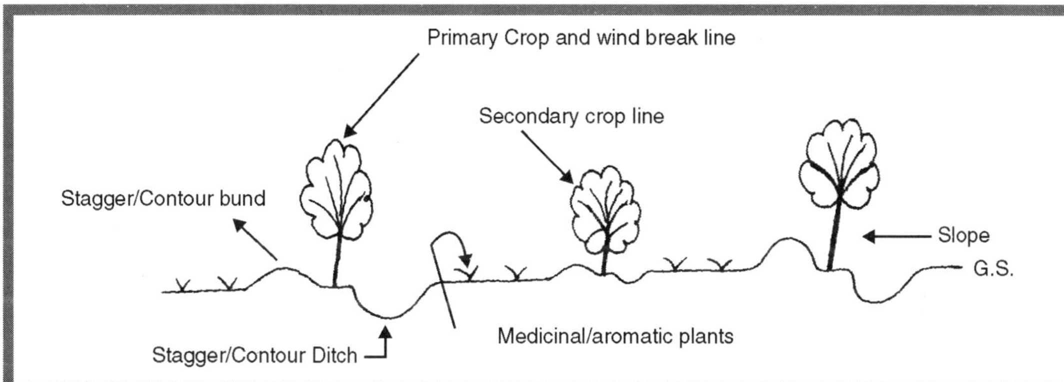

Fig. 6.3. Planting in Slopy Lands for Scheme-A

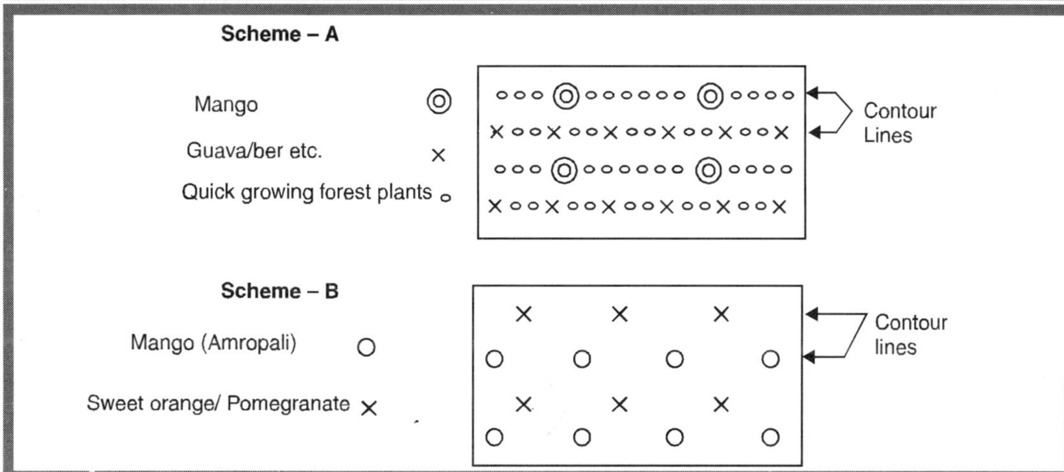

Fig. 6.4. Layout of Planting for scheme-B.

Projected gross Income/Year for Scheme-A

	Ist year	2nd year	3rd year	4th year	5th year	6th year	7th year	8th year	9th year	10th year
1. Primary crop	0	0	0	440	880	2200	3520	7040	13200	19360
2. Secondary crop.	0	1650	3300	5500	8250	11000	16500	22000	27500	33000
3. Medicinal/ aromatic crop	1980	3960	3960	1980	1980	1980	1980	900	900	900
4. Sale of fuel wood & fodder	0	0	200	500	800	1500	2500	0	0	0
Total	1980	5610	7460	8420	11910	16680	24500	29940	41600	53260

Cash Flow Analysis

Year	1	2	3	4	5	6	7	8	9	10
Cost	18880	4180	4660	5300	5300	5300	5300	5300	5300	5300
Benefit	1980	5610	7460	8420	11910	16680	24500	29940	41600	53260
Net	-16900	1430	2800	3120	6610	11380	19200	24640	36300	47960

Net present worth or NPW(At 15% DF)	**35,663**
Internal rate of return or IRR	**40.28%**
Benefit cost ratio or BCR (at 15% DF)	**1.96**
Present worth of cost or PWC (at 15% DF)	**37,140**
Present worth of benefit or PWB (at 15% DF)	**72,804**
Debt service coverage ratio or Av DSCR	**2.25**

Repayment Schedule

Year	Year wise Bank Loan disbursed	Bank loan Outst-anding	Interest @ 14%p.a.	Deferred Interest	Gross Income	Mainten-ance Cost	Surplus	Repayment of Bank Loan			Net Benefit +/–	DSCR
								Interest	Principal	Total		
1	2	3	4	5	6	7	8	9	10	11	12	
1	16992	16992	2379	2379	1980	B/L	1980	0	0	0	1980	
2	3762	20754	2906	1784	5610	B/L	5610	3,500	0	3500	2110	1.60
3	4194	24998	3493	0	7460	B/L	7460	5,277	0	5277	2183	1.41
4	4752	29700	4159	0	8420	B/L	8420	4158	0	4158	4262	2.03
5		29700	4158	0	11910	5300	6610	4158	0	4158	2452	1.59
6		29700	4158	0	16680	5300	11380	4158	5425	9583	1797	1.19
7		24275	3399	0	24500	5300	19200	3399	5425	8824	10377	2.18
8		18850	2639	0	29940	5300	24640	2639	5425	8064	16576	3.06
9		13425	180	0	41600	5300	36300	1880	5425	7305	28996	4.97
10		8000	1120	0	53260	5300	47960	1120	8000	9120	38840	
			30288		201360	31800	169560	30288	29700	59988	109572	2.25

6.2.2. SCHEME-B

Irrigated and well managed Fruit plantation inter cropped with medicinal & aromatic plants for 1 acre (on 2005 price)

Crops

1. Primary crop- Mango (Amropali)/Guava (L-49, Saffeda)- 110 No.
2. Secondary crop-Sweet Orange (Mosambi)/Pomegranate (Ganesh, Muscat, Jyoti)- 90 No.
3. Medicinal/Aromatic plants/Vegetables

   Safed Masli (Chlorophytum)

   Patchouli (Pogostamou Patchouli)

   All kinds of vegetables (mainly Cucurbits).

Cost Items Considered For Fruit Crops

Planting Materials (10% extra)

Farm Yard Manure (FYM)

Fertilizers

Plant protection and other Agro-chemicals-

Operation & Labour

Cost in ₹/Tree				
	1st yr.	**2nd yr.**	**3rd yr.**	**4th yr.**
(1) Primary crop-	83	34	44	47
(2) Secondary crop-	52	27	33	39

The cost of medicinal/Aromatic Plants/Vegetable cultivation is taken on lumpsum basis.

Project Cost				
	1st yr.	**2nd yr.**	**3rd yr.**	**4th yr.**
1. Primary crop (Amropali)	9,130	3,740	4,840	5,170
2. Secondary crop	4,680	2,430	2,970	3,510
3. Medicinal/Aromatic plants/vegetables	2,000	1,500	1,500	1,500
4. Land shaping	8,500	-	-	-
5. Live fencing with tree spp.	1,500	-	-	-
6. Irrigation-				
Dug well	16,000		-	-
Pump set	17,500	-	-	-
Drip pipe sys	15,000	-	-	-
Maintenance	1,500	1,500	1,500	1,500
Total	**75,810**	**9,170**	**10,810**	**11,680**

Total project cost	107470	Sale price	1. Mango (₹/Kg.)	5.5
Bank loan	96700		2. Mosambi (₹/Kg.)	8
Recovery period	10 years		3. Lemon grass (₹/Kg of Oil)	180
Grace period	5 years			
Rate of interest	14.00%			

Yield schedule	1	2	3	4	5	6	7	8	9	10
1. Mango (Amropali-(kg/tree)	0	0	3	7	15	20	30	35	35	35
2. Mosambi (kg/tree)	0	0	0	10	30	50	70	80	80	80
3. Medicinal/Aromatic Plant (kg of Lemongrass oil)	11	16	16	16	16	16	16	16	16	16

Projected Gross Income/Year for Scheme-B

	1st year	2nd year	3rd year	4th year	5th year	6th year	7th year	8th year	9th year	10th year
1. Primary fruit crop	-	-	1815	4235	9075	12100	18150	21175	21175	21175
2. Secondary fruit crop	-	-	-	7200	21600	36000	50400	57600	57600	57600
3. Medicinal, Aromatic crops	1980	2880	2880	2880	2880	2880	2880	2880	2880	2880
Total	**1980**	**2880**	**4695**	**14315**	**33555**	**50980**	**71430**	**81655**	**81655**	**81655**

Cash Flow Analysis

Year	1	2	3	4	5	6	7	8	9	10
Cost	75810	9170	10810	11680	11680	11680	11680	11680	11680	11680
Benefit	1980	2880	4695	14315	33555	50980	71430	81655	81655	81655
Net	**-73830**	**-6290**	**-6115**	**2635**	**21875**	**39300**	**59750**	**69975**	**69975**	**69975**

Net present worth or NPW(At 15% DF)	38,921
Internal rate of return or IRR	22.96%
Benifit cost ratio or BCR (at 15% DF)	1.35
Present worth of cost or PWC (at 15% DF)	111,914
Present worth of Benefit or PWB (at 15% DF)	150,836
Debt service coverage ratio or Av DSCR	1.88

Repayment Schedule

Year	Year wise Bank Loan disbursed	Bank loan Outst-anding	Interest @ 14%p.a.	Deferred Interest	Gross Income	Mainten-ance Cost	Surplus	Repayment of Bank Loan			Net Surplus +/–	DSCR
								Interest	Principal	Total		
1	2	3	4	5	6	7	8	9	10	11	12	
1	68229	68229	9552	9552	1980	B/L	1980	0	0	0	1980	
2	8253	76482	10707	19260	2880	B/L	2880	1,000	0	1000	1880	2.88
3	9729	86211	12070	28329	4695	B/L	4695	3,000	0	3000	1695	1.57
4	10489	96700	13538	28329	14315	B/L	14315	13,538	0	13538	777	1.06
6		96700	13538	22663	33555	11680	21875	19,204	0	19204	2671	1.14
7		77360	10830	11332	71430	11680	59750	16,496	19340	35836	23914	1.67
8		58020	8123	5666	81655	11680	69975	13,879	19340	33129	36846	2.11
9		38680	5415	0	81655	11680	69975	11,081	19349	39421	39554	2.30
10		19340	2708	0	81655	11680	69975	2,708	19340	22048	47927	3.17
Total			100019					100,019	96700		Average	1.88

6.2.3. Scheme of the Vermicompost unit

[A] Acpital investment:

1. 10 X 6 X 2.5 ft size tank (brick wall with plaster & concrete floor) ₹ 4,750.00
2. Roof structure made up of straw & bamboo ₹ 1,100.00
3. Vermin, soil, cuwdung, sand, bats, plastic etc. ₹ 750.00
4. Implements like fork-spade, bucket, wire mesh, shovel etc ₹ 700.00

[B] Working capital (for one cycle i.e. 100 days):

1. Collection & purchase of organic matter ₹ 600.00
2. Cost of Preliminary decomposition ₹ 150.00
3. Loading/filling, watering, turning, unloading, drying, packing etc. ₹ 550.00
4. Cleaning & maintenance cost (yearly) ₹1,000.00

TOTAL ₹ 10,000.00

[C] Gross return in one year i.e. in 3 cycles:

1. 1500 kg vermiconpost per cycle @ Rs3.50 / kg ₹ 15,750.00
2. Saleable vermin i.e. earthworm ₹ 750.00

[D] Yearly net return:

Year	Expenditure (₹)	Gross Return (₹)	Net Profit (₹)
1st year	12600.00	16500.00	3900.00
2nd yr. & Onwards	4900.00	16500.00	11600.00

Note:

(1) If this project runs in commercial basis then 75 to 80 days cycle per year can be achieved, i.e. 4 to 5 production cycle per year is achievable.

(2) A sketch of plan of this unit is attached herewith.

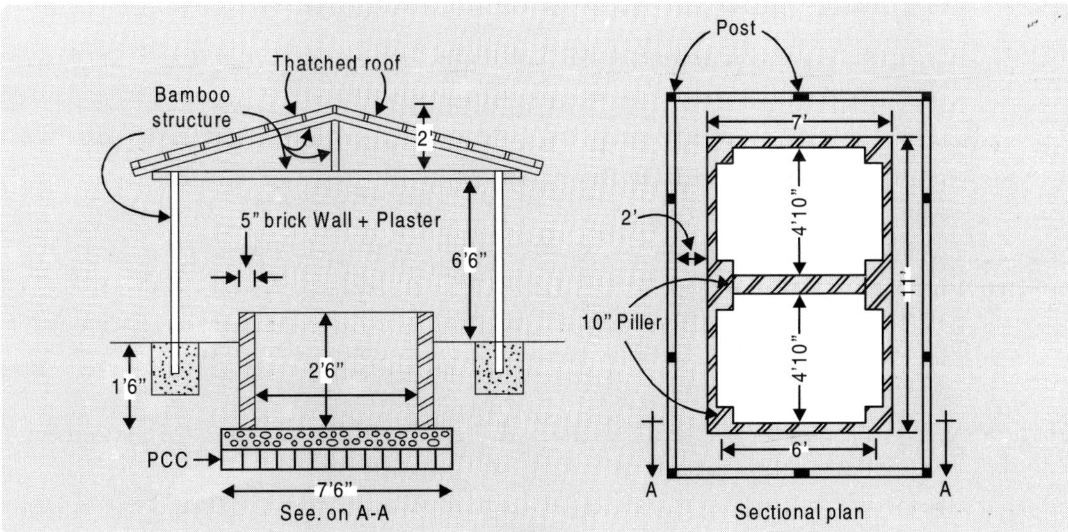

REFERENCES

Aina P. O. 1979. 'Soil changes resulting from long term management practices in Western Nigeria', J. Soil Sci. Soc. Amer. **43**, 173-7.

Beets W. C. 1982. 'Multiple Cropping and Tropical Farming System'. Gomer publishing Co., U.K.

Borst H. L. and Woodburn R. 1942. 'The effect of mulching and methods of cultivation on runoff and erosion from Muskingum silt loam', Agric. Engng. **23**,19-22.

Brady N.C. 1980. 'The Nature and Properties of Soils' 8th Edn. Macmillan, USA.

Burges A., and Raw F. (Ed.) 1967. 'Soil Biology', Academic press, New york.

Clarke G. R. 1971. "The study of the soil in the field" 5th Edn., Oxford, London.

Cogo N. P., Moldenhauer W. C. and Foster G. R. 1984. 'Soil loss reductions from conservation tillage practices', J. Soil. Sci. Soc. Amer. **48**, 368-73.

Cook R. V. And Doornkamp J. C. 'Geomorphology in Environmental Managment' Oxford University Press.

CRIDA perspective plan, 1995, - Vission 2020' Central Research Institute of Dryland Agriculture, Hydrabad.

Dakhinamurti C., Michael A.M. and Dastane N. G. 1971. 'Water resources and their optimum utilization in agriculture'. Proc. Sym. "Water resources", Indian Inst. Sci., Bangalore C_{6-1}, C_{6-14}.

DARE / ICAR IX Plan Document, 1996,' Indian Council of Agricultural Research, New Delhi.

Day P. R. 1965. "Methods of soil analysis", Agronomy, Vol. 9, Chapter 43.

De Vleeschauwer D., Lal R. and De Boodt M. 1978. 'Influence of soil conditioners on water movement through some tropical soils', In Lal and Greenland (Ed.), "Soil physical properties and crop production in tropics", Wiley, 149-50.

Demiruren A.S., Beheshti R.D., Sdlmi H., Salwh B.A. and Djaferi A. 1971.'Comparison of the reproductive and productive capacities of Sheep of the Kellakul, Kizil, Bakhtiari and Baluchi breeds' in Iran Tech. Rep. No. 1 Anim. Husb. Res. Ins.: Teheran

Dhruva N. V. V., Sastry G. and Patnaik U. S. 1990. 'Watershed Management' ICAR,Pusa, New Delhi.

Dixey F. (1950). 'A Practical Handbook of Water Supply' (2nd ed.), Thomas Marby, London.

Doglus J. T. and Goss M. J. 1982. 'Stability and organic matter content of surface soil aggregates under different methods of cultivation and in grass land', Soil and Tillage Research. **2**, 155-76.

Doorenbos, J., and W. O. Pruitt (1977). Crop Water Requirements. Irrigation and Drainage Paper 24. Food and Agriculture Organization, Rome.

Durning A. B. and Brough H. B. 1991. 'Takink stock: Animal farming and the environment'. Worldwatch Paper 103. Worldwatch Institute, Washington D.C.

Eberhart S. A. and Russel W. A. 1966. 'Stability parameters for comparing varieties.' Crop Sci., **6**, 36-40.

Ehrlich, P.R. and J. Roughgarden. 1987. 'The Science of Ecology'. Macmillan, New York.

Evans R. 1980. 'Mechanics of water erosion and their spatial and temporal controls: an empirical viewpoint' in kirkby and Morgan (Ed.), "Soil Erosion" Wiley, 109-28.

Evenari, 'The Challenge of the Negev', Ben-Gurion University of the Negev, P.O. Box 1025, Beer-Sheva 84110 ISRAEL).

Furon R. (1963) ' The problem of water: a world study' Faber and Faber, London.

F.A.O. 1990. 'Watershed Management Field Manual: Water shed survay and planning', Conservation Guid 13/6 Rome.

Findlay J. D., 1954. 'The climatic physiology of farm animals'. Met. Monogr., **2**, P-19-29.

Frscher R. A. and Maurer R., 1978. 'Draught resistance in spring wheat cultivars. I. Grain yield response. Aust. J. Agri. Res. **29**, 897-912.

Gillbertare S. M. 1949. 'Effect of hulching of Cofee with banana leaf", E. Afr. Agric.J. II. 75.

Gischler C., Jauregui C.F. (1984) 'Low cost technique for water conservation and management in Latin America' Nature and Resources, XX(3), 15-16.

Gong Y. S. and Jiang D. Q. 1977. 'Soil erosion and its control in small gully watersheds in the rolling loess area on the middle reaches of the Yellow River', *Paris Symp. Erosion and soil matter transport in inland waters,*

Gray T. R. G. and Parkinson D. (Ed.), 1967. 'Ecology of Soil Bacteria', Liverpool UP.

Haan C.T., H.P. Jhonson, and D.L. Braknsick (1982). Hydrologic modeling of small water shade. Amarican Society of Agriculture Engineers. St. Joseph.

H.C.Pereira, M. Dagg and P. H. Hosegood, 1964. *Emp.J.exp.Agric,* **32**, 31

Hanks R. J., Bowers S. A. and Bark L. D, 1961. *Soil Sci.,* **91**, 233.

Heitman H.Jr. and Hughes E.H. 1949. 'The effect of air temperature and relative humidity on the physiological wellbeing of Swine'. J. Anim. Sci., **8**, 171-81.

Hocking D. 1993. "Project planning Management and monitoring – Improving land management in Rajasthan". Intercooperation Coordination Office, Jaipur, India.

Hudson, Norman (1971). Soil Consrvation, B. T. Batsford Ltd., London.

Hudson N. W. and Jackson D. C. 1959. ' Results achived in the measurment of erosion and runoff in southern Rhodesia', Proc. Third inter-African soils conference, Dalaba, 575-83.

Hurni H. 1984. 'Soil conservation research project Ethiopia', Vol-4, Third progress report. Univ. Bern and United Nations Univ.

Hutchinson J.C.D. 1954. 'Heat regulation in Birds'. In Hammond J. (ed.), "Progress in the physiology of farm animals" Vol.1. pp- 299-362, Butterworths: London.

Jiang D. Q., Qi L. D. and Tan J. S. 1981. "Soil erosion and conservation in the Wuding River Vally, China" in 'Soil Conservation: problem and prospect' Wiley, 461-79.

Jolliet O., Hortitrans, (1994). A model for predicting and optimizing humidity and transpiration in greenhouse, J. Agric.Engg. Res. 57, 23-37.

Kehar N. D. I954. 'ICAR Silver Jubilee Souvenir', Govt. of India Press. New Delhi.

Khimina N. T. 1973. Bull. VNIIALM, **12**, 26-8.

Kirpich, Z.P. (1940). Time of concentration of small agricultural watershade.Civil Engg., 10: 362.

Klimentov P. P. 1983. 'General Hydrogeology'. Mir Publishers, Moscow.

Klimentov P. P. and Kononov V. M. 1973. Ground water dynamics. Mir Publishers, Moscow.

Krebs, C. J. 1985. Ecology 3rd ed., "The experimental Analysis of Distribution and Abundence". Harper & Row, New York.

Kumar Ashok and Adrol I.P., 1982. 'Relative performance of some forage grasses in relation to soil pH and Sodicity'. Forage Res. **8**, 127-31.

Laura E. Powers and R. Mcsorley. 1999. 'Ecological Principales of Agriculture'.Delmar Thomson learning, New york.

Laws W. D. and Evance D. D. 1949. "The effects of long-time culivation on some physical and chemical properties of two Rendzina soils," Soil. Sci. Soc.Amer.Proc.,**14**,15-19.

Lyon T. L., Bucman H. O. and Brady N. C. 1952. " The Nature and Properties of Soils", Macmillan, Inc., New york.

Martin J. P. and Aldrich D. G. 1955. 'Proc. Soil Sci. Soc. Amer., **19**, 52.

McCalla T. M. and Duley F. L. 1943. *J. Agron.,* **35,** 306.

Michael A. M. 1990. "Range of available water holding capacity of soils". Vikas Publishing House Pvt. Ltd., New Delhi.

Michael Evenari *et al* 1971, The Negev: The Challenge of a Desert, Oxford University Press, UK.

Morgan R. P. C. 1980. "Implications" in Soil Erosion, Wiley.

Moss A. J., Green P. and Hutka J. 1982. "Small channels: their formation, nature and significance", earth Surf. Proc. Landf. **7**, 401-15.

MYRADA and IIRR. 1997. 'Resource management in Rainfed Drylands – An information kit'

Nambiar P.T.C., Rupela O.P. and Kumar Rao J.V.D.K. 1988. 'Nodulation and nitrogen fixation in Ground nut, Chichpea and Pegeonpea' in "Biological Nitrogen Fixation: Recent development" Subba Rao N. S. (Ed.), Oxford and IBH Publishing Co. New Delhi.

Nimbalkar V.S., Umrani N.K. and Deokar A.B. 1986. 'Sina' and 'Man' two high yielding varieties of horsegram (kulthi) for Maharashtra. J. Maharashtra Agric. Univ. II (2), 148-150.

Nimje P.M. and Bhandarkar D.M. 1990. " Effect of suppliment irrigation through farm pond on crop productivity in vertisols' Ind. J. Sil Cons. **18** (1&2), 9 – 14.

Norman M. J. T. et.al. 1995. "The Ecology of Tropical Food Crops" Cambridge Uni. Press, UK.

Odum, E.P.1971. "Fundamentals of Ecology", Philadelphia: Sunders.

Ologe K. O. 1972. "Gullies in the Zaria area. A preliminary study of headscarp recession", Savanna, **1**, 55-66.

Oram P. 1980. 'What are the world resources and constrains for Dryland Agriculture?' P-17-78 in "Proceedings on the International congress for Dryland Farming", Adelaide, South Australia.

Pareek S. K. and Gupta R. 1993. 'Medicinal and aromatic improve profitability of cropping systems in India' in "Glimpse in plant Research" Vol-II, Govil J. W. et.al. (Ed.), Today and tomorrow printers and publishers, New Delhi, 325-35.

Parker C. F. 1990. "Role of animals in sustainable agriculture". In 'Sustainable Agricultural System'(Ed.) Edward, Madden, Miller and House, 238-45, Delray Beach, Fla.: St. Lucie Press.

Payne W. J. A. and Hancock J., 1957. 'The direct effect of tropical climate on the performance of European type cattle' II Production Emp. J. exp. Agric., **25**, 321-38.

Pedigo L. 1996. 'Entomology and pest management.' Upper Saddle River, N.J., Prentice-Hall.

Penman H. L. 1941. '— Reduction of evaporation—', J. Agric. Sci., 31, 454.

Pidgeon J. D. and Sone B. D, 1978. "Soil structure and strength relations following tillage, zero tillage and wheel traffic in Scotland", in 'Modification of soil structure, wiley, 371-8.

Pimentel D. (Ed.) 1991. 'Handbook of Pest Management', 2nd ed., Vol.II. Boca Raton, Fla.,CRC Press.

Powers,L.E. and Mcsorley, R. 1999. "Ecologycal principal of Agriculture" Delmar, New York.

R.B.Tukey and E.L.Schoff, *Proc. Amer. Soc. hort. Sci.*, 1963, **82**, 68.

Ragsdale A.C., Thompson H.J., Worstell D.M. and Bordy S. 1953. 'Environmental Physiology and Shelter Engineering with special reference to domestic animals'. XXI "The effect of humidity on milk production and composition of feed, and water composition and body weight of cattle". Res. Bull. Mo. Agric. Exp. Stn. No. 521

Randhawa N. S. and Venkateswarlu J. 1979. 'Indian experience in semi-arid tropics: prospects and retrospects'. In "Development and Transfer of Technology", Proc. Intl. Symp., Aug – Sept, 1979. Int. Crop Res. Inst. For Semi-Arid Tropics, Hyderabad, pp 207 – 220.

Rechcigl N. A. and Rechcigl J. E. (Ed.) 1997. 'Environmentally Safe Approaches to crop Disease Control', Boca Raton, Fla., CRC Lewis Publishers.

Reddy D.S., Raju A.P. and Redy M.S. 1985. 'Consolidated report of the co-ordinated soil physics project', Hyderabad Centre (1967–1983) A P Agric. Univ., Hyderabad.

Robart Mendelsohn, William D. Nordhaus, and Daigee Shaw (1994): The Impact of Global Worming on Agriculture: A Ricardian Analysis. The American Economic Review, Vol.84, No. 4 (sept.1994). Pp. 753-771

Rothamsted Rept., 1968. Part- I and II. Organic matter and manure.

Russel E. W. 1943. 'J. Agriculture Science,' **33**, 147.

Russel E. W. 1978. 'Soil Conditions and plant growth', 10th Edn. Longman, London.

Schofield R. K. and Penman H. L. 1941. J. Agric. Sci., 31, 74.

Schroth M.N. and Hancock J.G. 1981. 'Selected topics in biological control' in Annual Review of microdiol. **35**, 453-76.

Sheng T. C. 1981. 'The need for soil conservation structures for steep cultivated slopes in the humid tropics', in Lal, and Russel (eds), "Tropical Agricultural Hydrology", Wiley, 357-72.

Singer M. J. and Blackard J. 1978. 'Effect of mulching on sediment in runoff from simulated rainfall', J. Soil Sci. Soc. Amer. **42**,481-6.

Singh S. P. 1987. 'Efficient crops and cropping systems' in "Technological Advances in Dryland Agriculture", Sing S. P. (Ed.) CRIDA Hydrabad, 47-66.

Skidmore E. L. and Hagen L. J. 1977. 'Reducing wind erosion with barriers', Trans. Amer. Soc. Agric. Engnrs. **20**, 911-15.

Soil Survey Staff, 1960. "Soil Classification—A Comprehensive System—7th Approximation" (and its supplement in 1967). U S Dept. of Agric., Washington.

Sprent J. I. And Sprent P. 1990. 'Nitrogen Fixing Organisms', Chapman and Hall, London.

Steinbach J. 1972. 'Biodimatic influences on sexual activity in Boars'., Proc. 7th Int. Cong. Anim. Report and A. I., Munich. Summaries, P-427.

Stewart B. A. and Burnett E. 1985. Water conservation technology in Rainfed and Dryland Agriculture'. P-355-359. In "Water and water policy in world food supplies (ed.)" Procedings of the confarance. Texas.

Sthapit B. R., Joshi K. D. and Witcomber J. R. 1996. 'Farmer participatory Crop Improvement III', A case study for Rice in Nepal, Experimental Agriculture **32**, 479-96.

Subba Rao N. S. 1984. 'Biofertilizers in Agriculture'. Oxford and IBH Publishing Co. New Delhi.

Tandon H. L. S. (Ed.) 1992. "Fertilizers, Organic Manures, Recyclable Wastes and Biofertilizers", Fert. Dev. Con. Orga., New Delhi.

Tejwani K. G. 1981. 'Watershed management as a basis for land development and management in India' in "Tropical Agricultural Hydrology", Lal R. and Russell E. W. (Ed.), Wiely, 239-55.

Troeh, Hobbs, and Donahue (1980). 'Soil and water conservation for productivity and environmental protection'. Prentice Hall.

UNEP (1983) 'Rain ans storm water hervesting in rural areas: A report by the UN environment Programme', (Water Resources Series Vol-5), Tycooly International Publishing, Dublin.

Venkateswarlu J. and Singh R. 1982. ' Crop response to applied nutrients under limited water condition', in "Review of Soil Res. In Yndia", Part-I, 12th Int. Congress Soil Sci. 8-16 Feb, 1982. New Delhi.

Vijayalakshmi K. and Vittal K.P.R. 1981. 'Rainwater harvesting and re-cycling in semi-arid red and Sierozem soils' Ind. J. Soil. Cons. **9** (2&3), 87-99.

Vijayalaxmi K. 1987. 'Rainwater management in drylands' in Technological Advances in Dryland Agriculture" (Ed.) by Singh R.P. et.al. CRIDA, Hydrabad, P- 19-45.

Voetberg K. S. 1970. Erotion on agricultural lands, Agricultural University, Wageningen.

Wallwork J. A. 1970. 'Ecology of Soil Animals', MacGraw-Hill, London.

Williams C. N. and Joseph K. T. 1970. 'Climate, soil and crop production in humid tropics'. Oxford University Press, London.

Williamson G. and Payne W.J.A. 1978. "An Introduction to Animal Husbandry in the Tropics". 3rd Ed., Longman, London.

Wischmeier, W.H., and D.D. Smith (1978). Predicting Rainfall Erosion Losses, A Guide to Conservation Planning. Agriculture Hand book 537. U.S. Department of griculture.

Woodruff N. P. and Zingg A. W. 1952. ' Wind tunnel studies of fundamental problems related to windbreaks', US Soil Consurv. Serv. Pub. **SCP-TP-112**.

Worstell D. M. and Bordy S. 1953. 'Environmental Physiology and Shelter Engineering with special reference to domestic animals'. XX "Comparative physiological reactions of European and Indian cattle to changing temperature". Res. Bull. Mo. Agric. Exp. Stn. No. 515.

Yaniv Z. and Palevitch D. 1982. ' Effect of draught on the secondary metabolites of medicinal and aromatic plants.' In "Cultivation and utilization of medicinal and aromatic plants" Attal C. K. and Kapur B. M. (Ed.) RRL Laboratories, Jammu and Tawi, India.

GLOSSARY

- **Abscicic acid (ABA):** It is a plant metabolite. Its production and accumulation in plant body are altered under varying moisture stress situations. Amongst the various substances, accumulated under water stress, ABA is thought to be directly related to drought tolerance property of a plant.
- **Auxin:** It is a plant hormone that, in very small concentraion, is capable of producing elongation of plant shoots and to produce other specific growth effects.
- **Bar:** It is a unit for measurement of pressure and 1 bar is equal to one million $(10)^6$ dynes per square centimetre.
- **Biodiversity:** It is the state of an ecosystem where various forms of genetically diverse living materials are present.
- **Capillary water:** Water held vertically in the very small or microscopic tube like structures (capillary) present in the soil, usually with a suction force (surface tension) greater than 60 cm. fo standing water.
- **Cation exchange capacity (CEC):** It is the sum total of exchangeable positively charged ions (cations) that a soil can absorb. It sometimes called as 'base exchange capacity' and is expressed in milliequivalant per 100 grams of soil or any such adsorbing material.
- **Consumptive use:** The total quantity of water used by a plant for transpiration and growth, plus water that evaporates from the adjacent soil and plant surface in specific time. It is usually expressed as equivalent depth of free water layer per unit time.
- **Contour:** It is an imaginary line connecting points of equal elevation of the land surface. It runs at right angle to the direction of slope of the land.
- **Control plot:** In a field trial with various treatments a plot is left untreated for comparing the performance of the treated plots against it.
- **Crust:** It is a thin but hard and compact dry layer formed on the surface of soil with a relatively moist layer beneath it.
- **Diffusion:** It is the transmission of matter as a result of movement of molecules between two intermingling fluids or gases when they come in contact.
- **Drought susceptibility index:** It is the index (see stability index) of capability of a crop to withstand water stress.

- **Eluviation:** The removal of soil particles, which are in suspension or solution, from a particular layer of soil. Leaching is a kind of eluviation.
- **Exchangeable sodium percentage (ESP):** The extent to which the adsorption complex of a soil is occupied by exchangeable sodium. It is normally expressed in a formula as under :

  ESP = exchangeable sodium (meq./100gm. of soil)/CEC (meq./100gm. of soil) × 100.
- **Granulation:** It is the process of formation of small globular structure out of very fine soil particles. This structure of soil causes better air circulation in the soil. This condition is helpful for good plant growth.
- **Gross Domestic Production (GDP):** It is the assessment of total production of a country in terms of money. It broadly includes Agricultural, Industrial and other non-agricultural production as well as the growth of service sector.
- **Humidity:** It is the amount of water vapour present in the layer of atmosphere attached to the earth's surface.
- **Hybrid:** It is the sexual progeny obtained by crossing two individuals. Normally, the selection of parents is done in such a way that good genetic materials from each parent accumulate in the new progeny to make it superior to either of the parents but this superiority does not get transmitted to the next generation.
- **Hydraulic conductivity:** It is an expression by which water flows (vertically or horizontally) through soil or any other porous medium in response to a given potential gradient. It is expressed in meter/day and referred as 'K'.
- **Hydrosphere:** It is the water present on the surface of the earth in oceans, lakes, rivers, reservoirs, glaciers, underground water, and the water contained in the tissues of all living flora and fauna.
- **Igneous rock:** Rocks formed by cooling and solidification of magma. This type of rock has suffered little change (due to weathering) since its formation.
- **Illuviation:** It is the process of formation of a soil layer by deposition of materials from solution or suspension from the overlaying layer of that soil.
- **Infiltration:** It is the downward penetration of water through the surface layer of any medium (soil). It is expressed as equivalent height of water passing through the surface in unit time.
- **In-situ Biomass:** It is the total mass of living matter and un-decomposed materials produced from any living organism present in a particular area. It may be expressed in fresh or dry wright basis.
- **Keolinite, Illite & Montmorilonite:** The Ist and 3rd are the aluminosilicate clay minerals having 1:1 and 2:1 lattice structure (silicon tetrahedral and aluminium octahedral layers) respectively of which 2:1 is of expanding type. The 2nd one is silicate clay with 2:1 lattice structure of indefinite chemical composition.
- **Leaching:** It is the downward removal of soluble materials from a medium (here soil) with the help of a percolating agent like water.

- **Manning's equation:** It is the relationship among the flow depth or hydraulic radius (r), slope (S), and roughness of the surface (n) that determine the velocity of water flow (v). It is defined as $v = (r^{2/3} S^{1/2})/n$.
- **Marls:** It is the soft and unconsolidated calcium carbonate, usually mixed with varying amounts of clay or other impurities.
- **Metamorphic rock:** Rock that has undergone great changes from its original state through the combined action of heat and pressure. Examples are marbles, gneiss and slate, which are produced from limestone, granite and shale.
- **Nitrogen assimilation:** Conversion of nitrogen into organic cell substances by living organisms.
- **Nitrogen fixation:** It is the collection of atmospheric elemental nitrogen (N_2) principally through the effort of the soil biota.
- **Nomadic:** The habit of moving from place to place without staying in a permanent abode.
- **Percolation:** It is the process of downward movement of water through a specific mass of a saturated or nearly saturated porous medium (soil). This process precedes infiltration and seepage and is driven by gravity to flow along the hydraulic gradients of the order of 1.0 or less.
- **pH of the soil:** It is the unit of measurement of acidity or alkalinity of the soil. Mathematically, it is the negative logarithm of the hydrogen-ion concentration of the soil.
- **Photoperiod:** It denotes the length of the day measured from sunrise to sunset. It controls the flowering response of many plants.
- **Porosity:** It is the total void, i.e. space occupied by air, present in a particular porous medium (like soil) and expressed as percentage of volume.
- **Proline:** It is the solute, which has been found to accumulate in plant cells to withstand drought, i.e. water stress and to maintain the turgor of the cells.
- **Residual toxicity:** It is the remains of pesticides, used to save the crop plants from the attack of pests and diseases. If it is present in the produce at a concentration higher than a specified limit residual toxicity is dangerous to the consumer.
- **Reynold's number (Re):** It is the hydraulic characteristics of the overland flow, which is the relationship among hydraulic radius (r), velocity (v) of the flow and kinematic viscosity of water (u). It is an index of the turbulence of the flow and is expressed as $Re = vr/u$.
- **Shift cultivation:** It is a cropping system in which a land is cleared of its natural vegetation for growing of crops. After a few years, when the yield declines, the growers shift to a new land for cultivation leaving the old one fallow allowing it to regain its natural vegetation.
- **Soil aggregate:** It is a soil mass, usually of globular shape, formed by the aggregation of miniature soil particles being held together. This structure of soil improves air circulation inside the soil thus helping the root of the crop plant to grow better.

- **Solarization:** It is the process of heating of soil under clear plastic sheet utilising the solar radiation to kill soil borne pests and pathogens.
- **Stability index:** It is the index to adopt a crop plant in a specific situation, and is assessed on the basis of stability in yield of that crop over the years. Different physiological aspects of a crop contribute the stability in yield, which may be estimated statistically.
- **Salinization:** It is the process of accumulation of salt in the soil generating salinity and alkalinity in it.
- **Time of concentration:** It is the time taken by a drop of runoff water to travel from the farthest point to reach the outlet point of a catchment area or the point of accumulation of a watershed.
- **Topography:** It is the height of a land surface in respect to that of a reference level, generally the mean sea level.
- **Viscosity:** It is the property (thickness) of any fluid by virtue of which it resists its flow while the fluid is in motion. It is basically the internal friction developed within the imaginary layers of a moving fluid.
- **Water harvesting:** It is the system of trapping, collecting and storing of precipitation, mostly the rainfall. This system predominantly consists of mechanism of trapping and conveyance of water and a storing facility/structure. Addition of a few auxiliary components can make the system more sustainable.
- **Watershed:** It is a well-defined area, enclosed by an imaginary boundary line, within which every drop of runoff water is supposed to accumulate in a particular structure like pond, lake, reservoir etc., natural or manmade. It is nearly synonymous to the term 'catchment area' which, in principle, is related to any waterway like rive, stream, channel etc. that *carry away* the run off water.
- **Xerophytic:** The ability of some plants to stand extreme water stress conditions.

INDEX

F

G

H

I

L

M

N

O

P